COMPUTATIONAL MOLECULAR BIOLOGY

COMPUTATIONAL MOLECULAR BIOLOGY

By

Dr. Rajiv Tyagi

Department of Zoology
M.M. College
Modi Nagar (U.P.)
(India)

Published by:

DISCOVERY PUBLISHING HOUSE
4383/4B, Ansari Road, Darya Ganj
New Delhi-110 002 (India)
Phone: +91-11-23279245; 23253475; 43596065
Mobile: +91 9811179893 / +91 9871656464
E-mail: discoverybooksindia@gmail.com
orderdphbooks@gmail.com
namitwasan9@gmail.com
web: www.discoverypublishinggroup.com

First Published: 2009
Reprinted: 2025

ISBN: 978-81-8356-465-6

Computational Molecular Biology

Printed at:
Infinity Imaging Systems
Delhi

Preface

The present title "*Computational Molecular Biology*" is an exciting new field bringing together life scientists, mathematicians, computer scientists and engineers to explore a new and deeper understanding of biological systems. The present text has been compiled for computer science and mathematics graduate and level undergraduate students. It will also be of interest to molecular biologists interested in bioinformatics.

The rationable of the book is to present algorithmic ideas in computational biology and to show how they are connected to molecular biology and to biotechnology. To achieve this goal, the book has a substantial "computating biology without formulas" component that presents biological motivation and computational ideas in a simple way. This simplified presentation of biology and computing aims to make the book accessible to computer scientists entering this new areas and to biologists who do not have sufficient background for more involved computational techniques. Every chapter has an introductory section that describes both computational and biological ideas without any formulas. Thus book concentrate on computational ideas rather than details of the algorithms and makes special efforts to prevent there ideas in a simple way.

To make the work more comprehensive and informative, the author has consulted many authoritative books, research journals, abstracts, monographs etc. He is grateful to all those great scholars whose work are cited or substantially reproduced.

There can be no claim to originality except in the manner of treatment and much of the information has been obtained from the books and scientific journals available in the different libraries.

The author expresses his thanks to his friends and colleagues whose continue inspirations have initiated him to bring out this book.

The author expresses his gratitude to Mr. Wasan and staff of M/s Discovery Publishing House Pvt. Ltd. for their whole hearted cooperation in the publication of this book.

In the mean time, the author will remain sincerely responsible for any shortcomings of the book and be grateful to the readers for their suggestions and constructive criticism for the continuous betterment of the book. He takes this opportunity to appeal to the readers to send their suggestions straightaway to his Publisher.

Author

Contents

Introduction

Survival machines began as passive receptacles for the genes, providing little more than walls to protect them from the chemical warfare of their rivals and the ravages of accidental molecular bombardment. In the early days they *'fed'* on organic molecules freely available in the soup.

In the past, living organisms were grouped into two distinct life forms or *domains* :

(*i*) *prokaryotes,* represented by *cyanobacteria* (blue-green algae) and common bacteria such as *Escherichia coli,* not having a nuclear membrane to separate genomic DNA from the cytoplasm; and

(*ii*) *eukaryotes,* represented by unicellular organisms such as the flagellum *Trypanosoma brucei* (one species belonging to the *T. brucei* complex causes sleeping sickness in humans) and multicellular organisms, all of which have a nuclear membrane.

However, in August 1996, the genome sequence of the methane-generating archaebacterium *Methanococcus jannaschii* was published by Bult et al. [BWO+96]. A small initial portion of the DNA sequence, consisting of over 1.6 million characters, is given as follows :

```
TACATTAGTGTTTATTACATTGAGAAACTTTATAATTAAAAAAGATTCATGTAAATTT
CTTATTTGTTTATTTAGAGGTTTTAAATTTAATTTCTAAGGGTTTGCTGGTTTGATTG
TTTAGAATATTTAACTTAATCAAATTATTTGAATTTTTGAAAATTAGGATTAATTAGG
TAAGTAAATAAAATTTCTCTAACAAATAAGTTAAATTTTTAAATTTAAGGAGATAAAA
ATACTCTGTTTTATTATGGAAAGAAAGATTTAAATACTAAAGGGTTTATATTATGAAG
TAGTTACTTACCCTTAGAAAAATATGGTATAGAAAAGCTTAAATATTAAGAGTGATGA
AGTATATTATGT...
```

Analysis of this sequence provided solid evidence for a startling hypothesis advanced two decades earlier by Carl Woese: there is a third domain of life called *Archaea,* which is distinct from *Prokarya* and *Eukarya:*

(*i*) *archaebacteria,* representing life domain *Archaea,* share cellular organization features with prokaryotes (such as lack of a nuclear membrane), but seem closer to eukaryotes in transcription and translation mechanism.

M. jannaschii was discovered in 1982 in a sediment sample collected by the research submarine *Alvin* from a *white smoker* on a hot spot of the sea floor of the Pacific Ocean (21° N on the East Pacific Rise) at a depth of 2600 meters. This methanogenic *archaeon* is *thermophilic,* living in a temperature range from 48°–94°C, with an optimum temperature around 85°C, and obtains its energy needs from high-energy bonds in certain inorganic compounds near thermal vents. Producing methane from carbon dioxide, *M. jannaschii* represents a potentially non-polluting energy source. How can one determine the (hypothetical) genes of *M. jannaschii* from its 1.66 megabase pair genome?

How can one compute that 56% of its 1738 genes are completely new. unlike any genes found in prokaryotes or eukaryotes? *M. jannaschii* has only one kind of DNA polymerase, while all other organisms studied have several kinds. Could this archaeon have an additional, completely different mechanism for DNA replication? If so, how can one begin to find the genes involved? How can one assert with confidence that *Archaea* is sufficiently different from *Protokarya* and *Eukarya* to warrant that it constitute a third domain of life? How can one devise a likely phylogenetic (ancestral) tree for life, which has *Protokarya* separating first from a common trunk, while *Archaea* separates somewhat later from the trunk leading to *Eukarya*? With genome sizes measuring into the millions and billions, only through the use of computers can one store and analyze genomic data.

Computational biology, also called *bioinformatics,* is concerned with the development of efficient algorithms, statistical analysis, and mathematical modeling in order to answer such questions. In this chapter, we'll give a brief survey of those concepts of molecular biology, which are essential to computational biology. The *'central dogma'* of information flow in biology states that information flows from DNA to RNA to protein; since a protein's functionality is determined by its unique three-dimensional structure, it follows that the one-dimensional sequence information in DNA determines the three-dimensional structure of the corresponding protein.

The central dogma states that once 'information' has passed into a protein it cannot get out again. The transfer of information from nucleic acid, or from nucleic acid to protein, may be possible, but transfer from protein to protein, or from protein to nucleic acid, is impossible. Information here means the precise determination of sequence, either of bases in the nucleic acid or of amino acid residues in the protein. Thus we have the following picture of the information flow in biology:

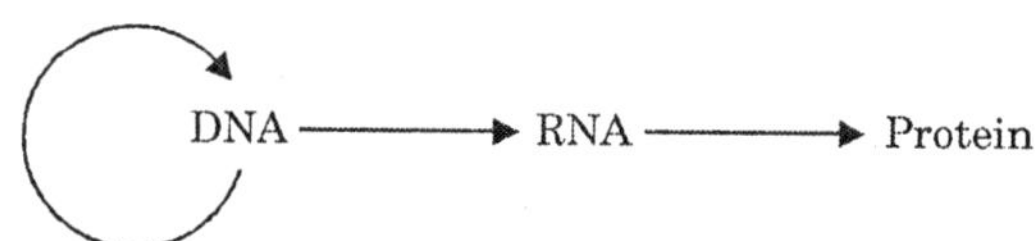

Following this diagram, the remainder of the chapter is organized as follows. We begin with an overview of organic chemistry needed in the sequel, then move on to DNA, RNA, amino acids and proteins. Finally, we describe the transcription and translation machinery that nature uses to process information in DNA into proteins.

SOME ORGANIC CHEMISTRY

Organic chemistry is centered around the chemistry of the carbon atom. The reason for this is that carbon has properties that distinguish it from most other elements: These properties are

(*i*) carbon has a small size,

(*ii*) carbon has 4 covalent bonds.

(*iii*) several carbon atoms can form rings and chains.

Thus, it is possible to build large, stable molecules using carbon.

One distinguishes two types of bonds, namely covalent and non-covalent bonds. Non-covalent bonds are much weaker than covalent bonds (typically 30-300 times weaker in aqueous environment). The average energy of a non-covalent bond is normally not much stronger than the energy of thermal movement (at room temperature). A *covalent bond* is formed between two atoms that share electrons.

Molecules are built using covalent bonds. *Non-covalent bonds* are subdivided into three classes, namely hydrogen bonds, ionic bonds, and van der Vaals interactions. Because of their importance, we concentrate on hydrogen bonds. Hydrogen bonds are used for base-pairing, which holds together the two strands of the DNA double helix. Hydrogen bonds also play an important role in RNA (which is normally single-stranded), as well as in stabilizing protein structure.

A *hydrogen bond* can be formed if there are two electronegative atoms (such as oxygen or nitrogen) that share a hydrogen atom, *e.g.*,

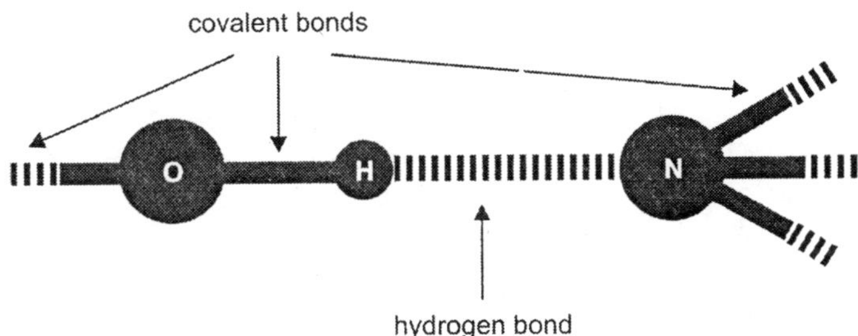

The hydrogen bond is caused by the polarity of the participating molecules. These bonds are in the range of 3-5 kJ/mole, and are easily disassociated by heat or certain chemicals. By comparison, a covalent C-C bond has 380 kJ/mole.

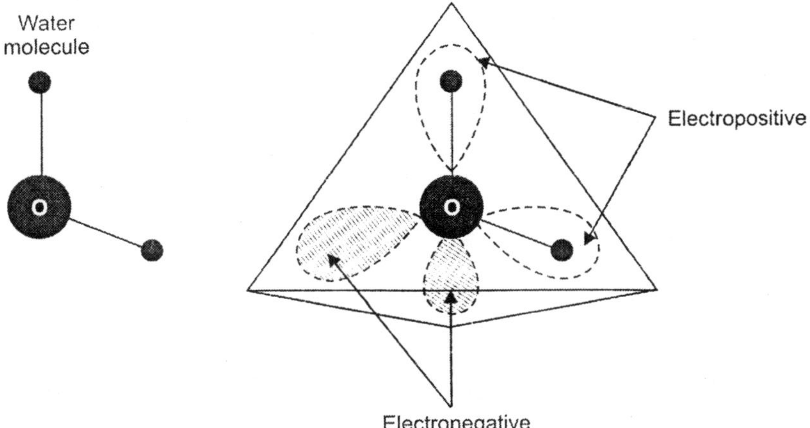

Fig. 1.1. Polarity of the water molecule. The positive and negative charged regions, which span a tetrahedron, are shown in the left part.

An important example of a polar molecule is the water molecule. The electronegative oxygen attracts the electrons from both hydrogens, thus yielding electropositive and electronegative regions as depicted in the figure. Due to this polarity and the form of the region, water molecules are able to form hydrogen bonds with four other water molecules. At room temperature, 15% of the water molecules have (short-duration) hydrogen bonds with four other water molecules. Other molecules that are polar can also form hydrogen bonds with water molecules. Since the formation of hydrogen bonds is energetically advantageous, polar molecules combine well with water and are therefore called *hydrophilic* molecules. On the other hand, non-polar groups cannot form hydrogen bonds.

Since non-polar molecules disturb the network of water molecules connected via hydrogen bonds, such molecules do not combine well with water and hence are called *hydrophobic*. To minimize the effects of disturbing the (energetically favored) network of water molecules, hydrophobic molecules are forced together by the water. This force is called the *hydrophobic force.*

Small Molecules

A cell is made up of a small number of elements, where C, N, O and H make up 99% of the mass. A *hydrocarbon* molecule consists of only carbon and hydrogen. An *acid* (resp. *base*) has the property that in water it donates (resp. accepts) a free hydrogen ion, or proton, H^+. For instance, hydrochloric acid, HCl, disassociates in an aqueous solution to form H^+ and Cl^-, hence yielding the hydronium ion H_3O^+. The base sodium hydroxide, NaOH, disassociates in water to form Na^+ and OH^-. Some specific simple groups such as the *hydroxyl group* –OH, the *carboxyl group* —COOH (which characterizes an organic acid), and the *amino group* —NH_2 commonly occur in organic molecules.

Another important class of C–O compounds are generated by an *ester* linkage. *Small molecules* are defined as organic molecules with up to 30 carbon atoms. A macromolecule, which is composed of a number of equal or similar smaller molecules (called *monomers),* is called a *polymer.* The most important polymers are DNA/RNA (composed of nucleotides) and proteins (composed of amino acids).

One distinguishes four groups of small molecules:

Sugar: A sugar molecule is a small carbohydrate, where a *carbohydrate* has the generic formula $C_x(H_2O)_y$. Larger carbohydrates are called *polysaccharides.*

Fatty acids: For example, components of lipid molecules, which make up the cell membrane.

Nucleotides: Building blocks for DNA and RNA. The nucleotide ATP, adenoside triphosphate, is used for the storage and transportation of energy in cells, while GTP, guanine triphosphate, plays a role as neurotransmitter in brain.

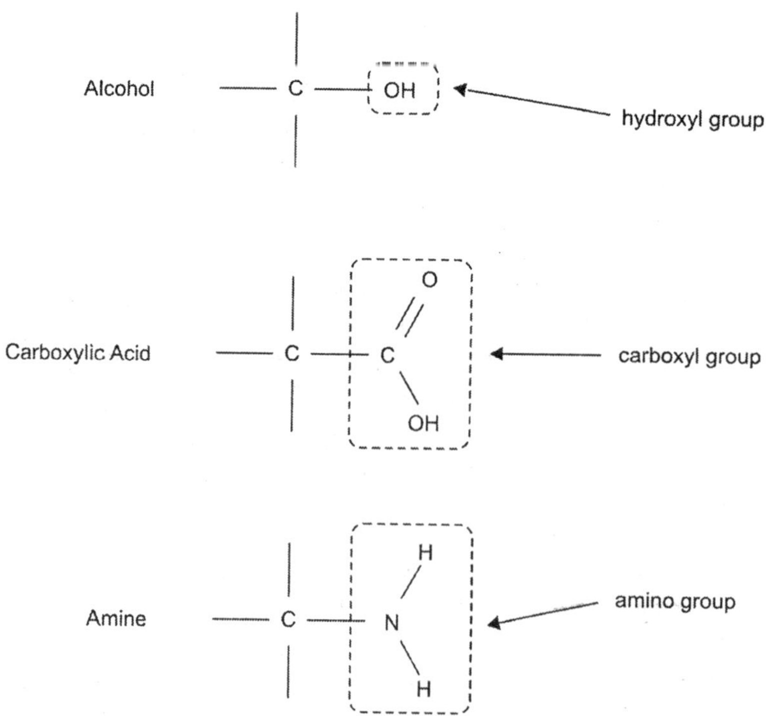

Fig. 1.2. Groups of C–O and C–N compounds. Common groups are the hydroxyl group, the carboxyl group, and the amino group.

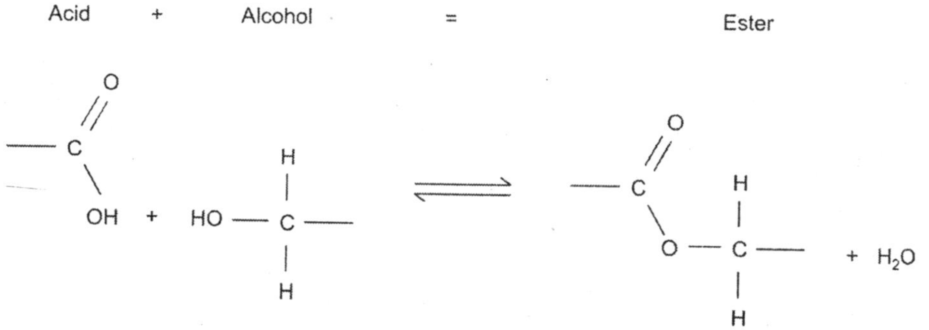

Acid + Alcohol = Ester

$$-C(=O)-OH + HO-CH_2- \rightleftharpoons -C(=O)-O-CH_2- + H_2O$$

Fig. 1.3. An ester is the result of a reaction of an alcohol with an acid (such as carboxylic acid). The sign ⇌ indicates that the reaction can proceed in both direction. Proceeding from left to ritht, an ester and a water molecule are products, while from right to left, a water molecule is broken down, a process called *hydrolysis.*

Amino acids: Proteins are composed of amino acids (where the amino acids are connected in a linear chain by a peptide bond).

Sugars

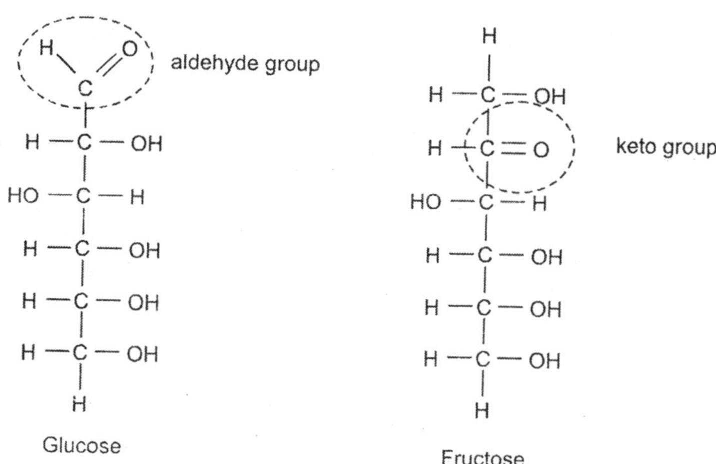

Fig. 1.4. Two common 6-carbon sugars.

Sugars can be distinguished with respect to their number of carbon atoms *and* with respect to their molecular structure. Important classes with respect to the number of carbons are *pentoses* (having 5 Cs) and *hexoses* (having 6 Cs). With respect to structure, there are two forms of sugar, namely *aldehyde sugars* (also called *aldoses*) and *ketone sugars* (or *ketoses*). Figure 1.4 shows two common hexoses. Since the keto or aldehyde group can react with a hydroxyl group, a sugar can occur in cyclic form. The carbons in a sugar are usually numbered, where one starts either with the aldehyde group, or with the end that is nearest to the keto group. Of particular significance for molecular biology is the 5-carbon sugar (or pentose) called *ribose,* which is a ketose. There are two different forms of ribose that occur, namely ordinary ribose, and *deoxyribose,* where one oxygen atom is missing.

Nucleic Acids

Nucleotides: Both DNA and RNA are polymers, which are composed of nucleotides (nucleotide polymers are also called *oligonucleotides*). A *nucleotide* is a molecule consisting of a base, a ribose sugar (in DNA, deoxyribose), and a phosphate molecule. The base is bound to the 1′ carbon, wherease the phosphate is bound to the 5′ carbon. A *nucleoside* consists of a base and a ribose sugar only. The *base* is a carbon ring molecule containing nitrogen atoms.

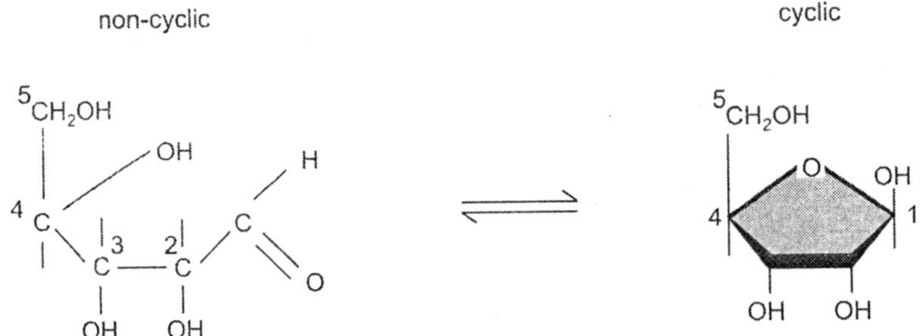

Fig. 1.5. The cyclic and non-cyclic form of ribose. We have suppressed most of the H-atoms attached directly to C.

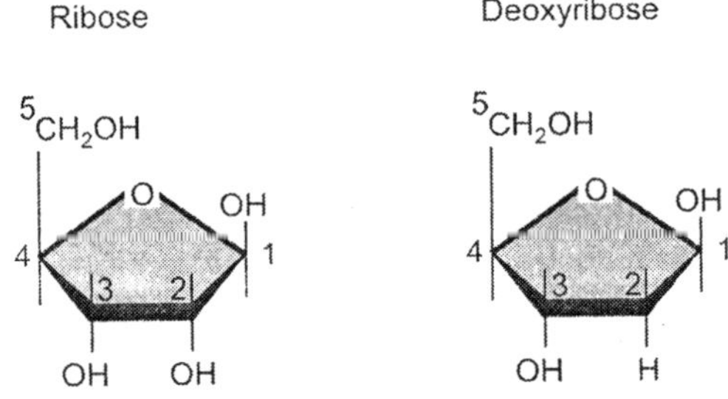

Fig. 1.6. Ribose and deoxyribose.

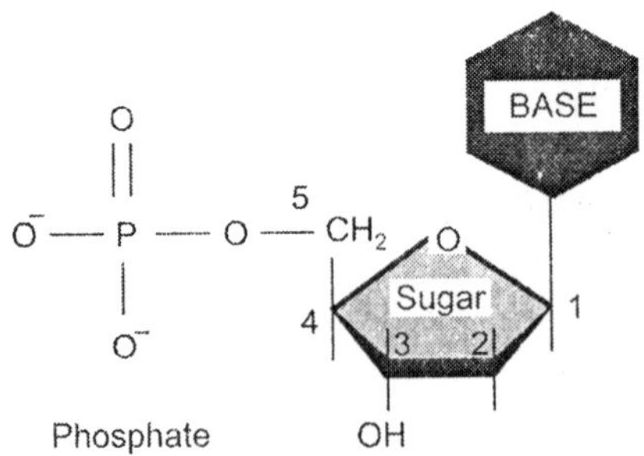

Fig. 1.7. General picture of a nucleotide.

In DNA, we have the four bases *adenine* (A), *cytosine* (C), *guanine* (G), and *thymine* (T). In RNA, the thymine is replaced by *uracil* (U). The *purines* adenine and guanine have a 2-ring structure, and the *pyramidines* cytosine, thymine and uracil have a 1-ring structure. Note that adenine and thymine have 2 hydrogen bond sites whereas cytosine and guanine have 3 hydrogen bond sites.

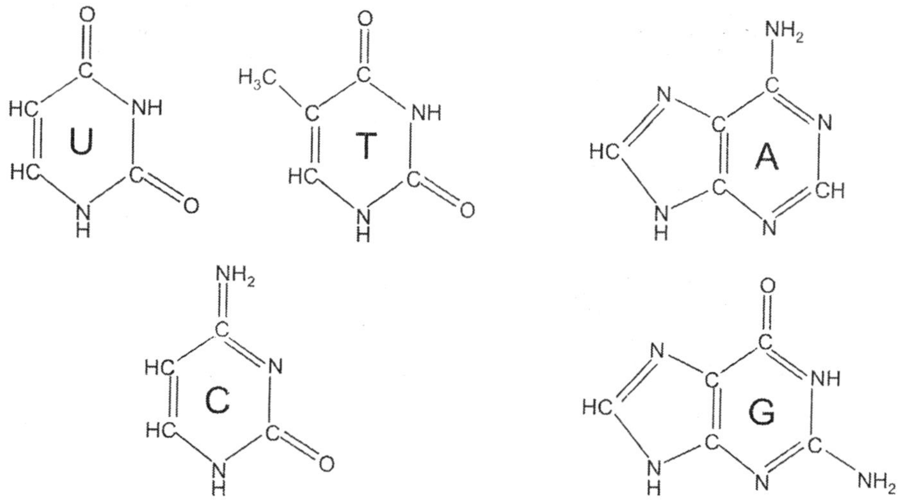

Fig. 1.8. Chemical forms of the bases.

Nucleotides occur in three different forms, namely as *monophosphate, diphosphate,* and *triphosphate* nucleotides, having 1, 2 and 3 phosphate groups attached, respectively. The additional phosphate groups in diphosphate and triphosphate nucleotides can be split off, yielding energy that can be used for another process. Thus, the diphosphate and triphosphate nucleotides are used to transport energy in the cell.

DNA

At room temperature, DNA exists as a double-stranded molecule, formed by hydrogen bonds between *complementary* bases: *A with T, and C with G,* the so-called *Watson–Crick rules.*[6] These rules explain an experimental observation made by Chargaff in 1951 (two years before publication of the Watson–Crick model of DNA): adenine and thymine (resp. cytosine and guanine) appear in equal quantities in DNA.[7] In double-stranded B-DNA, the A-T and G-C base pairs are stacked in a planar fashion (with slight tilt) with about 3.3 Å per base pair.

A single strand of DNA is generated by chaining together nucleotides via a *phosphodiester bond.* In this bond, the phosphate molecule of the first nucleotide is attached to the hydroxyl group at the 3′ carbon of the next nucleotide. The different ends of the DNA strand are numbered by the carbon atom position, where the next nucleotide can be attached. The 5′ *end* contains a phosphate group, attached to the 5′ carbon of the pentose sugar. The 3′ *end* is characterized by the fact that the 3′ position of the corresponding nucleotide is free for appending the next nucleotide.

Theoretically, there could be two directions for extending a single strand, namely by extending the 5′ end, or by extending the 3′ end. In nature, DNA is always, extended at the 3′ end, which implies that DNA grows in the 5′ → 3′ direction. The reason is that the energy for the extension of the DNA is given by the new nucleotide containing a triphosphate group. If the extension were in the 3′ → 5′ direction, then the strand would have to provide the energy for extension (by having a triphosphate attached to the 5′ end.) But then it would be difficult to remove the last extended nucleotide from a strand, since this would remove the group providing the energy for extension. Thus, the error correction mechanism would be more difficult (which is one of the possible reasons why nature uses the 5′ → 3′ direction for DNA).

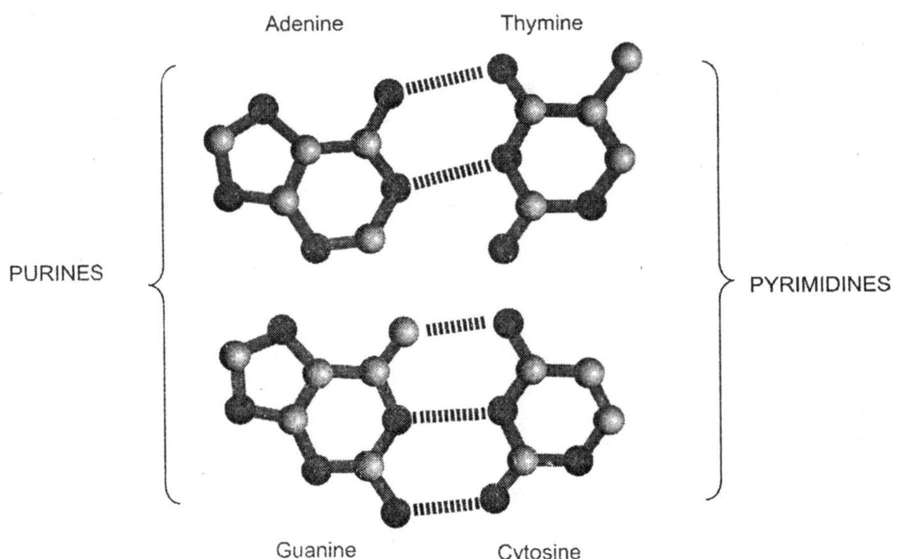

Fig. 1.9. The four DNA bases together with their hydrogen bonds. The relative location of the bases is shown as they occur in DNA.

Double-stranded DNA forms a *helix*, where one strand goes in the direction from 5′ to 3′, while the second goes from 3′ to 5′; thus the second strand is the *reverse complement* of the first strand. This mechanism allows one strand of DNA to serve as a template for producing the reverse complement strand, thus explaining how DNA can replicate. In many viruses and bacteria, DNA exists in a circular form. For instance, the genome of *M. jannaschii* consists of one large circular chromosome consisting of 1664976 base pairs (bp), together with a large circular *extrachromosomal element,* or ECE, consisting of 58407 bp, and one small circular ECE consisting of 16550 bp. Locally, DNA is often tightly wound around *histone* proteins in a structure called a *nucleosome,* thus causing a bunching-up or banding, visible when using certain color stains under a light microscope-hence the name *chromosome.*

Nucleosomes appear to be an essential structure, as manifested by the extreme conservation of the histone complex throughout organisms. On the order of 150-200 bp lie on the nucleosome, forming about 2 superhelices. Moreover, the DNA within nucleosomes is more tightly wound – roughly 10 Å per helix, as contrasted with 10.4 and perhelix in B-DNA. In higher organisms, DNA may appear in linear chromosome pairs – one chromosome from each parent.

Circular DNA (and even linear DNA, since the latter can be locally considered circular in the region of a nucleosome) has *linking, winding*, and *twist* numbers. Think of an unbuckled belt that you twist several times and then buckle. Twist is the number of times you twisted the belt before buckling it. Now make a figure-of-eight with the belt, and notice how the twist number is reduced by increasing the writhe. Formal mathematical definitions of linking number (L), twist (T), and writhe (W) can be given, for which the important topological invariance

$$L = T + W \tag{1.1}$$

holds a result due to J. White. In a later chapter, we shall present a method due to C. Benham *et al.,* which incorporates equation (1.1) to predict where double stranded DNA begins to separate in replication and transcription events. *In vivo*, DNA is usually negatively supercoiled, and it is known that certain enzymes, called *topoisomerases,* actually cut the DNA (topoisomerase I cuts a single strand while topoisomerase II cuts both strands).

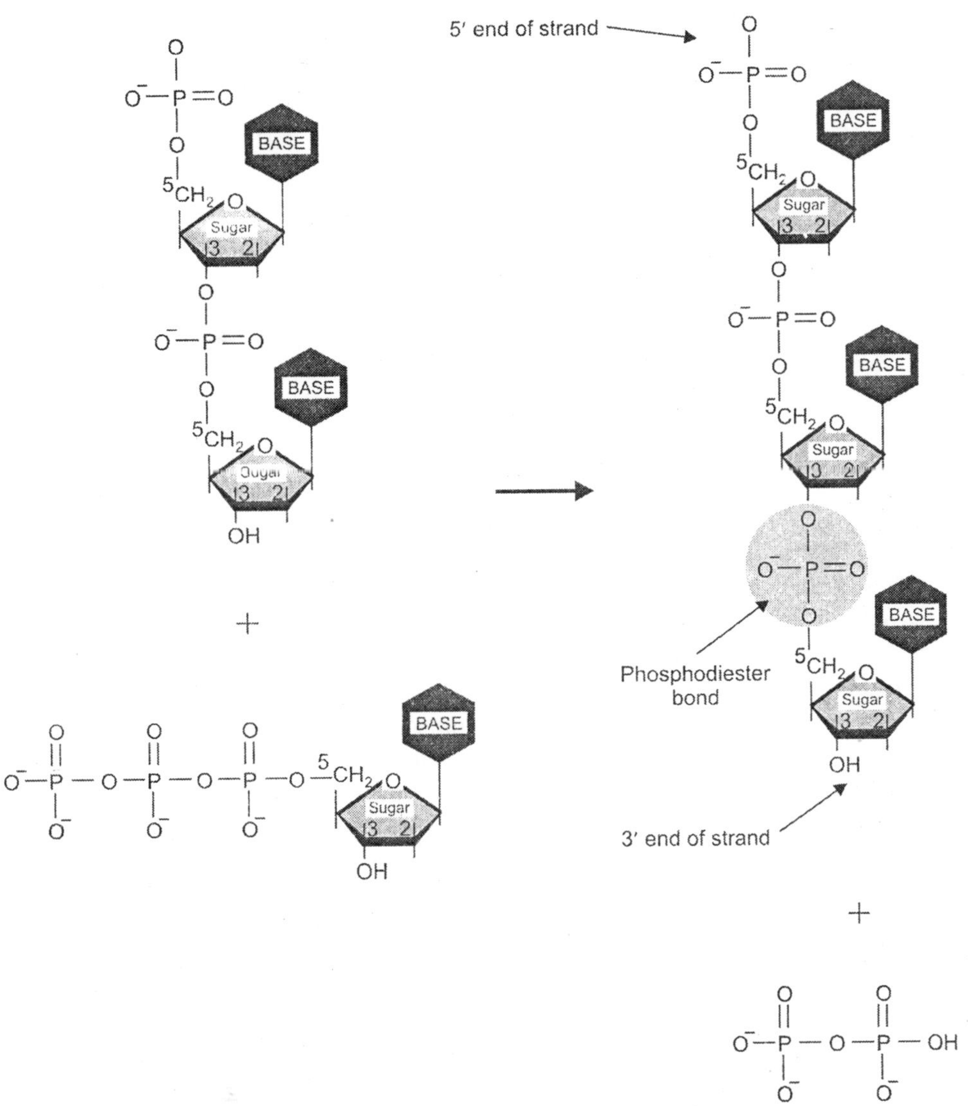

Fig. 1.10. Extension of a single DNA strand, where the new nucleotides has a triphosphate group attached. The gray circle shows the 3′ → 5′ phosphodiester bond connecting two nucleotides.

When determining the entire nucleotide sequence, or *genome*, of an organism (such as *M. jannaschii* above), laboratory uncertainties lead at times only to a partial determination of certain bases – for instance, that either A or C occurs at a certain position. The nucleotide code displayed in Table 1.1 includes a symbol for each subset of {A, C, G, T}. A segment of DNA that codes for a protein (as explained below) is called a *gene*. In *diploid* cells, where chromosomes donated from each parent are paired, the corresponding genes may be identical (*homozygous*) or non-identical (*heterozygous*). In the latter case, the corresponding, but slightly different genes are called *alleles*. Heterozygous alleles may be

Table 1.1. Nucleotide codes.

Code	*Meaning*	*Complement*
A	A	T
C	C	G
G	G	C
T	T	A
M	A or C	K
R	A or G	Y
W	A or T	W
S	C or G	S
Y	C or T	R
K	G or T	M
V	A or C or G	B
H	A or C or T	D
D	A or G or T	H
B	C or G or T	V
X/N	A or C or G or T	X
•	not A, C, G, T	•

dominant or *recessive.* Non-diploid germ cells are called *haploid.* These are formed by *meiosis,* where *crossing-over* between corresponding chromosomes happens randomly. In meiosis (germ cell formation), the cell's chromosomes are duplicated (with possible crossing-over), but the cell quadruples, leading to 4 haploid cells. This is contrasted with *mitosis,* where the cell's chromosomes are duplicated (without crossing-over) and the cell is then doubled. While the human genome consists of 23 pairs of chromosomes amounting to a total of roughly 3×10^9 base pairs, biologists estimate that there are only 10^5 genes.

From the genetic code, explained later, there are 3 bases that correspond to an amino acid, and most proteins consist of several hundred amino acids. Assuming that a protein consists of at most 1000 amino acids, a back-of-the-envelope calculation indicates that less than $3 \cdot 10^5 \cdot 10^3 \cdot = 3 \cdot 10^8$ bp of the genome correspond to genes, while $\frac{3\times10^9 - 3\times10^8}{3\times10^9} \approx 90\%$ of the genome is non-coding. Coding regions encode proteins. In general, the percentage of the genome consisting of coding regions *decreases* with increasing complexity of life.

For instance, in the *polyoma* virus genome, consisting of 5297 base pairs, coding regions comprise over 90% of the genome. Certain viruses even encode two distinct proteins within roughly the same genomic region by a shift in the reading frame.[9] Indeed, for lower organisms, the reproduction rate depends essentially only on the DNA replication time, which in turn depends on genome length (bacteria in ideal conditions can reproduce every 20-30 minutes). However, in eukaryotes, genomic length seems to have little to do with either the complexity of the organism or time between mitosis. The so-called TATA box is a consensus sequence consisting of TATA, or close relative thereof, which is part of the promotor sequence for a gene. The proximity of a TATA box upstream from the beginning of coding region is an indicator for a coding region, as well as a terminating hairpin structure and poly-A region at the end. A-T has two hydrogen bonds as compared with the three hydrogen bonds of G-C, so poly-A regions are less stable, and especially after a hairpin structure can lead to termination of transcription.

An *open reading frame* or ORF is a region thought to encode a protein, but for which the functionality is not currently known. Humans have 22 pairs of *autosomal,* or non-sex-related chromosomes, together with either an X,Y pair (in males) or a X,X pair (in females). DNA is a very thin (width about 10 Å) and long molecule (human DNA, if concatenated from all chromosomes and stretched out, would measure about 2 meters). Generally, more complex organisms have longer DNA, though: for instance, the lungfish genome is almost 50 times as large as the human genome, and the *Amoeba dubia* genome is over 600 Mbp. Some sample genome sizes:

(*i*) *Polyoma virus:* one circular chromosome of size 5297 bp.

(*ii*) *λ-phage virus:* one circular chromosome of size 48502 bp.

(*iii*) *E. coli:* one circular chromosome of size 4.6×10^6 bp.

(*iv*) *S. cerevisiae* (yeast): 16 linear chromosomes of total size 13×10^6 bp.

(*v*) *Homo sapiens:* 23 pairs of linear chromosomes of total size 3×10^9 bp.

RNA

Ribonucleic acid (RNA) consists of the bases adenine, cytosine, guanine, and uracil (in place of thymine) attached in a similar manner to the pentose sugar ribose. Ribose has an extra *hydroxyl* (OH) group at the 2′ position, and so is able to form more hydrogen bonds than DNA. In deoxyribose, this oxygen atom is missing, hence the name deoxyribose. Because of the ability to easily form hydrogen bonds, RNA manifests a number of catalytic properties, including self-splicing activity in certain cases." It is generally believed that an RNA world first existed, where RNA played both an information storage and an enzymatic role.

It is known that cytosine spontaneously deaminates to uracil, and since there is no known RNA repair mechanism (unlike the various repair mechanisms of DNA, such as SOS, etc.), RNA cannot be reliably used as a carrier of genetic information. In replicating DNA, uracil is *always* removed. Moreover, it is speculated that this spontaneous deamination is the mechanism for one form of RNA editing, to be explained later. Thus it is believed that, once it evolved, the more stable DNA gradually replaced RNA in its information storage role, while enzymatic functions of RNA were gradually replaced by more efficient proteins. There are certain enzymatic functions still carried out by RNA. It is thought that such instances are *molecular fossils.*

In contrast to DNA, RNA is single-stranded, though it can form Watson-Crick hydrogen bond pairs (A-T, C-G) and even the weaker pairs A-G with itself, thus forming hairpin loops and more complicated structures.

Though DNA serves only the function of information storage, RNA serves certain catalytic functions through its complex 3-dimensional forms. Before describing transcription and translation, here is an overview of the different kinds of RNA (undefined concepts are explained in the next section):

1. *mRNA:* messenger RNA, transcribed from DNA where introns have been spliced out and pointwise editing performed.

2. *tRNA:* transfer RNA generally consisting of 70-80 bases having a cloverleaf secondary struction, and an L-shaped tertiary structure. The anticodon occurs at the end of a hairpin loop. The corresponding amino acid is bonded to the tRNA by a reaction involving aminoacyl-tRNA synthetase.
3. *rRNA:* ribosomal RNA, main component of the ribosomes. Mammals have on the order of 10^6 ribosomes per cell. Sometimes rRNA is described using Svedberg units (denoted by S), which are sedimentation rates in an ultracentrifuge; for instance, 16 S rRNA.
4. *snRNA:* small nuclear RNA, a molecular fossil which assists in splicing nuclear mRNA.
5. *gRNA:* guide RNA, used to control the editing of mRNA (pointwise nucleotide insertions or deletions in RNA).

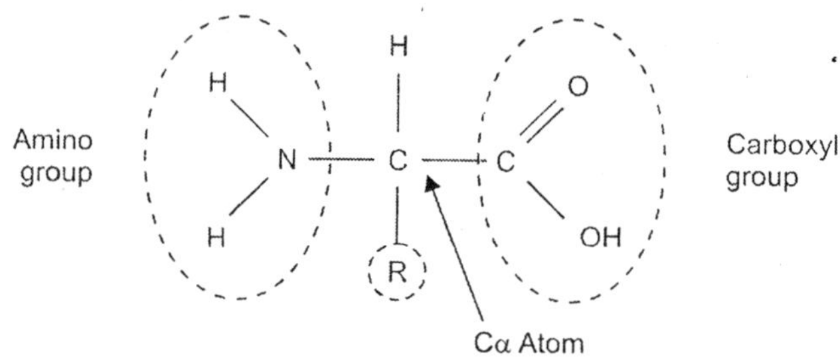

Fig. 1.11. General form of an amino acid.

PROTEINS

Amino Acids: Proteins are sequences composed of an alphabet of 20 amino acids. An amino acid is a chemical group of the form, where R is a chemical group (called *chain residue)* specifying the type of the amino acid. The central carbon atom is the ∝ *carbon* (Cα), the left NH_2 group is the *amino group,* and the left COOH the *carboxyl group.* There are 20 different chain residues, which have different chemical properties. Residues can be hydrophobic or hydrophilic, small or large, charged or uncharged.

Two amino acids can be connected via a *peptide bond,* where the carboxyl group of the first amino acid reacts with the amino group of the second. The result is a group of the form using the peptide bond, long linear chains of amino acids (*i.e.,* proteins) can be generated The ends are characterized by the fact that they have an amino group (resp. carboxyl group) that is not part of the peptide bond. Thus, one speaks of the *N terminus* (resp. *C terminus)* of the amino acid chain.

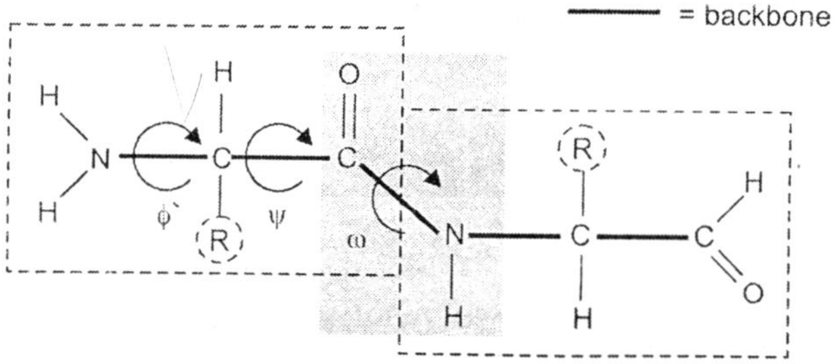

= backbone

...(1.2)

The peptide bond itself (indicated with a gray rectangle in (1.2)) is usually planar, which means that there is no free rotation around this bond. There is more flexibility for rotation around the N-Cα-bond (called the ϕ-angle), and around the C∝-C bond (called the ψ-angle).

Nevertheless, the allowed values of combinations of ϕ and ψ angles are restricted to small regions in natural proteins (which are displayed on so-called Ramachandran plots, Using this freedom of rotation, the protein can *fold* into a specific three-dimensional structure (called conformation). In natural proteins, the final structure that is achieved is uniquely determined by the sequence of amino acids. For this reason, one speaks of the *native structure* of a given amino acid sequence. The native structure of a protein is believed to be the conformation that is a free energy global minimum. Terms in the energy function include van der Waals attraction, electrostatic (Coulomb) forces, hydrogen bonds between different residues (*e.g.* between H bound to N and the O in polypeptides; such bonds are responsible for the secondary structures α helices and β strands), salt bridges (ionic attractions), disulfide bridges (disulfide bond between two cysteines), and hydrogen bonds with the solvent water.

PROTEIN STRUCTURE

The function of a protein is determined by its 3-dimensional structure – for instance, antibodies in the human immune system recognize antigens by having a complementary surface to that of the antigen *(key and lock* paradigm). Enzymes are proteins, which, by forming a substate to lower the reaction energy, facilitate or accelerate chemical reactions, which would otherwise might never take place. Because of the importance of the immune system, immunoglobulins have been widely studied, and hence form a large portion of current protein databases.

Proteins may roughly be divided into three classes: *globular (enzymes), fibril* (collagen, elastin), and *membrane* proteins. Transmembrane proteins, responsible for pores/channels in cell membranes, have in part been much less studied because of technical difficulties in their X-ray crystallography. The Brookhaven Protein Database (PDB) contains the 3-dimensional coordinates of 10310 proteins and 788 nucleic acids, as determined by X-ray crystallography or nuclear magnetic resonance (NMR). See Appendix B.2 for a more detailed discription of an example PDB file.

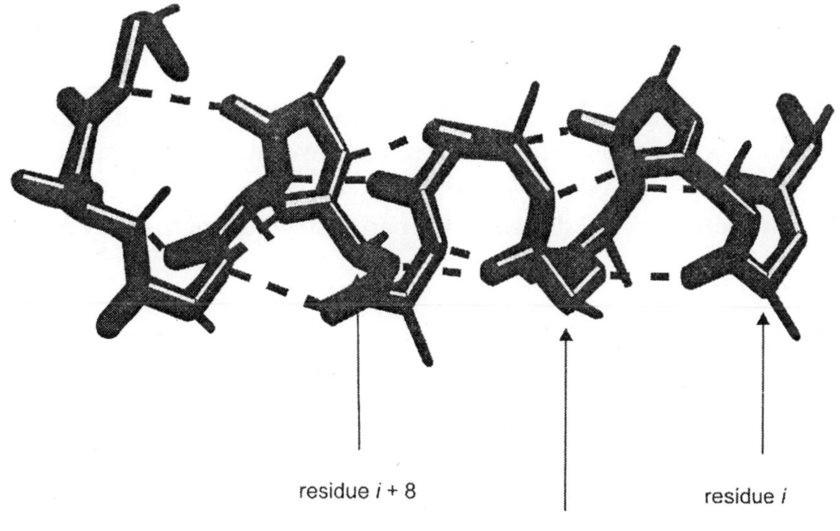

Fig. 1.13. An ∝ helix. The stabilizing hydrogen bonds are shown with dashed lines.

Concerning the 3-dimensional structure of proteins, there are some architectural features that are commonly used in natural proteins. These structural pieces are called *secondary structure elements.* The two main motifs are α helices and β sheets. An α helix is generated by stacking amino acids in a helix, forming a fixed cylinder. See Fig. 1.13 Note that residues 1, 5, 9,... appear aligned, with residue n+4 stacked above residue n in the enclosing cylinder. A β *sheet* is a structure where parts of the amino acid chain are stacked onto each other in a linear manner). Since a protein has a direction (from the N terminus to the C terminus), one distinguishes *parallel β sheets* and *antiparallel β sheets.* In the parallel form, the folded parts have the same direction, whereas in the antiparallel form, they have alternating directions.

Certain combinations of α helices and β sheets have been identified as *motifs* (sometimes called *supersecondary structures)* common to several proteins. An example is the helix-turn-helix (H-T-H) motif, often signaling a DNA-binding site.

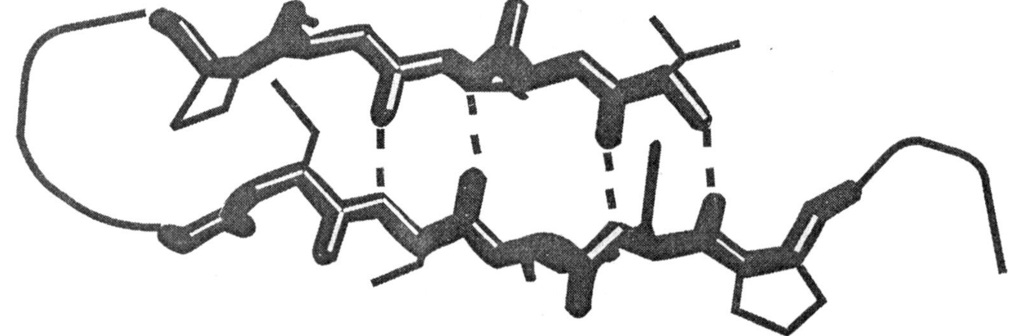

Fig. 1.14. An antiparallel β sheet. The stabilizing hydrogen bonds are shown with dashed lines.

The structure of a protein is described on different levels:

Primary Structure : the amino acid sequence of the protein.

Secondary Structure : describes the regions in the primary structure where secondary structure elements (such that α helix, β sheets, etc.) occur.

Tertiary Structure : the 3-dimensional structure of a protein domain in the native structure. If a protein is dimeric (*i.e.,* is composed of several subproteins), then the tertiary structure describes the structure of a subprotein.

Quarternary Structure : the 3-dimensional, native structure of the fully functional protein.

From DNA to Proteins

Amino Acids and Proteins : Amino acids have numerous different chemical properties, including molecular size, electric charge, etc. It is held that one of the most important driving forces in protein folding is the *hydrophobic force,* or tendency for hydrophobic residues to avoid contact with water molecules and hence form a compact inner core in the protein. *Salt bridges* (attractions between oppositely charged residues) and *disulfide bonds* between the sulfur atoms of cysteine residues also play a role in the conformation determination. The 20 amino acids found in proteins along with their 1-letter and 3-letter codes, together with a hydrophobicity value technically called *polar requirement* and the designation 'H' (hydrophobic) or 'P' (polar, *i.e.,* hydrophilic).

The genetic code is now known to be a triplet, non-overlapping, *comma-free*[16] code, where successive *codons* consisting of 3 successive RNA nucleotides encode one of the 20 amino acids or the signal to stop translation. For instance, the oligonucleotide codes for the amino acid

1	2	3	4	5	6	7	8	9	10	11	12
G	U	U	U	U	A	A	G	U	C	C	U

Table 1.2. Amino acid codes.

Amino Acid	*Code (1 Ch)*	*Code (3 Ch)*	*Polar Req.*	*H/P*
Alanine	A	Ala	7.0	H
Arginine	R	Arg	9.1	P
Asparagine	N	Asn	10.0	P
Aspartic acid	D	Asp	13.0	P
Asparagine or aspartic acid	B	Asx		P
Cysteine	C	Cys	4.8	P
Glutamine	Q	Gln	8.6	P
Glutamic acid	E	Glu	12.5	P
Glutamine or glutamic acid	Z	Glx		P
Glycine	G	Gly	7.9	P
Histidine	H	His	8.4	P
Isoleucine	I	Ile	4.9	H
Leucine	L	Leu	4.9	H
Lysine	K	Lys	10.1	P
Methionine	M	Met	5.3	H
Phenylalanine	F	Phe	5.0	H
Proline	P	Pro	6.6	H
Serine	S	Ser	7.5	P
Threonine	T	Thr	6.6	P
Tryptophan	W	Trp	5.2	H
Tyrosine	Y	Tyr	5.4	P
Valine	V	Val	5.6	H

sequence Val `Leu Ser Pro` if the *reading frame* begins at position 1, while it codes for `Phe Stop` if the reading frame begins at position 2, and `Phe Lys Ser` if the reading frame begins at position 3. To complicate matters, in double-stranded DNA there are three additional reading frames possible for the reverse complement strand. In prokaryotes and eukaryotes, the reading frames of different proteins do not overlap. This is different in some viruses, where the need to compactify the genetic information may lead to a situation where different reading frames for viral proteins overlap greatly. There is much interesting speculation about the origin of the genetic code.

It is thought that in the protobiotic RNA world, different genetic codes may have existed, in which case pressure from natural selection would have optimized both the codes and translation machinery, until, as F. Crick has suggested, *a frozen accident* occurred with the emergence of life, meaning that further random mutations would most likely be detrimental to living organisms. Two rather different views for the emergence of the genetic code have been proposed. In the late 1960s, Crick suggested that few amino acids were initially encoded, that with time, more amino acids were encoded using similar codes to those that then existed, and that tRNA and aminoacyl-tRNA synthetases were derived from those already existent.

In 1973, C. Woese *el al.,* proposed that stereochemical affinities between codons and respective amino acids exist, that early translation mechanisms were imprecise, and that ancestors of tRNA were capable of recognizing only classes of codons, rather than unique codons. In particular, Woese *et al.,* introduced a measure of hydrophobicity of amino acids, and noted that amino acids having *A* (resp. *U*) as middle codon are hydrophilic, (resp. hydrophilic) while the nucleotide adenine (resp. uracil) is chemically hydrophobic (resp. hydrophobic). As later explicitly stated in middle anti-codon positions with *A* (resp. *U*) are thus associated with hydrophobic (resp. hydrophilic) amino acids. It has also been suggested by M. Eigen and P. Schuster that the original code was of the form *RNY,* meaning purine, followed by any nucleotide, followed by pyrimidine.

Table 1.3. Genetic code.

	U	*C*	*A*	*G*	
	Phe	Ser	Tyr	Cys	U
	Phe	Ser	Tyr	Cys	C
U	Leu	Ser	Stop	Stop	A
	Leu	Ser	Stop	Trp	G
	Leu	Pro	His	Arg	U
	Leu	Pro	His	Arg	C
C	Leu	Pro	Gln	Arg	A
	Leu	Pro	Gln	Arg	G
	Ile	Thr	Asn	Ser	U
	Ile	Thr	Asn	Ser	C
A	Ile	Thr	Lys	Arg	A
	Met	Thr	Lys	Arg	G
	Val	Ala	Asp	Gly	U
	Val	Ala	Asp	Gly	C
G	Val	Ala	Glu	Gly	A
	Val	Ala	Glu	Gly	G

TRANSCRIPTION AND TRANSLATION

A segment of a chromosome, corresponding to a *coding region* (a region that gives rise to a protein), is often preceded by a *promotor* sequence, usually consisting of a rough consensus pattern of TATAA.... (called a TATA box) as well as other sequence signals. RNA polymerase is capable of recognizing the promotor sequence, and after DNA strand separation in the coding region occurs, RNA polymerase assists in the *transcription* of DNA into a pre-edited form of messenger RNA called the RNA *transcript.*

The transcription occurs in the 5′ to 3′ direction, where the DNA nucleotides A, C, G, T are respectively transcribed into RNA nucleotides U, G, C, A. There are different models for the transcription and translations in prokaryotes (*i.e.,* bacteria), and eukaryotes (non-bacteria; *e.g.* unicellular paramecia, and all plants and animals). *Prokaryotes* are cells that have no compartment to separate genomic DNA from the cytoplasm and no membraneous organelles, and have one circular chromosome.

The reading frame for the transcription of a proteins is continuous, *i.e.,* not interupted. *Eukaryotes* on the other hand, do have a compartment, called a *nucleus*, to separate DNA from the cytoplasm, have membraneous organelles (*e.g.* mitochondria), and generally have more than one linear chromosome. The reading frame for a protein is not continuous, but rather is disrupted by *introns* (or *interveaning sequences*). Phillip A. Sharp and Richard J. Roberts discovered introns in 1977, for which they later later received the Nobel Prize. The remaining segments between the introns are called *exons*. Either by self-catalysis, or facilitated by enzymes, the introns in the RNA transcript are excised and removed. In proteins, a similar splicing occurs, where the removed segments are called *inteins* and the remaining segments *exteins*. Moreover, it is an important algorithmic problem in bioinformatics to determine intron/ exon splice sites. After removal of introns, it may occur that nucleotides are inserted or deleted (pointwise, non-consecutive insertions and deletions). This phenomenon is called *RNA editing*, and, at least in the case of kinetoplastid protozoa, is directed by *guide RNA* (gRNA).

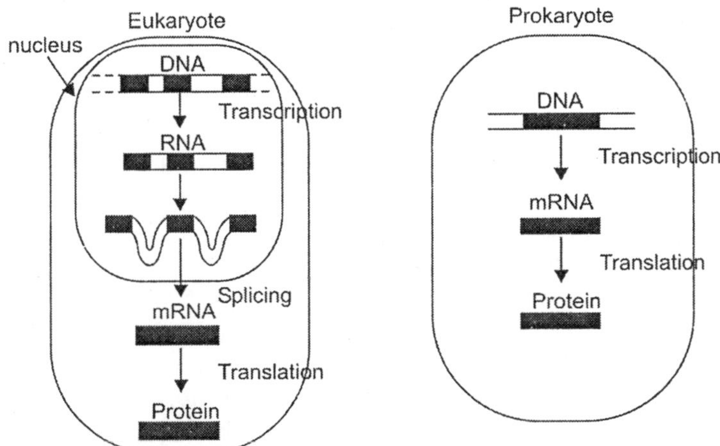

Fig. 1.15. The different models of transcrption and translation in prokaryotes and eukaryotes.

That editing occurs in most living organisms is truly remarkable, since the protein coded by mRNA after editing may have nothing to do with the protein that would have been coded by mRNA before editing occurred.

This is because the genetic code is a triplet, non-overlapping code, and an insertion of one nucleotide effectively shifts the original reading frame, producing codes for possibly completely unrelated amino acids.

Transfer RNA (tRNA), whose existence was first postulated by F. Crick in his *adapter molecule* hypothesis, is a small (around 70 bp) RNA, having a cloverleaf secondary structure and an L-shaped tertiary structure. At the extremity of the middle lobe of the cloverleaf lies an *anticodon, i.e.*, three nucleotides, whose Watson-Crick reverse complement is a triplet codon for an amino acid, as given in table 1.3 *Aminoacyl-tRNA synthetase* assists in the attachment of the *correct* amino acid to tRNA, *i.e.,* the amino acid whose codon is the reverse complement of the anticodon in the tRNA. Despite the fact that, with the exception of mitochondria and certain bacteria, the genetic code is essentially universal, different organisms may prefer the usage of certain codons for particular amino acids. In particular, an organism may have no tRNA for particular codon, or may produce no aminoacyl-tRNA synthetase for that codon. For instance, leucine aminoacyl-tRNA synthetase was not found in *M. jannaschii. Ribosomes* are cytoplasmic complexes constituted from ribosomal RNA (rRNA) together with certain proteins, and form a surface for mRNA (after intron removal and possible editing) and tRNA to assemble amino acids together in a linear chain, thus forming a protein. This process is called *translation.* Eukaryotic translation usually begins with the codon for the amino acid *methionine*, which serves as a start codon.

Methionine may later be removed from a functional protein. It is known that certain proteins, called *chaperones,* are found in the vicinity of ribosomes, and are responsible for refolding incorrectly folded proteins, or ensuring their destruction, when correct refolding does not occur. Chaperones recognize incorrectly folded proteins, because of hydrophobic residues on the protein's surface (rather than in the core). At least for one particularly well-studied class, chaperones are known to have a cylindrical 3-dimensional structure with hydrophobic-rich interior walls. When a protein with a hydrophobic surface lies within the chaperone's walls, the chaperone undergoes a conformational change causing the inner walls to become hydrophilic and the 'lid' of the cylinder to clap shut, thus providing an inert 'test tube' medium, allowing the protein to refold correctly. DNA replication occurs in a similar fashion as RNA transcription, but facilitated with DNA polymerase, as opposed to RNA polymerase.

After DNA strand separation begins to occur (*i.e.* as the 'zipper' proceeds to unzip), the adjoining of complementary nucleotides to the template strand always occurs in the 5′ to 3′ direction, because of the high-energy phosphate group at the 5′ end of the nucleotide. This occurs without major difficulty along one template strand, but is more complicated along the other template strand, since the 5′ direction lies at the border between the 'zipped' and 'unzipped' portion of the partially separated double strands. The replication error rate along this latter strand is significantly higher. Various correction mechanisms for DNA replication, including the SOS system, reduce the final replication error rate to less than 1 in 10^9. The above description of DNA replication holds for *mitosis* (cell division) as well as *meiosis* (germ cell formation). In mitosis, though DNA can in principle be replicated forever (eternal life), the accumulation of replication errors is thought to lead to dysfunctional cells, cancerous cells, etc. In meiosis, in addition to *pointwise mutation* caused by DNA replication errors as just explained, another source of mutation, essential for evolution, consists of *inversions* and *transpositions* as well as *crossover.*

Finding Genes

2

The more comprehensive and accurate initial computational analysis performed for new genomic sequences, the less time-consuming and costly experimental work will have to be done to determine their functions. For this reason, computational gene identification is an issue of obvious importance as a tool of identifying biologically relevant features (protein coding sequences) that often cannot be found by the traditional sequence database searching technique. This chapter describes statistically based methods for the recognition of eukaryotic genes. We review the structure and significant characteristics of gene components, and discuss recent advances and open problems in gene-finding methodology and its application to sequence annotation of long genomic sequences. Finally, we consider the application of gene expression data for large-scale verification of predicted genes.

GENERAL FEATURES OF EUKARYOTIC GENES

Genes carry and express the hereditary information encoded by segments of nucleic sequence involved in producing protein or RNA molecules. Genetic organization and packaging of eukaryotic genes is fundamentally different from those of prokaryotes. The major differences are large proportion of noncoding DNA (regulatory sequences, introns, repeats, pseudogenes) and the existence of interruptions (introns) that separate different parts of protein coding region in DNA. A typical DNA fragment of protein coding gene includes noncoding regulatory sequences, exons, and introns.

Gene Expression Steps

The gene is expressed by a several stage process comprising transcription and translation. (Figure 2.1) Transcription (or pre-mRNA synthesis on DNA template) involves initiation, elongation, and termination steps. RNA polymerase catalyzing RNA synthesis binds a special region (promoter) at the start of the gene and moves along the template, synthesizing RNA, until it reaches a terminator sequence. Post-transcriptional processing of messenger RNA precursors includes capping, 3'-polyadenilation, and splicing. The processing events of mRNA capping and poly A addition take place before pre-mRNA splicing and result in producing the mature mRNA. The mRNA consist of sequences (called exons) that encode the protein product (according to the rules of the genetic code). The gene sequence often includes noncoding regions, called introns, that are removed from the primary transcript during RNA splicing.

Eukaryotic pre-mRNA is processed in the nucleus and then transported to the cytoplasm for translation. The sequence of mRNA contains a series of triplet codons that interact with the anticodons of aminoacyl-tRNAs (carrying the amino acids) so that the corresponding series of amino acids is incorporated into a polypeptide chain. The small subunit of eukaryotic ribosome binds to the 5'-end of mRNA and then migrates to the special sequence on mRNA (preceding to the start codon) called the ribosome binding site, where it is joined by a large ribosome subunit, forming a complete ribosome. The ribosome initiates protein synthesis at the start codon (AUG in eukaryotes) and moves along the mRNA synthesizing polypeptide chain until it reaches a stop codon sequence (TAA, TGA, or TAG), where release of polypeptide and dissociation the ribosome from the mRNA take place. After that, many proteins undergo post-translational processing (*i.e.*, covalent modifications such as proteolytic cleavage, attachment of carbohydrates and phosphates) before they become functional.

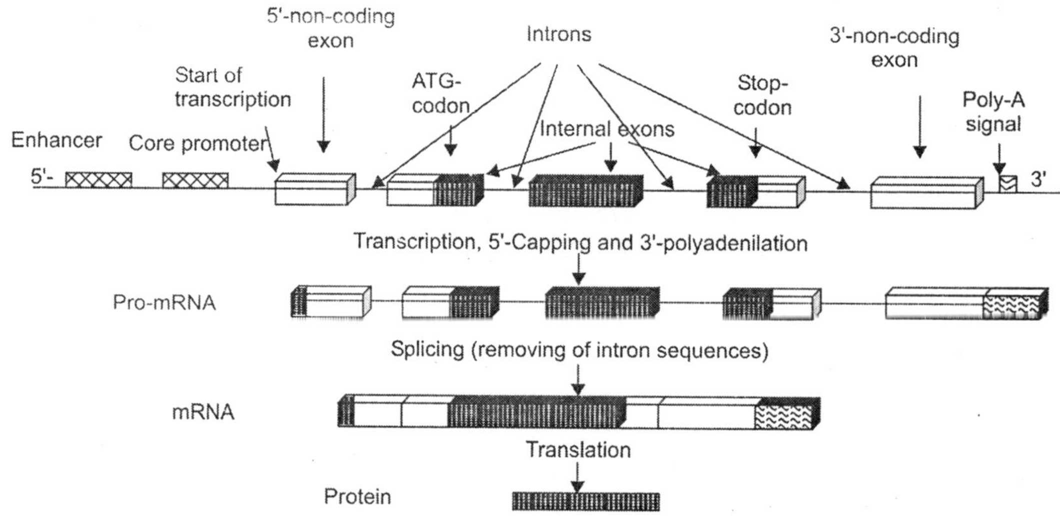

Fig. 2.1. Expression stages and structural organization of typical eukaryotic protein-coding gene including associated regulatory regions.

Structural Characteristics

Information about gene structure is accumulated in GenBank and EMBL nucleotide sequence databases. These databases contain annotations of contiguous sequences; therefore, one gene can be described in dozens of entries with partially sequenced gene regions, alternative splicing forms, or mRNA. The gene-centric database InfoGene contains descriptions of known genes and their basic functional signals extracted from GenBank. InfoGene also includes all predicted genes for human and *Drosophila* draft genomes and several chromosomes of the *Arabidopsis* genome. InfoGene is realized under JAVA interactive environment system that provides visual analysis of known information about complex gene structure and search for different gene component and signals.

Table 2.1. Structural characteristics of human genes deposited in Genbank (Release 119)

Gene features	*Numbers from the Infogen database*
CDS/partially sequenced CDS	53435/29404
Exons/partial sequenced exons	83488/21342
Genes/partial sequenced genes	20791/16141
Alternative splicing	2167, 10.4%
Pseudogenes	8.5%
Genes without introns	1552, 7.4%
Number of exons (maximal, average)	117, 5.7
Exon length (range, average)	1 – 1088, 201.6
Intron length (maximal, average)	259776, 2203.5
Gene length (maximal, average)	401910, 9033
Repeats in genome	41% of total DNA
DNA occupied by coding exons	3%
Donor sites	58707, 98.0%
Acceptor sites	58112,98.53%

Major organisms are presented in the InfoGene separate divisions. The human division (based on GenBank 119 release) contains about 21,000 genes, 53,000 coding regions, 83,000 exons, and about 58,000 donor and acceptor splice sites. Table 9.1 shows the major structural characteristics of human genes. About 41 percent of sequenced human DNA consists of different kinds of repeats. Only about 3 percent of the genome sequence contains protein coding exon sequences. Table 2.2 presents the characteristics of genes in major model organisms such as mouse, *D. melanogaster, C. elegans, S. cerevisiae,* and *Arabidopsis.* The gene sizes are often larger in vertebrates, and especially in primates. The average size of an exon is about 190 bp, which is close to the DNA length associated with the nucleosome particle.

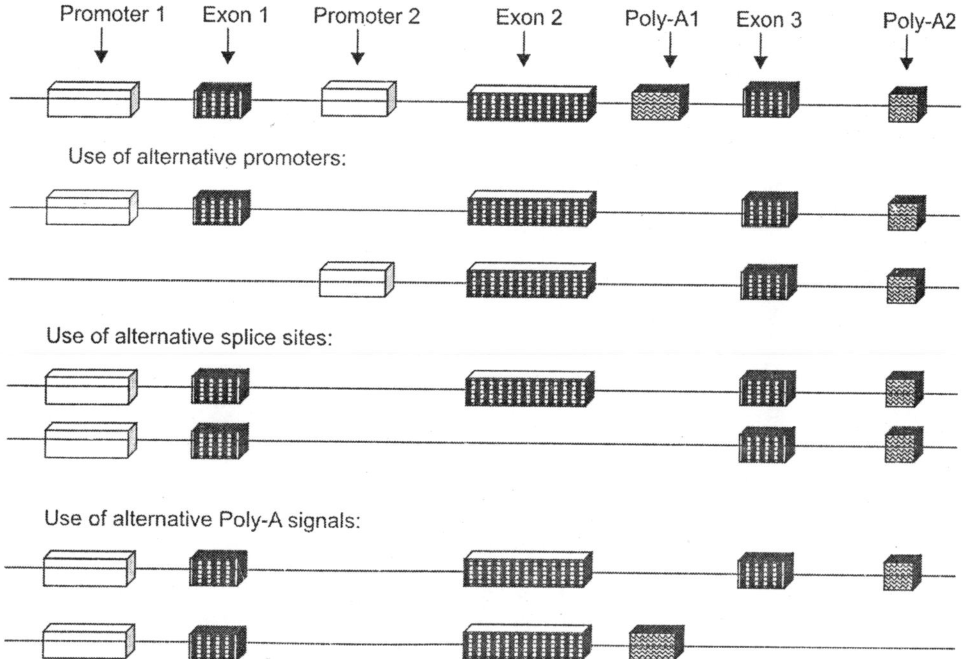

Fig. 2.2. DNA region coding alternative gene products.

Table 2.2. Structural characteristics of genes in eukaryotic model organisms.

	Mus musculus	*Drosophila melanogaster*	*C. elegans*	*S. cerevisiae*	*Arabidopsis thaliana*
CDS/partial	24527/13060	20314/1510 (20622)	20634/526	12635/1016	31194/461
Exons/partial	24508/7913	66960/19343 (75935)	122951/38293	13572/13127	145942/42844
Genes/partial	7428/5573	17435/1154 (20622)	19658/1263	12513/1098	28346/1023
Alt. splicing	749, 10%	1785, 10%	1194, 6.1%	598, 4.7%	227, 0.1%
No introns genes	748, 10%	3583, 20%	669, 3%	11070, 88.5%	5776, 19.7%
Number of exons	64, 4.72	50, 3.88	52, 6.1	3, 1.03	78, 5.1
Exons length	1–6642, 207.1	6–10785, 419.5	1–14975, 221.5	1–7471, 1500.0	3–75916, 192.0
Intron length	42573, 818.3	205244, 613.7	19397, 244.0	7317, 300	118637, 174.4
Gene size	150523, 3963	155515, 2854	45315, 2624	14733, 1462	170191, 2027
Donor sites	6225, 96.9%	49592, 98.0%	102872, 99.5%	471, 93.0%	117658, 99.2%
Acceptor sites	5627, 97.5%	49602, 97.9%	102933, 99.7%	475, 95.6%	121917, 96.9%

Human exon sizes are significantly smaller than the gene sizes. There are many exons as short as several bases. Computational identification of small exons (1–20 bp) cannot be done using the composition based methods that were successful for predicting prokaryote coding regions. Eukaryotic gene prediction approaches should be based on recognition of functional signals encoded in the DNA sequence. Figure 2.2 illustrates how the same DNA sequences may code several different proteins due to alternative promoters or terminators and alternative splicing. These processes also can significantly complicate computational gene finding.

FUNCTIONAL SIGNALS DESCRIPTION AND IDENTIFICATION

In this section, we focus on several approaches for gene functional signal recognition and some features of these signals used in gene identification. We describe the application of different weight matrices, which usually contain more information about the structure of functional signal than the corresponding consensus sequences, and later elaborate their implementation gene prediction approaches to score potential functional signals.

Position Specific Discrimination

The consensus sequence consists of the most common base at each position of an alignment of binding sites of a particular type. Often it uses a special letters (IUPAC) to indicate the potential presence of more than one nucleotide at a given position. Position weight matrices usually provide better representation of functional signals including quantitative information. We can consider the weight matrix as a simple model based on a set of position-specific probability distributions $\{p_s^i\}$, that give the probability of observing a particular type of nucleotide in a particular position of a functional signal (S) sequence. The probability of generating the signal sequence sequence $X(x_i,\ldots, x_k)$ under this model is

$$p(X/S) = \sum_{i=1}^{k} p_{x_i}^{i} \qquad \ldots(2.1)$$

where nucleotides of the signal are generated independently. The corresponding model can be constructed for nonsite (N) sequences, $\{\pi_s^i\}$, with the same probability distribution in each position. A discriminative function based on these models is the log likelihood ratio:

$$\text{LLR}(X) = \log P(X/S)/P(X/N) \qquad \ldots(2.2)$$

It can be written in weight matrix notation, $w(i, s) = \{\log(p_s^i/\pi_s^i)\}$, and

$$\text{Score} = \text{LLR(X)} = \frac{1}{k}\sum_{i=1}^{k} w(i, x_i) \qquad \ldots(2.3)$$

The other types of weight functions can be used to score the sequence of signal. For example, weights can be generated by some optimization procedures such as perceptron or neural network; different position-specific probability distributions $\{p_s^i\}$ might also be considered.

More general types of weight matrix uses position-specific probability distributions $\{p_s^i\}$ of oligonucleotides (instead of nucleotides). Oligonucleotide frequencies are successfully used in Markov chain models, where the probability to generate a particular nucleotide x, of the signal sequence depends on $k_0 - 1$ previous bases (*i.e.*, depends on oligonucleotide [$k_0 - 1$ base long] ended at the position $i - 1$). Then the probability of generating the signal sequence X is :

$$P(X/S) = p_0 \prod_{i=k_0}^{k} P_{s_{i-1},x_i}^{i-1,i} \qquad \ldots(2.4)$$

where $p^{i-1,i}_{s_{i-1},x_i}$ is the conditional probability of generating nucleotide x_i in position i given that oligonucleotide s_{i-1} ends at position $i - 1$; p_0 is the probability of generating oligonucleotide $x_1 \ldots x_{k_0-1}$. A simple weight matrix represents an independent mononucleotide model (or 0-order Markov chain), where $k_0 = 1$, $p_0 = 1$ and $p^{i-1,i}_{x_{i-1},x_i} = p^i_{x_i}$. When we use dinucleotides (first order Markov chain) $k_0 = 2$, $p_0 = p^1_{x^1}$, and $p^{i-1,i}_{x_{i-1},x_i}$ is the conditional probability of generating nucleotide x_i in position i given nucleotide x_{i-1} at position $i - 1$. The conditional probability can be estimated from the ratio of observed frequency of oligonucleotide k_0 bases long ($k_0 > 1$) ending at position $i(s_{i-1}, x_i)$ to the frequency of the oligonucleotide $k_0 - 1$ bases long ending at position $i - 1(s_i - 1)$ in a set of aligned sequences of some functional signal :

$$p^{i-1,i}_{x_{i-1},x_i} = f(s_{i-1}, x_i)/f(s_{i-1})$$

A model for nonsite sequences for computing $P(X/N)$ is usually based on a 0-order Markov chain with genomic base frequencies (or even equal frequencies [0.25]).

A log likelihood ratio (9.3) with Markov chains was used in a description of promoter, splice sites, and start and stop of translation signals in gene finding programs such as Genscan (Burge and Karlin 1997), Fgenesh (Salamov and Solovyev 1998, 2000) and GeneFinder (Green and Hillier 1998).

A useful discriminative measure taking into account a priori knowledge can be based on computing Bayesian probabilities as components of position-specific distributions $\{p^i_s\}$:

$$P(S/o^i_s) = P(o^i_s/S)P(S)/(P(o^i_s/S)P(S) + P(o^i_s/N)P(N) \quad \ldots(2.5)$$

where $P(o^i_s/S)$ and $P(o^i_o/N)$ can be estimated as position specific frequencies of oligonucleotides o^i_s in the set of aligned sites and nonsites; $P(S)$ and $P(N)$ are the a priori probabilities of site and nonsite sequences, respectively, and o^i_s is a type of the oligonucleotide starting (or ending) in r'th position (Solovyev and Lawrence 1993). The probability of a sequence X to belong to a signal, if one assumes independence of oligonucleotides in different positions, is

$$P(S/X) = \sum_{i=1}^{k} P(S/o^i_s)$$

Another empirical discriminator, called "Preference," uses average positional probability to belong to a signal:

$$Pr(S/X) = 1/k \sum_{i=1}^{k} P(S/o^i_s) \quad \ldots(2.6)$$

This measure was used in constructing discriminant functions for the Fgenes gene finding program (Solovyev 1998).

Content Specific Discrimination

To take into account general oligonucleotide composition of a functional region (such as GC-rich promoter sequences) we can use probability distributions and their estimations by oligonucleotide frequencies computed on the whole set of functional signal sequences. Then the Markov chain based probability formula (9.4) of generating the signal sequence X is:

$$P(X/S) = p_0 \sum_{i=k_0}^{k} p_{s_{i-1},x_i} \quad \ldots(2.7)$$

Frame Specific Discrimination

The best discrimination of coding and noncoding sequences in gene prediction approaches was achieved by frame specific recognizers (Claverie and Bougueleret 1986; Claverie et al. 1991; Fickett and Tung 1992). The coding sequence is a sequence of triplets (codons) read continuously from a fixed starting point. Three different reading frames with different codons are possible for any nucleotide sequence (six, if a complementary chain is also considered). It was noted that nucleotides are distributed very unevenly relative to the positions within codons. Therefore, the probability of observing a specific oligonucleotide in coding sequences depends on its position relative to the coding frame (three possible variants), as well as on neighboring nucleotides (Shepherd 1981; Borodovsky et al. 1986; Borodovsky and McIninch 1993). Asymmetry in base composition between codon positions arises due to uneven usage of amino acids and synonymous codons, in addition to the particular structure of genetic code. In Markov chain approaches, the frame dependent probabilities (f = {1, 2, 3}) are used to model coding regions. The probability of generating a protein coding sequence X is

$$P(X/C) = p_0 \sum_{i-k_0}^{k} p^{f}_{s_{i-1}, x_i} \qquad \text{...(2.8)}$$

where f is equal 1, 2, or 3 for oligonucleotides ending at codon position 1, 2, or 3, respectively.

Prediction Performance Measures

Sensitivity and specificity measures are widely used to characterize the accuracy of an algorithm or a recognition function. Let us have S sites (positive examples) and N nonsites (negative examples). By applying the recognition function, we identify correctly T_p sites (true positives) and T_n nonsites (true negatives). At the same time, F_p (false positives) sites are wrongly classified as nonsites and F_n (false negative) nonsites are wrongly classified as sites. Note that $S = T_p + F_n$ and $N = T_n + F_n$. Sensitivity (S_n) measures the fraction of the true positive examples that are correctly predicted: $S_n = T_p/(T_p + F_n)$. Specificity (S_p) measures the fraction of the predicted examples that are correct: $S_p = T_p/(T_p + F_p)$. When we see only one value of accuracy estimation, it means the average accuracy of sites and nonsites is a true prediction: $AC = 0.5(T_p/S + T_n/N)$. The more general single measure (correlation coefficient) takes into account a possible difference in the sizes of site and nonsite sets :

$$CC = (T_pT_n - F_pF_n)/\sqrt{(T_p + F_p)(T_n + F_n)(T_p + F_n)(T_n + F_p)}$$

Fisher's Linear Discriminant

The linear discriminant analysis approach provides a method to select a "best" set of an objects features and combine them in a discriminant function that yields an output that is an estimate of the class membership of this object. We assume that each given sequence fragment can be described by a vector $\dot{x}$ of p characteristics ($x_1, x_2, .., x_p$), which can be measured. The procedure of linear discriminant analysis is to find a linear combination of the measures (called the linear discriminant function or LDF) that provides maximum discrimination between sites sequences (class 1) and nonsite examples (class 2). The LDF

$$Z = \sum_{i=1}^{p} a_i x_i$$

classifies (X) into class 1 if $Z > c$ and into class 2 if $Z < c$. The vector of coefficients $(\alpha_1, \alpha_2, .., \alpha_p)$ and the threshold constant c are derived from the training set by maximizing the ratio of the between-class variation of z to within-class variation (or minimizing expected probability of misclassification) and are equal to

$$a = s^{-1}(\dot{m}_1 - \dot{m}_2)$$

$$c = \dot{a}(\dot{m}_1 + \dot{m}_2)/2 + \text{In}\frac{p_2^0}{p_1^0}$$

where $\dot{m}_i$ are the sample mean vectors of characteristics for class 1 and class 2, respectively; s is the pooled covariance matrix of characteristics,

$$s = \frac{1}{n_1 + n_2 - 2}(s_1 + s_2)$$

s_i is the covariation matrix, p_i^0 is the prior probability, and n_i is the sample size of class i. Based on these equations, we can calculate the coefficients of LDF and the threshold constant c using the characteristics of site and nonsite sequences from the training sets, and we can then test the accuracy of LDF on the test set data. This classification actually assigns a feature vector x to the category of the nearest mean measure, the squared Mahalanobis distance

$$\dot{D}^2 = (\dot{x} - \dot{m}_i)s^{-1}(\dot{x} - \dot{m}_i)$$

from $\dot{x}$ to each of the mean vectors m_i. The significance of a given characteristic or a set of characteristics can be estimated by the Mahalonobis distance between two classes:

$$\dot{D}^2 = (\dot{m}_1 - \dot{m}_2)s^{-1}(\dot{m}_1 - \dot{m}_2)$$

which is computed based on values of the characteristics in the training sequences of classes 1 and 2. To find discriminating sequence features, many possible characteristics, such as score of weigh matrices, distances, oligonucleotide preference at different subregions, and so on, are generated. Selection of the subset of significant characteristics q (among the tested p) is performed by a step-wise discriminant procedure including only characteristics, which significantly increases the Mahalonobis distance. The procedure to test this significance uses the fact that the quantity:

$$F = \frac{n_1 + n_2 - p - 1}{p - q} \frac{n_1 n_2 (D_p^2 - D_q^2)}{(n_1 + n_2)(n_1 + n_2 - 2) + n_1 n_2 D_q^2}$$

has an $F(p - q, n_1 + n_2 - p - 1)$ distribution when testing hypothesis H_0: $\Delta_p^2 = \Delta_q^2$, where Δ_m^2 is the population Mahalonobis distance based on m variables (Afifi and Aizen 1979).

Quadratic Discriminant Analysis

Classical linear discriminant analysis often assumes the probability model in which the observations for classes have different means, but a common covariation matrix. The feature space is partitioned by hyperplane optimally separating observations of different classes. To classify groups having different covariation matrices, one can use the quadratic discriminant analysis (QDA). Quadratic discriminant analysis provides a curved boundary in multidimensional feature space. Maximum discrimination between the two classes is achieved with the quadratic discriminant function QDF:

$$QDF = \log\frac{p_1^0}{p_2^0} - \frac{1}{2}(D_1^2 - D_2^2) - \frac{1}{2}\log\left|\frac{S_1}{S_2}\right|$$

where D_i^2 is Mahalonobis distance from an object to the mean and Si is the covariation matrix of class i (i = 1, 2). Quadratic discriminant function might provide a more effective discrimination, but will require a larger learning set of observations to accurately define its larger set of parameters. Such an approach was used in exon prediction method developed by Zhang (1997) to improve the accuracy of the linear discriminant exon predictor (Solovyev et al. 1994)..

Splice Sites Conservative Features

The precise removal of introns from mRNA precursors is mainly defined by the highly conserved sequences near the ends of introns Breatnach and Chambon 1981; Wieringa et al. 1983). The donor (or 5′-splice site) is characterized by a sequence of eight nucleotides AG|GTRAGT. The acceptor (or 3′-splice site) possesses a sequence of four nucleotides preceded by a pyrimidine rich region: YYTTYYYYYYNC|AGG. (Senapathy et al. 1990). The third less conserved intron sequence (branch site), of about 5–8 nucleotides and containing an adenosine residue, usually lies between 10 and 50 nucleotides upstream of the acceptor splice site. The vast majority of introns contains invariant GT and AG dinucleotides at their termini excised from pre-mRNA by the spliceosome, including Ul, U2, U4/U6, and U5 snRNPs.

A rare type of splice pair, the AT- AC, has also been discovered. It is processed by related but different splicing machinery . For the AT-AC group, different conserved positions have been noticed: |ATATCCTTT for the donor site and YAC | for the acceptor site. Burset *et al.* (2000) have recently done a comprehensive investigation of canonical and noncanonical splice sites. They applied ESTs and high-throughput genomic (HTG) sequences to analyze 43,437 pairs of exon-intron boundaries and their sequences from InfoGene database, including all annotated genes in mammalian genomic sequences. Of the 43,437 pairs of donor and acceptor splice sites (splice pairs), 1,215 were annotated as nonstandard donor sites (2.80 percent), and 1,027 were annotated as nonstandard acceptor sites (2.36 percent). Forty-one thousand seven hundred and sixty-seven splice pairs (96.18 percent) contained the standard splice site pair GT-AG. Analysis showed that of 1,615 non-canonical pairs, 441 were supported by EST (27.3 percent) and just 290 (18 percent) were supported by EST after removing potential annotation errors and examples with ambiguity in the position of the splice junction. Analysis of human

Table 2.3. Splice sites sequences presented in the SpliceDB (Burset, Seledtsov, and Solovyev 2001)

Sequences of splice pairs	*Canonical*	*Noncanonical*
Mammals		
Original from GenBank	41722 (96.27%)	1615 (3.73%)
Est supported	22374 (98.07%)	441 (1.93%)
EST supported + corrected	22199 (98.71%)	290 (1.29%)
Human		
Original from GenBank	27486 (96.55%)	982 (3.45%)
EST supported	15384 (98.33%)	261 (1.67%)
EST supported + corrected	15263 (98.89%)	171 (1.11%)
HTG supported		156

noncanonical splice pairs that have corresponding EST and HTG sequences shows that all human EST-supported GC-AG cases having HTS matches were supported (39 cases).

Thirty-one errors were found damaging the standard splice pairs (seven cases had one or both intronic GenBank sequences completely unsupported by HTS, whereas eight cases had intronic GenBank sequences supported; there was a gap between exonic and intronic parts, and in the end, 16 cases had small errors, such as insertions, deletions, or substitutions). Three of five observed AT-AC pairs were correctly annotated in the original noncanonical set; two were recovered from errors. Two more cases were annotated as introns, but in HTS, the exonic parts were continuous (accession numbers: U70997 and M13300). This analysis shows that the overwhelming majority of splice sites contain conserved dinucleotides GT-AG (99.2 percent).

The other major group includes GC-AG pairs (0.62 percent), the alternative splicing machine group AC-AT (about 0.08 percent), and a very small number of other noncanonical splice sites (about 0.03 percent). Therefore, gene-finding approaches using just standard GT-AG splice sites can potentially correctly predict 97 percent genes (if we assume four exons per gene, on average). Including the GC-AG splice pair will increase this level to 99 percent. Twenty-two thousand two hundred and fifty-three verified examples of canonical splice pairs were presented in a SpliceDB database, which is available at http:// genomic.sanger.ac.uk. It also includes 1,615 annotated and 292 EST-supported and shift-verified noncanonical pairs. The weight matrices and consensus sequences for the major group of splice sites are presented in figure 2.4.

(*a*) CT-AG group (canonical splice sites) : 22199 examples

$$\mathbf{M_{70}A_{60}G_{80}|GTR_{95}A_{71}G_{81}T_{46}}$$

	–3	–2	–1	1	2	3	4	5	6
A	34.0	60.4	9.2	0.0	0.0	52.6	71.3	7.1	16.0
C	36.3	12.9	3.3	0.0	0.0	2.8	7.6	5.5	16.5
G	18.3	12.5	80.3	100	0.0	41.9	11.8	81.4	20.9
U	11.4	14.2	7.3	0.0	100	2.5	9.3	5.9	46.2

$$\mathbf{Y_{73}Y_{75}Y_{78}Y_{79}Y_{80}Y_{79}Y_{78}Y_{81}Y_{86}Y_{86}NC_{71}AG|G_{52}}$$

	–14	–13	–12	–11	–10	–9	–8	–7	–6	–5	–4	–3	–2	–1	1
A	9.0	8.4	7.5	6.8	7.6	8.0	9.7	9.2	7.6	7.8	23.7	4.2	100	0.0	23.9
C	31.0	31.0	30.7	29.3	32.6	33.0	37.3	38.5	41.0	35.2	30.9	70.8	0.0	0.0	13.8
G	12.5	11.5	10.6	10.4	11.0	11.3	11.3	8.5	6.6	6.4	21.2	0.3	0.0	100	52.0
U	42.3	44.0	47.0	49.4	47.1	46.3	40.8	42.9	44.5	50.4	24.0	24.6	0.0	0.0	10.4

(*b*) GC-AG group : 126 examples

$$\mathbf{M_{83}A_{89}G_{98}|GCA_{87}A_{84}G_{97}T_{71}}$$

	–3	–2	–1	1	2	3	4	5	6
A	40.5	88.9	1.6	0.0	0.0	87.3	84.1	1.6	7.9
C	42.1	0.8	0.8	0.0	100	0.0	3.2	0.8	11.9
G	15.9	1.6	97.6	100	0.0	12.7	6.3	96.8	9.5
U	1.6	8.7	0.0	0.0	0.0	0.0	6.3	0.8	70.6

$$Y_{75}Y_{72}Y_{78}Y_{79}Y_{77}Y_{80}Y_{66}Y_{78}Y_{85}Y_{84}NC_{68}AG|G_{63}$$

	–14	–13	–12	–11	–10	–9	–8	–7	–6	–5	–4	–3	–2	–1	1
A	11.1	12.7	3.2	4.8	12.7	8.7	16.7	16.7	12.7	9.5	26.2	6.3	100	0.0	21.4
C	36.5	30.9	19.1	23.0	34.9	39.7	34.9	40.5	40.5	36.5	33.3	68.2	0.0	0.0	7.9
G	9.5	10.3	15.1	12.7	8.7	9.5	16.7	4.8	2.4	6.3	13.5	0.0	0.0	100	62.7
U	38.9	4.13	58.7	55.6	42.1	40.5	30.9	37.3	44.4	47.6	27.0	25.4	0.0	0.0	7.9

(*c*) AT-AC group : 8 annotated exmaples + 2 examples recovered from annotation erroes.

$$S_{90}|ATA_{100}T_{100}C_{100}C_{100}T_{100}T_{90}T_{70}$$

$$T_{70}G_{50}C_{70}NC_{60}AC|A_{60}T_{60}$$

AC002397	**TGCCAAGATG**|atatccttgtgt	ctgtctgctcac|**CTTGGAGAAG**
AC004976	**GAAAGAACCC**|atatcctttctg	actacttcatac|**AAAACAGTCA**
AF136179	**TATGGTAGAG**|atatcctttact	actgtttcggac|**ATTGACCAAA**
AL021578	**ACGCTGAACC**|atatcctttggg	ttaaccgctcac|**TGGCCCAGCT**
L10295	**ATTGGTGAAG**|atatcctttttag	aatcattactac|**ATGTGAATCC**
U39892	**AGATTAGAGA**|atatcctttctt	aactgccagcac|**ATTTTGTCAG**
U47924	**TGCCAAGATG**|atatccttctgc	aaccctcctcac|**CTTGGAGAAG**
U53004	**GGAAGTGGTC**|atatccttcctg	aactctgcacac|**GAAGCTCACG**

Fig. 2.3. Consensus sequences and weight matrices for major groups of splice site pairs. Numbering splice site positions is provided relative to the splice junction along the gene sequence.

Table 2.4. Significance of various characteristics for discimination of donor splice sites.

Characteristics	*1*	*2*	*3*	*4*	*5*	*6*	*7*
Individual D^2	9.3	2.6	2.5	0.1	1.5	0.01	0.4
Combined D^2	9.3	11.8	13.6	14.9	15.5	16.6	16.8

1, 2, 3 are the triplet preference (13) of consensus (_4 __ +6), intron G rich (+7 __ +50) and coding regions (_30 __ _5), respectively. 4 is the number of significant triplets in the consensus region. 5 and 6 are the octanucleotide preference for being coding 54 bp region on the left and for being intron 54 bp region on the right of donor splice site junction. 7 is the number of G bases, GG doublets, and GGG triplets in +6 __ +50 intron G rich region.

Computational Recognition of Splice Sites

Computational analysis of splice site sequences demonstrates that their consensuses are slightly specific for distant classes of organisms (Senapathy et al. 1990; Mount 1993) and that some important information is encoded by the sequences outside the short conserved regions. There were several attempts to develop accurate splice site identification algorithms based on consensus sequences or weight matrices, which take into account the information about open reading frames, free energy of base pairing of RNA with snRNA, and other sequence features. These approaches reached the accuracy of about 80 percent for the prediction splice site positions (Nakata et al. 1985; Gelfand 1989). More accurate prediction was achieved by neural network algorithms. To demonstrate what sequence features can help identify authentic splice sites, we will describe a simple method using a linear discriminant function.

Donor Splice Site Recognition

To test the significance of different sequence features by liner discriminant approach, described in section 2.2.5, seven characteristics were selected for donor splice site identification. In table 9.4, we can see the Mahalonobis distances showing the significance of each characteristic. The strongest characteristic of donor sites is triplet composition in consensus region (D^2 = 9.3), in the adjacent intron region (D^2 = 2.6), and in coding region (D^2 = 2.5). Other significant characteristics are a number of significant triplets in conserved consensus region; the number of G bases, GG doublets, and GGG triplets; and the octaplet composition of the coding and intron regions. A rigorous testing of several splice site prediction programs on the same sets of new data demonstrated that the linear discriminant function provides the most accurate local donor site recognizer.

Table 2.5. Comparing the prediction accuracy of local donor splice site recognizers.

Method / program	*False positives*	*False negatives*	*CC*	*Reference*
Wegiht matrix	5.0%	20%	0.22	Guigo *et. al.* 1991
Neural network (NETGENE)	16.3%	6.7%	0.35	Brunak *et. al.* 1991
Discriminant analysis (SPL)	22.0%	2.3%	0.51	Solovyev *et. al.* 1994

The accuracy is averaged for three tested sets.

A single weight matrix provides less accurate recognition than more sophisticated approaches, but it can be easily recomputed for new organisms and is very convenient to use in probabilistic HMM gene prediction methods. Using maximal dependence decomposition procedure (Burge 1998), we constructed five donor recognition weight matrices for different subsets of splice site sequences. The subclassification of donor signals and the matrices constructed, based on 22,306 EST supported splice sites, are presented in Figure 2.4. Performance of these matrices comparing with the other methods was estimated on the Burset and Guigo (1996) data set (figure 2.6). It shows that several weight matrices provide better splice site discrimination than just one. However, their discriminative power is similar to the triplet matrix and lower for most levels of sensitivity than the liner discriminant function of the SPL program.

Acceptor Splice Site Recognition

The performance of acceptor site recognition by different computational methods is presented in table 2.6. We can see that acceptor site recognition accuracy is lower than the accuracy of predicting donor sites. The linear discriminant function (Solovyev et al. 1994) implemented in the SPL program demonstrates the higher accuracy.

Burge (1998) demonstrated that the first order Markov chain model formula based on dinucleotide frequencies of [–20, +3] acceptor site region gives slightly better discrimination than the simple weight matrix model. Such a model was implemented in the Genscan gene prediction method. Recently, Thanaraj (2000) evaluated several splice site recognitions. Among them, the SPL program remains the best local recognizer. Of course, complex gene prediction systems (HMM gene, Genscan, Fgen^sh, Fgenesh, and some intermediate approaches such as Net-Gene2) using a lot of global information about optimal exon (or splice site) combination will have a better accuracy level. However, they cannot be applied to study possible alternative splice sites in a particular gene. Local recognizers might be useful for such tasks.

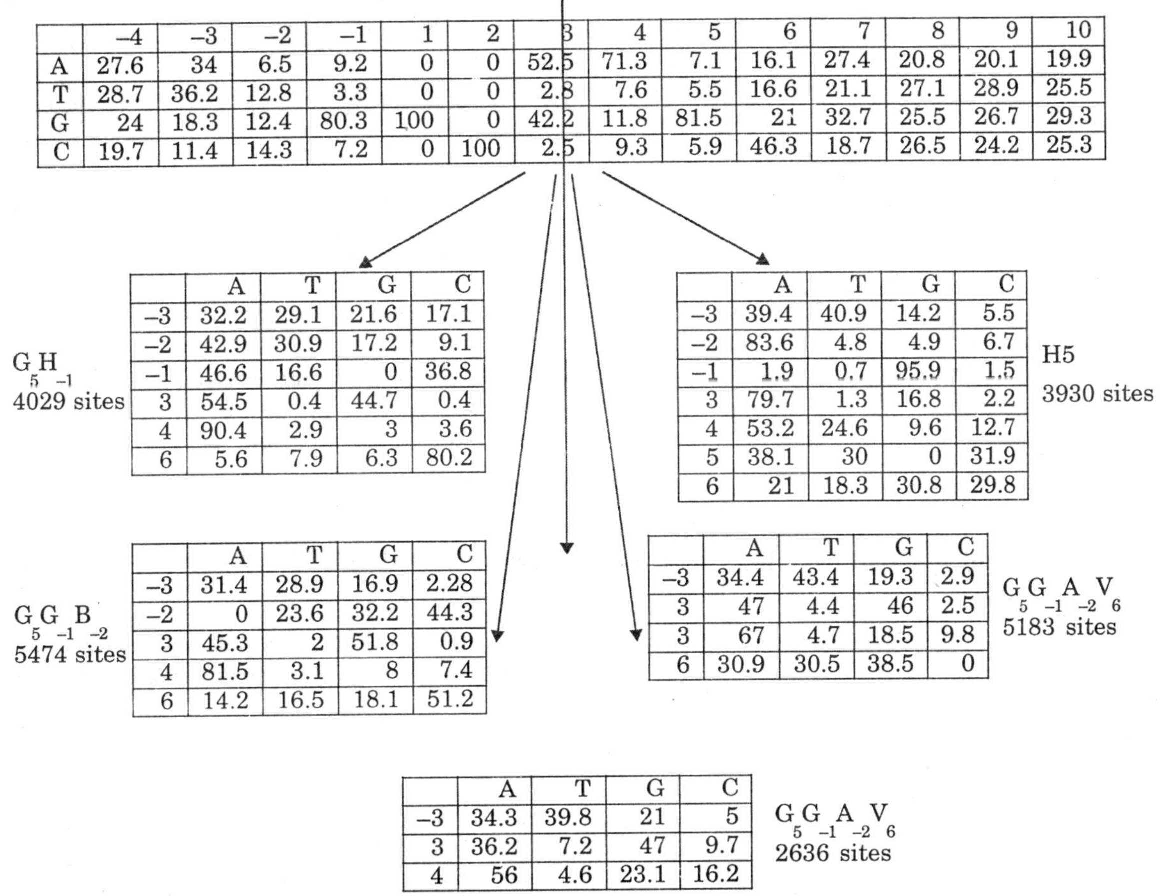

	–4	–3	–2	–1	1	2	3	4	5	6	7	8	9	10
A	27.6	34	6.5	9.2	0	0	52.5	71.3	7.1	16.1	27.4	20.8	20.1	19.9
T	28.7	36.2	12.8	3.3	0	0	2.8	7.6	5.5	16.6	21.1	27.1	28.9	25.5
G	24	18.3	12.4	80.3	100	0	42.2	11.8	81.5	21	32.7	25.5	26.7	29.3
C	19.7	11.4	14.3	7.2	0	100	2.5	9.3	5.9	46.3	18.7	26.5	24.2	25.3

G_5H_{-1}
4029 sites

	A	T	G	C
–3	32.2	29.1	21.6	17.1
–2	42.9	30.9	17.2	9.1
–1	46.6	16.6	0	36.8
3	54.5	0.4	44.7	0.4
4	90.4	2.9	3	3.6
6	5.6	7.9	6.3	80.2

H_5
3930 sites

	A	T	G	C
–3	39.4	40.9	14.2	5.5
–2	83.6	4.8	4.9	6.7
–1	1.9	0.7	95.9	1.5
3	79.7	1.3	16.8	2.2
4	53.2	24.6	9.6	12.7
5	38.1	30	0	31.9
6	21	18.3	30.8	29.8

$G_5G_{-1}B_{-2}$
5474 sites

	A	T	G	C
–3	31.4	28.9	16.9	2.28
–2	0	23.6	32.2	44.3
3	45.3	2	51.8	0.9
4	81.5	3.1	8	7.4
6	14.2	16.5	18.1	51.2

$G_5G_{-1}A_{-2}V_6$
5183 sites

	A	T	G	C
–3	34.4	43.4	19.3	2.9
3	47	4.4	46	2.5
3	67	4.7	18.5	9.8
6	30.9	30.5	38.5	0

$G_5G_{-1}A_{-2}V_6$
2636 sites

	A	T	G	C
–3	34.3	39.8	21	5
3	36.2	7.2	47	9.7
4	56	4.6	23.1	16.2

Fig. 2.4. Classification of donor splice sites by several weight matrices reflecting different splice site groups. Classification computed on 21,252 verified splice sites from SpliceDB. Each cell represents the frequency (in percent) of a particular base in some position of donor site. H is (A, C, or T), B is (C, G, or T), V is (A, C, orG).

POLII Promoter Recognition

Because each eukaryotic polymerase II promoter has a unique selection and arrangement of regulatory elements providing a unique program of gene expression, the computational identification of promoter sequences in genomic DNA is an extremely difficult problem (see chapter 2). Here we consider a version of promoter recognition program TSSW, several modules of which are implemented in the gene prediction program FGENES (Solovyev 1997). In the last version of TSSW (Solovyev and Salamov 1997), it was assumed that TATA+ and TATA— promoters have very different sequence features, so these groups were analyzed separately. Potential TATA+ promoter sequences were selected by the value of the score of Bucher TATA box weight matrix (Bucher 1990) with the threshold close to the minimal score value for the TATA+ promoters in the learning set. Such a threshold divides the learning sets of known promoters into approximately equal groups. Selected significant characteristics of both groups found by discriminant analysis are presented in table 2.7. This analysis demonstrated that TATA- promoters have much weaker general features than TATA+ promoters. Probably TATA–- promoters possess more gene specific structure; they will be extremely difficult to predict by any general-purpose methods.

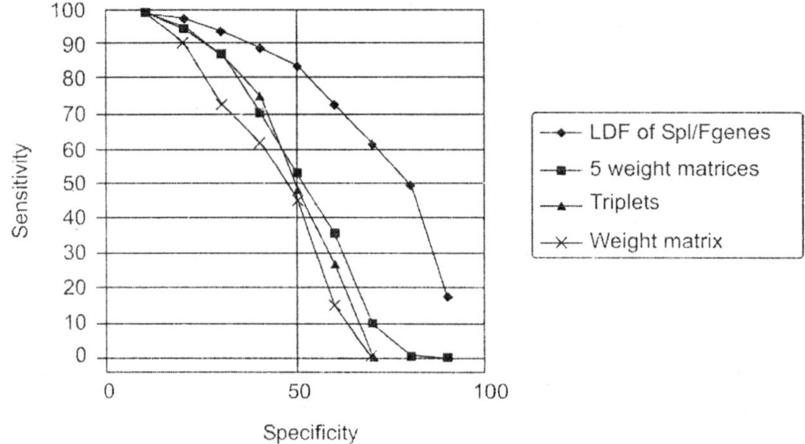

Fig. 2.5. Comparing accuracy of donor splice site recognizers: single weight matrix, five weight matrices, matrix of triplets, linear discriminant function.

Table 2.6. Comparing the prediction accuracy of local acceptor splice site recognizers.

Method / program	*False positives*	*False negatives*	*CC*	*Reference*
Weight matrix	2.3%	53%	0.13	Guigo *et al.* 1992
Consensus MAG/GURAGU	6.0%	18%	0.27	Mount 1982
Consensus MAG/GURAGU	6.0%	18%	0.27	Mount 1982
Five consensuses	4.2%	15%	0.31	Rogozin and Milanesi 1997
Neural network (NETGENE)	25.0%	2.7%	0.51	Brunak *et. al.* 1991
Discriminant analysis (SPL)	10.0%	3.0%	0.56	solovyev *et. al.* 1994

The accuracy is averaged for three tested sets.

For each position of a given sequence, the TSSW program evaluates the occurrence of TSS using two linear discriminant functions (for TATA+ and TATA-promoters) with characteristics computed in the (-200, +50) region around the given position. If we find a TATA-box (using TATA-box weight matrix) in this region, then we compute the value of LDF for TATA+ promoters, otherwise the value of LDF for TATA-less. Only one prediction with the highest score of LDF and greater than some threshold will be selected within any 300-bp region. If we observe a lower scoring promoter predicted by the TATA- LDF near a higher scoring promoter predicted by TATA+ LDF, then the first prediction is also displayed as a potential enhancer region.

Figure 2.5 shows an example of TSSW program results for the sequence of the human connexin 32 (GJB1) gene (GenBank accession number L47127). The TSSW predicts one enhancer at position 246 and one potential TSS at position 428, with corresponding TATA-box at the position 388. GenBank annotation based on experimental data shows real TATA-signal in positions 389-394. TSSW also optionally lists all potential TF binding sites around predicted promoters or enhancers (Fig. 2.6). It outputs the position, the strand (±), TRANSFAC identifier, and the consensus sequences of found sites. The information about these sites may be of interest for researchers studying the transcription of a particular gene. Due to a high false positive rate of promoter prediction in long genomic sequences, they are more useful when we can remove some predictions about the positions of coding regions.

Table 2.7. Characteristics of promoter sequences used by TSSW programs for identification of TATA+ and TATA– promoters.

Characteristics	*D^2 for TATA + promoters*	*D^2 for TATA– promoters*
Hexaplets –200 – – 45	2.6	1.4 (–100 – –1)
TATA box score	3.4	0.9
Triplets around TSS	4.1	0.7
Hexaplets +1 – +40		0.9
Spl-motif content		0.9
TATA fixed location	0.7	
CpG content	1.4	0.7
Similarity –200 – –100	0.3	0.7
Motif Density (MD) –200 – +1	4.5	3.2
Direct/Inverted MD – 100 – +1	4.0	3.3 (–100 – –1)
Total Mahalonobis distance	11.2	4.3
Number promoters/nonpromoters	203/4000	193/74000

The TSSW was additionally tested on the several GenBank entries that have information about experimentally verified TSS and were not included in the learning set (table 2.8).

```
tssw Wed Dec 20 20 : 02 : 08 PST 2000
> Homo sapiens connexin 32 (GJB1) gene
 Length of sequence- 950
 Thresholds for TATA+ promoters - 0.45, for TATA-/enhancers - 3.70
     2 promoter/enhancer(s) are predicted
 Enhancer Pos: 246 LDF-14.25
 Promoter Pos: 428 LDF- 0.64 TATA box at     388   18.68
 Transcription factor binding sites:
for position - 246
179   (+)   CHICKSACRA       CCGCCC
241   (-)   CHICKSACRA       CCGCCC
220   (-)   CHICKSACRA       CCGCCC
188   (+)   MAIZESADH1       CGTGG
```

```
 26  (+)  YSADH2_01     TCTCC
168  (+)  YSADH2_01     TCTCC
 42  (-)  YSADH2_01     TCTCC
215  (+)  HSSAPOE_08    GGGCGG
236  (+)  HSSAPOE_08    GGGCGG
184  (-)  HSSAPOE_09    GCCCGCCC
243  (-)  HSSAPOE_09    GCCCGCCC
146  (+)  MOUSESMCK     CCCaaCACCTGCtgcctgagcc
 87  (-)  RATSEAI_09    GTCAG
 14  (+)  RATSAFEP_0    TGTCCT
 50  (+)  YSGALI_10     AGCCT
116  (+)  YSGAL1_10     AGCCT
226  (-)  HSSBG_01      CCaCACCCg
214  (+)  CHICKSBAG     GGGGCGGG
185  (-)  CHICKSBAG_    GGGGCGGG
.............................  .  .
for promoter at position - 428
179  (+)  CHICKSACRA    CCGCCC
423  (-)  CHICKSACRA    CCGCCC
312  (-)  CHICKSACRA    CCGCCC
241  (-)  CHICKSACRA    CCGCCC
220  (-)  CHICKSACRA    CCGCCC
389  (+)  CHICKSBAC_    TATAA
188  (+)  MAIZESADH1    CGTGG
168  (+)  YSADH2_01     TCTCC
215  (+)  HSSAPOE_08    GGGCGG
236  (+)  HSSAPOE_08    GGGCGG
307  (+)  HSSAPOE_08    GGGCGG
418  (+)  HSSAPOE_08    GGGCGG
184  (-)  HSSAPOE_08    GGGCGG
243  (-)  HSSAPOE_09    GCCCGCCC
357  (+)  PBV1SBPV1_    ACCagtaatGGT
368  (-)  BPV1SBPV1_    ACCagtaatGGT
357  (+)  BPV1SBPV1_    ACCgttgccGGT
368  (-)  BPV1SBPV1_    ACCgttgccGGT
357  (+)  BPV1SBPV1_    ACCgtcttcGGT
368  (-)  BPV1SBPV1_    ACCgtcttcGGT
417  (+)  MOUSESA21C    gcccagccctcccATTGGtggagacg
146  (+)  MOUSESMCK_    cccaaCACCTGCtgcctgagcc
381  (+)  ADSE3_06      gggcagggTATAActcacctga
.............................  .  .
```

Fig. 2.6. Results of promoter prediction by TSSW program in human connexin 32 (GJB1) gene. (GenBank accession number L47127).

Table 2.8. The TSSW predictions on some GenBank entries with experimentally verified TSS.

Gene	*GenBank accession number*	*Length (bp)*	*True TSS*	*Predicted TSS*	*Number of false postives*
CXCR4	AJ224869	8747	2632	2631	4
HOX3D	X61755	4968	2280	2278	2
DAF	M64356	2003	733	744	1
GJB1	L47127	950	404	428	0
ID4	AF030295	1473	1066	1081	1
C inhibitor	M68516	15571	2200	2004	4
MBD1	AJ132338	2951	1964	1876	1
Id –3	X73428	2481	665	663	0

The lengths of sequences varied from 950 to 28,438 bp, with a median length of 2,938 bp. All true TSS in these sequences can be considered as correctly predicted, with an average of 1.5 false positives per sequence or 1 false positive per 3,340 bp. The distances between true TSS and the correctly predicted ones varied from exact matching to 196 bp, with the median deviation of about 15 bp, which means that half of the predictions are close to the experimental mapping of TSS with the estimated precision of ± 5 bp (Perier et al. 2000).

The above prediction algorithm uses the propensities of each TF binding site independently, not taking into account their mutual orientation and positioning. At the same time, the transcription regulation is a highly cooperative process, involving the simultaneous binding of several transcription factors to their corresponding sites. In future algorithms we should analyze patterns of regulatory sequences, where mutual orientation and location of individual regulatory elements are necessary.

Recognition of Poly A Signals

A 3′-untranslated region (3’UTR) has a diversity of cytoplasmic functions affecting the localization, stability, and translation of MrNAs (Decker and Parker 1995). Practically all eukaryotic mRNAs undergo 3′-end processing, which involves endo-nucleotide cleavage followed by polyadenylation of the upstream cleavage product. The formation of large RNA-protein complexes is essential for 3′-end processing. RNA sequences directing the binding of specific proteins are usually poorly conserved and often recognized in a cooperative fashion. Therefore, the approaches for poly-A signal identification use statistical characteristics of poly-A regions, which can reflect some unknown functional sequences.

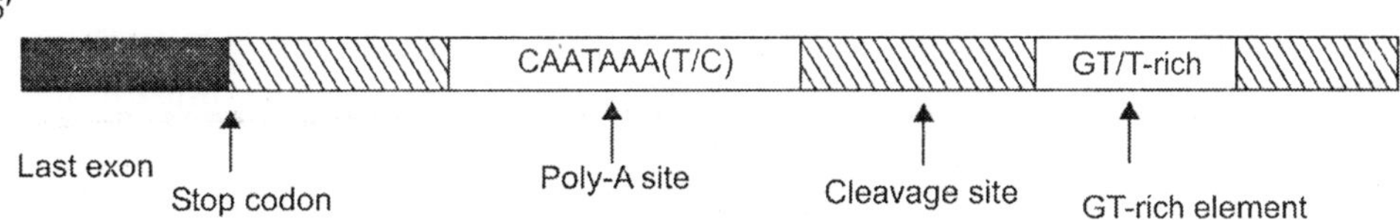

Fig. 2.7. Basic structure of poly-A signal sequences.

There are three types of basic RNA sequences defining a 3′-processing site (Wahle 1995; Proudfoot 1991) (Fig. 2.7). The most conserved is the hexamer signal AAUAAA (polyA signal), situated 10-30 nucleotides upstream of the 3′-cleavage site. About 90 percent of sequenced mRNAs have a perfect copy of this signal. Two other types, the upstream and the downstream elements, are poorly conserved and characterized. Downstream elements are frequently located within approximately 50 nucleotides 3′ of the cleavage site and often GU or U rich (Wahle and Keller 1992). Comparing their sequences, McLachlan et al. (1985) suggest a possible consensuse of one downstream element: YGUGUUYY. The efficiency of polyadenylation in a number of genes can be also increased by generally U-rich sequences upstream of AAUAAA (Whale 1995). A few computer programs were developed to identify 3′-processing.

Yada *et al.* (1994) analyzed human DNA sequences in the vicinity of the poly-A signal, trying to distinguish them from other AATAAA sequences nonactive in polyadenylation (pseudo poly A signals). They found that C frequently appears on the upstream side of the AATAAA signal and T or C often appears on the downstream side, generating an extended consensus of poly-A signal: CAATAAA(T/C). Kondrakhin *et al.* (1994) constructed a generalized consensus matrix using 63 sequences of cleavage/ polyadenylation sites in vertebrate pre-mRNA. The matrix elements were the absolute frequencies of triplets at each site position. Using this matrix for recognition of polyadenylation regions produces a very high number of false positive predictions. A LDF recognition function for poly-A signal was developed by Salamov and Solovyev (1997). The prediction algorithm was realized in the POLY AH program.

It searches for the AATAAA pattern by using weight matrix. After it finds the pattern, it computes the value of the linear discriminant function, defined by seven sequence characteristics around this position. The POLY AH program has been tested on the sequence of Ad2 genome, where for eight correctly identified sites, it predicts only four false sites. Further improvement of poly-A recognition was reached in using a pair of quadratic discriminant function in the Polyadq program (Tabaska and Zhang 1999). This program outperformed the POLY AH detection method and is the first that can detect significant numbers of ATTAAA-type signals.

ORF, EXON, AND SINGLE GENE PREDICTION APPROACHES

The first generation of computational gene finding programs searched for open reading frames with organism-specific codon usage (Staden and McLachlan 1982). These approaches worked successfully for bacterial genes (Staden 1984; Borodovsky et al. 1986), but short eukaryotic exons and spliced eukaryotic genes require algorithms taking into account additional information about functional signals. One application of such approaches is useful for predicting coding regions or coding ORF in partially or completely sequenced mRNA(EST) sequences. Several HMM-based predictors developed recently, such as BESTORF, (Solovyev and Salamov 1999a) and ESTscan (Iseli et al. 1999), significantly improve the accuracy of earlier approaches. The internal exon prediction program SORFIND (Hutchinson and Hayden 1992) was designed based on codon usage and Berg and von Hippel (1987) discrimination energy for intron-exon boundary recognition.

The accuracy of exact internal exons prediction (at both 5′- and 3′-splice junctions and in the correct reading frame) by the SORFIND program reaches 59 percent, with a specificity of 20 percent. Snyder and Stormo (1993) applied a dynamic programming approach (an alternative to the rule-based approach) to internal exon prediction in the GeneParser algorithm. It recognized 76 percent of internal exons, but the structure of only 46 percent of the exons was exactly predicted when tested using entire GenBank entry sequences. The HEXON (Human

EXON) program, (Solovyev et al. 1994a) based on linear discriminant analysis, was one of the most accurate in exact internal exon prediction. It was recently upgraded to predict all type of exons (and renamed FEX—find exon) (Solovyev et al. 1994b).

The FEX program can be useful to analyze a possible set of alternatively spliced exons in a given sequence in addition to the optimal variant of gene structure produced by exon assembling programs. In an effort to improve the accuracy of exon prediction, Zhang (1997) applied quadratic discriminant technique (in the MZEF program) as a direct extension of the classical liner discriminant approach used in the HEXON program. The statistical evaluation of MZEF predictions on 473 genes (partially included in MZEF training) demonstrated a better performance than the HEXON program. Later, a number of single-gene prediction programs were developed to assemble potential eukaryotic coding regions into translatable mRNA sequences, selecting optimal combinations of compatible exons (Field and Soderlund 1990; Gelfalnd 1990; Guigo el. al. 1992; Dong and Searls 1994).

Dynamic programming was suggested as a fast method to find an optimal combination of pre-selected exons (Gelfand and Roytberg 1993; Solovyev and Lawrence 1993b; Xu el. at. 1994), which is different from the approach in the GeneParser algorithm suggested by Snyder and Stormo (1993) to recursively search for exon-intron boundary positions. The FGENEH (Find GENE in Human) algorithm incorporated 5′-, internal, and 3′-exon identification linear discriminant functions and a dynamic programming approach (Solovyev et al. 1994 1995). Burset and Guigo (1996) conducted a comprehensive test for it and the other gene finding algorithms; the FGENEH program was one of the best in the tested group, having an exon prediction accuracy 10 percent higher than the others and the best level of accuracy on the protein level. A novel step in gene prediction approaches was the application of generalized hidden Markov models implemented in the Genie algorithm (Kulp et al. 1996). Genie is similar in design to GeneParser, but is based on a rigorous probabilistic framework. It is similar to FGENEH in performance. (Kulp et al. 1996).

MULTIPLE GENE PREDICTION BY DISCRIMINATIVE AND PROBABILISTIC APPROACHES

Whole genome sequencing projects were initiated for a number of organisms, from bacteria to higher eukaryotes. They require gene-finding approaches that are able to identify many genes encoded in the genomic sequences. The most accurate multiple gene prediction programs include such HMM-based probabilistic approaches as Genscan (Burge and Karlin 1997) and Fgenesh (Salamov and Solovyev 2000), Fgenes (discriminative approach) (Solovyev 1997), and Genie (generalized HMM with neural network splice site detectors) (Reese *et al.* 2000). In the next section, we will describe a general scheme of HMM-based gene prediction that was initially realized in the works of Dr. Haussler's group (Krogh et al. 1994; Kulp et al. 1996). This pattern-based approach can also be considered as a variant in which transition probabilities are not taken into account.

HMM-Based Eukaryotic Gene Identification

Exons, introns, 5′-, and 3′-untranslated regions are different components (states) of gene structure that occupy k non-overlapping subsequences of a sequence $X = \bigcup_{i=1,\,k} x_i$. There are 35 states constituting an eukaryotic gene model, considering direct and reverse chains as possible gene locations (Figure 2.8). The absence of protein coding characteristics reduces significantly prediction accuracy of noncoding 5′- and 3′-exons (and introns; therefore, they are not considered in the current gene prediction algorithms. The other 27 states consist of six exon states (first, last, single, and three types of internal exons due to three possible reading frames) and seven non-coding states (three intron, noncoding 5′- and 3′-, promoter, and poly A) in each chain, plus the noncoding intergenic region.

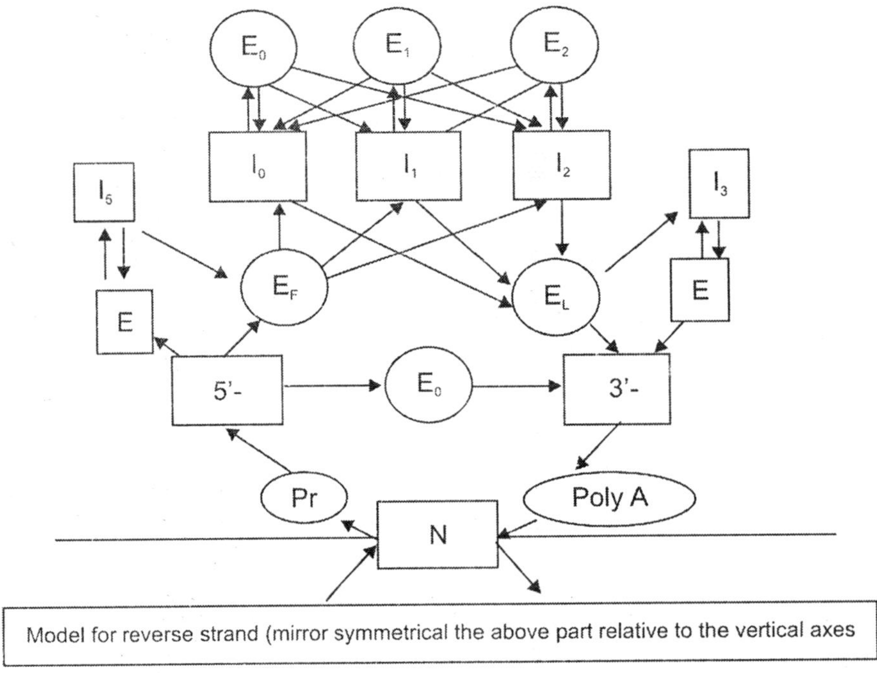

Fig. 2.8. Different states and transitions in eukaryotic gene model.

A gene structure can be considered as the ordered set of state/subsequence pairs, $\phi = \{(q_1, x_1), (q_2, x_2), \ldots, (q_k, x_k)\}$,called the parse. We call the predicted gene structure to be such a parse ϕ that the probability $P(X, \phi)$ of generating X according to ϕ is maximal over all possible parses (or when some score is optimal in some meaningful sense, *i.e.*, best explains the observations. This probability can be computed using statistical parameters describing a particular state and generated from the training set of known gene structures and sequences. Successive states of this HMM model are generated according to the Markov process with the inclusion of explicit state duration density. A simple technique based on the dynamic programming method for finding the optimal parse (or the single best state sequence) is called the Viterbi algorithm. The algorithm requires $o(N^2D^2L)$ calculations, where N is the number of states, D is the longest duration, and L is the sequence length. Burge (1997) introduced a helpful technique to reduce the number of states and simplify computations by modeling noncoding state length with a geometrical distribution.

We will shortly consider the algorithm of gene finding using these technique, which was initially implemented in the Genscan program (Burge 1977; Burge and Karlin, 1997) and used later in the Fgenesh program. As any valid parse will consist of only alternating series of noncoding and coding states, NCNCNC, . . . , NCN, we need only 11 variables, corresponding to the different types of N states. For each sequence position (starting from 1), we select the maximum joint probability to continue the current state or to move to another noncoding state defined by a coding state (from a pre-computed list of possible coding states) that terminates at the analyzed sequence position. The parse probability is

$$P(X,\ddot{o}) = P(q_1)\sum_{i=1}^{k-1} P(x_i \mid l(x_i),q_i)\,P(l(x_i)\mid q_i)\,(P(q_{i+1},q_i))\,P(X_i \mid l(x_k),q_k)\,P(l(x_k)\mid q_k)$$

where $P(q_1)$ denotes the initial state probability; $P(x_i \mid l(x_i), q_i)\ P(l(x_i) \mid q_i)$ and $P(q_{i+1}, q_i)$ are the independent joint probabilities of generation of the subsequence x_i of length l in the state q_i and transitioning to the q_{i+1} state. $P(x_i \mid l(x_i), q_i)\ P(l(x_i) \mid q_i)$ is a production of a probability of generation l-length sequence x_i and the probability of observing such an l-length sequence in the state q_i, which are computed using the sequence of x_i and the statistical data from a training set of known genes. To compute $P(x_i \mid l(x_i), q_i)$ for an internal exon state, we use donor and acceptor site models based on position specific weight matrices and frame-specific Markov models based on hexaplet frequencies in exons and introns.

Search for Optimal Parse : Let us define the best score (the highest joint probability $y_i[j]$ of the optimal parse of the subsequence $S_{1,j}$ $[s_1, s_2, ..., s_j]$,which ends in state q_i at position j). Assume a set A_j of coding states $\{c_k\}$ of lengths $\{d_k\}$, starting at positions $\{m_k\}$ and ending at position j, which have the previous states $\{b_k\}$. The length distribution of state c_k is denoted by $f_{ck}(d)$. The searching procedure can be stated as follows:

Initialization

$$y_i(1) = \pi_i P_i(s_1)\, p_i \text{ and } Z_i(1) = 0,\ i = 1, \ldots 11$$

Recursion

$$y_i(j+1) = \max\begin{Bmatrix} y_i(j)p_i\,P_i(s_{j+1}), \max\limits_{ck|A_j}\{y_i(m_k-1)(1-p_{b_k})t_{b_k,c_k} f_{c_k} \\ (d_k)P(S_{m_k,j}) \times t_{c_k,} ipi P_i(s_{j+1})\} \end{Bmatrix} i = 1,\ldots 11, j = 1,\ldots, L-1.$$

Termination

$$y_i(L+1) = \left\{ y_i(L), \max_{c_k|A_j}\{y_i(m_k-1)(1-p_{b_k})t_{b_k,c_k} f_{ck}(d_k)P(S_{m_k,j})t_{c_k,j} \right\}$$
$$i = 1,\ldots 11$$

On each step we record the location and type of transition maximizing the functional to restore the optimal set of states (gene structure) by a backtracking procedure. Most parameters of these equations can be calculated from the learning set of known gene structures. Instead of scores of coding states $P(S_{m_k,j})$, it is better to use log-likelihood ratios, which do not produce scores below the limits of computer precision. This technique to predict multiple eukaryotic genes was initially implemented in the Genscan algorithm (Burge and Karlin 1997).

Several other HMM-based gene prediction programs were developed later: Veil (Hederson et al. 1997), HMMgene Krogh 1997), Fgenesh (Salamov and Solovyev 1999, 2000), a variant of Genie (Kulp et al. 1996), and GeneMark (Lukashin and Borodovsky 1998). Fgenesh (Find GENES using Hmm) is currently the most accurate program. It is different from Genscan in computing the coding scores of potential exons, where a priori probabilities of exons were taken into account according to the Bayes theorem. As a result, the coding scores of potential exons are generally lower than in Genscan. Some minor differences exist in the functional signal description and preparing of training sets to compute specific parameters for each model organism, such as human, Drosophila, nematode, yeast, Arabidopsis, monocotolydons, and so on. Coding potentials were calculated separately for four isochores (human) and for two isochores (other species). The run time of Fgenesh is practically linear; the current version has no practical limit on the length of analyzed sequence. Prediction of about one thousand genes in 34.5 MB of chromosome 22 sequence takes about 1.5 minutes with a Dec-alpha processor EV6.

Discriminant Analysis—Based Gene Prediction

The Fgenes (*Find GENES*) program predicts multiple genes using dynamic programming and discriminant classifiers to generate a set of exon candidates. The following major steps describe analysis of genomic sequences by the Fgenes algorithm:

1. Create a list of potential exons by selecting: ATG . . . GT, AG-GT, AG. Stop sequence regions having exons scores higher than the specific thresholds depending on GC content (four groups);
2. Find a set of compatible exons with the maximal total score. Guigo (1999) described an effective procedure for finding such a set. Fgenes uses a simpler variant of this algorithm: we order all exon candidates according to their 3′-end positions. Then, going from the first to the last exon, we select for each exon the maximal score path (compatible exons combination) terminated by this exon using a dynamic programming approach. Include in the optimal set either this exon or an exon with the same 3′-splicing pattern ending at the same position or earlier, whichever has the higher maximal score path.

Table 2.9. Significance of internal exon characteristics selected by LDA.

	Characteristics	*1*	*2*	*3*	*4*	*5*
a	Individual D^2	15.0	12.1	0.4	0.2	1.5
b	Combined D^2	15.0	25.3	25.8	25.8	25.9

Characteristics 1 and 2 are the values of donor and acceptor site LD recognition functions. 3 gives the octanucleotide preference for being coding of potential exon region. 4 gives the octanucleotide preference for being intron 70-bp region on the left and 70-bp region on the right of potential exon region.

3. Take into account promoter or polyA scores (if predicted) in terminal exon scores. The run time of the algorithm grows approximately linearly with the sequence length. Fgenes is based on usage linear discriminant functions developed for identification of splice sites, exons, promoter, and polyA sites Solovyev et al. 1994; Solovyev and Salamov 1997). We consider these functions for internal exons to demonstrate what sequence features are important to exon identification.

Internal Exon Recognition : We consider as potential internal exons all open reading frames in a given sequence flanked by AG (on the left) and GT (on the right). The structure of such exons is presented in figure 2.10. The values of five exon characteristics were calculated for 952 authentic exons and for 690,714 pseudo-exon training sequences from the set. Table 2.9 gives the Mahalonobis distances, showing the significance of each characteristic. We can see that the strongest characteristics are the recognition functions of flanking donor and acceptor splice sites (D^2 = 15.04 and D^2 = 12.06, respectively). The preference of ORF as a coding region has D^2 = 1.47 and adjacent left intron region has D^2 = 0.41 and right intron region has D^2 = 0.18. The accuracy of the discriminant function based on these characteristics was estimated on the recognition of 451 exon and 246,693 pseudo-exon sequences from the test set.

The sensitivity of exact internal exon prediction is 77 percent, with a specificity of 79 percent. At the level of individual nucleotides, the sensitivity of exon prediction is 89 percent, with a specificity of 89 percent; the sensitivity of intron positions prediction is 98 percent, with a specificity of 98 percent. This accuracy is better than that demonstrated by dynamic programming and neural network based methods (Snyder and Stormo 1993), which have a 75 percent sensitivity of the exact internal exons prediction, with a specificity of 67 percent. 5′_coding exon recognition LDF uses the average value of positional triplet preference in the (_15, 10) region around ATG codon (instead of donor splice site score). 3′-exon coding region recognition LDF includes the average value of positional triplet preference in the (_10, 30) region around the stop codon (instead of the acceptor site score). The recognition function of single exons combines corresponding characteristics of 5′- and 3′-exons (Figure 2.9) (Solovyev et al. 1994; Solovyev and Salamov 1997). Features describing sequences near initial and stop codons have much less discriminative power than the splice site characteristics; therefore, terminal and short single exons have a lower accuracy of recognition.

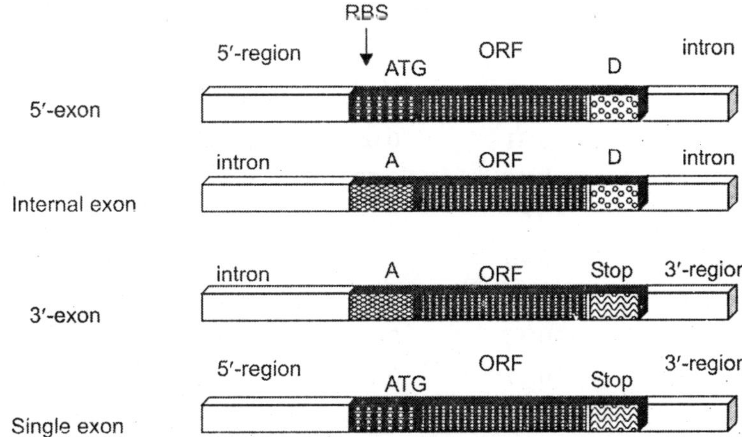

Fig. 2.9. Different functional regions of the first (*a*), internal (*b*), last (*c*) corresponding to components of recognition functions. Single exons include left and right characteristics of first and last exons, respectively.

ACCURACY OF GENE GENE IDENTIFICATION PROGRAMS

Burset and Guigo (1996) specially selected a set of 570 single-gene sequences of mammalian genes, which they used to evaluate the performance of many gene finding approaches. The results of this test are presented in table 2.10. Of course, some of these data have only a historical value to show the progress in gene finding development, and some of these programs have been improved since the test. We can see that the best programs on average predict accurately 93 percent of exon nucleotides (Sn = 0.93), with just 7 percent false positive predictions. However, the accuracy on the nucleotide level does not completely reflect the quality of gene structure prediction because missing small exons and the imperfect location of exon ends will not much affect its value. Therefore, it is important to provide the accuracy of exact exon prediction level, which is usually lower than at the nucleotide level.

The table clearly demonstrates that the recent multiple gene prediction programs such as Fgenesh, Fgenes, and Genscan significantly outperform the older approaches. The exon identification rate is actually even higher than the presented data because the overlapped exons were not counted as true predictions in exact exon accuracy evaluation. Yet there is still room for significant improvement. The accuracy of exact gene prediction is only 59 percent for Fgenesh, 56 percent for Fgenes, and 45 percent for Genscan programs computed on this relatively simple test with single gene sequences. A more practical task is to identify multiple genes in long genomic sequences containing genes in both DNA strands.

We selected a test set of 19 long genomic sequences of 26,000-240,000 bp and 19 multigene sequences with 2-6 genes from GenBank to compare performance of gene-finding programs in analyzing genomic DNA. Table 2.11 demonstrates the results of gene prediction for these data. The results show that the accuracy is still pretty good on the nucleotide and exon level, but exact gene prediction is lower than for the test with short single gene sequences. Sensitivity for exact internal exon prediction is 85-90 percent, but 5′-, 3′-, and single exons have a prediction

Table 2.10 Characteristics of accuracy for the gene prediction programs on single gene sequences of Burset and Guigo 1996 dataset)

Algorithm / program	*Sn (exons)*	*Sp (exons)*	*Sn nucleotides*	*Sp nucleotides*	*Reference*
Fgenesh	0.84	0.86	0.94	0.95	Salamov & Solovyev 1998
Fgenes	0.83	0.82	0.93	0.93	Solovyev 1997
Genscan	0.78	0.81	0.93	0.93	Burge & Karlin 1997
Fgeneh	0.61	0.64	0.77	0.88	Solovyev *et. al.* 1995
Morgan	0.58	0.51	0.83	0.79	Salsberg *et. al.* 1998
Veil	0.53	0.49	0.83	0.79	Henderson *et al.* 1997
Genic	0.55	0.48	0.76	0.77	Kulp *et. al.* 1996
GenLang	0.51	0.52	0.72	0.79	Dong & Searls 1994
Sorfind	0.42	0.47	0.71	0.85	Hutchinson & Hyden 1992
GeneID	0.44	0.46	0.63	0.81	Guigo *et. al.* 1992
Grail2	0.36	0.43	0.72	0.87	Xu *et. al.* 1994
GeneParser2	0.35	0.40	0.66	0.79	Snyder & Stromo 1995
Xpound	0.15	0.18	0.61	0.87	Thomas & Skolnick 1994

Sn (sensitivity) = number of exactly predicted exons/number of true exons (or nucleotide); Sp (specificity) = number of exactly predicted exons/number of all predicted exons. (Accuracy data for programs developed before 1996 were estimated by Burst and Guigo (1996). The other data were produced by the authors of the corresponding programs.

sensitivity of about 50-75 percent, which can partially explain relatively low level of exact gene prediction. As a result, we observe the splitting of some actual genes and/or joining some other multiple genes into a single one. Another limitation of current gene-finding programs is that they cannot detect the nested genes, that is, genes located inside introns of other genes. This is one of the future directions for improvement of gene-finding software.

Although this is probably a rare event for the human genome, for organisms like Drosophila, it presents a real problem. For example, annotators identified 17 examples of such cases in the Adh region. (Ashburner *et al.* 1999). Masking repeats is important. It significantly increases (~10 percent) the specificity of prediction.

Table 2.11. Performance of gene-finding for 38 genomic sequences.

Program	Sequences / Genes	Accuracy per nucleotide			Accuracy per exon				Gens / Entries
		Sn	Sp	CC	Sn / Sn_0	Sp	Me	We	
Fgenes	38/77 M_r	0.94	0.87	0.90	0.85/0.93	0.80	0.08	0.14	0.36/0.11
		0.94	0.78	0.85	0.84/0.92	0.75	0.08	0.21	0.34/0.08
Genescan	38/77 M_r	0.93	0.82	0.87	0.80/0.90	0.74	0.10	0.18	0.29/0.03
		0.92	0.70	0.79	0.79/0.90	0.66	0.11	0.30	0.29/0.03
Fgenes	38/77 M_r	0.91	0.80	0.84	0.84/0.92	0.72	0.08	0.21	0.36/0.18
		0.92	0.76	0.83	0.84/0.93	0.68	0.07	0.30	0.39/0.21

Me, missing exons; WE, wrong exons. M_r lines provide predictions on sequences with masked repeats. Sn_o, exon prediction accuracy including overlapping exons.

KNOWLEDGE OF SIMILAR PROTEIN OR EST CAN IMPROVE GENE PREDICTION

Automatic gene prediction approaches can take into account some information about exon similarity with a known protein or EST (Gelfand et al. 1996l Xu and Uberbacher 1996; Krogh 2000; Birney and Durbin 2000). Fgenesh+ (Salmove and Salvyev 2000) is a modification of the Fgenesh algorithm, which uses additional information from available similar prcteins. These proteins can be acquired by running Fgenesh on a given sequence. Then the predicted proteins (or amino acid fragments translated from predicted exons) are used to select similar proteins in some protein database. After that, we can use selected proteins to improve prediction accuracy. Fgenesh+ reads the protein homolog sequence and aligns all predicted potential exons with that protein using the Smith-Waterman algorithm, as implemented in the *Sim* program (Huang et. at. 1990) or the *Lial* (local iterative alignments) algorithm developed by Seledtsov and Solovyev (1999).

To improve the computational time, all overlapped exons in the same reading frame are combined into one sequence and aligned only once. Fgenesh+ includes two major additions to the Fgenesh algorithm: augmentation of the exon score (for exons having detected similarity) by an additional term proportional to the alignment score and imposing a penalty for the adjacent exons in a dynamic programming procedure, when the distance between their corresponding protein segments is significantly different from the distance between the corresponding fragments of a similar protein. We tested Fgenesh+ on a set of 61 GenBank human sequences, which have imperfect ab initio Fgenesh predictions and known protein homologs from other organisms with identity varying from 99 percent to 40 percent. The results of applying Fgenesh+ to these sequences show (table 2.12) that when the alignment covers the entire lengths of both proteins, the accuracy increases (relative to Fgenesh) and the improvement does not depend significantly on the level of percent identity (for ID > 40 percent).

Table 2.12. The accuracy of Fgenesh and Fgenesh+ on the same set of human genes with known protein homologs from another organisms.

Program	CG	Sn_e	Sp_e	Sn	Sn	CC
Fgenesh	0	63	68	86	83	0.74
Fgenesh+	46	82	85	96	98	0.95

The set includes 61 genes and 370 exons. CG—percent of correctly predicted genes; Sn_e, Sp_e—sensitivity and specificity at the exon level (in %); Sn, Sp—sensitivity and specificity at the nucleotide level (in %); CC—correlation coefficient.

This feature makes valuable the proteins from distant organisms for improving the accuracy of gene identification. Having a sequence of the human genome, we can find where in the genomic sequence a given protein is located using Blast-like search in all predicted proteins of this genome. Then we use Fgenesh+ for a prediction of the full-length mRNA (its coding part) for a given protein using its sequence and the selected genomic sequence. Recently we have developed a Fgenesh++ script, which initially predicts genes using the Fgenesh program and then selects from NR (nonredundant protein database) similar proteins for predicted genes using the Dbscan program (a Blast-like program, but about 10 times faster). The Fgenesh++ script uses found protein sequences to improve initial gene prediction and can automatically generate annotation of the entire chromosome. Similar to the Fgenesh+ algorithm scheme of exploiting known EST/cDNA information to improve accuracy of gene identification is the program Fgenesh_c (Salamov and Solovyev 2000a). Fgenesh+ and Fgenesh_c are very fast programs. For example, gene prediction by Fgenesh+ for a sequence of 80,000 bp with a protein of eight hundred amino acids takes about 1 second on an EV6 processor in a Dec-alpha computer.

ANNOTATION OF GENOMIC SEQUENCES

GenBank (Benson et al. 1999) and EMBL (Stoesser et al. 1999) databases have for many years collected information about sequences of different genomes. A sequence-based structure of these databases often produces annotations of one gene in many different records when several gene fragments are sequenced independently. Last year, the vast amount of sequence information was produced by genome sequencing projects. Absence of experimental information about genes in a major part of these sequences makes valuable a presentation of computationally identified genes to provide positional cloners, gene hunters, and others with the gene candidates contained in finished and unfinished genomic sequences. Using these predictions, the scientific community can experimentally work with most real genes, because gene finding programs usually predict correctly most exons in a gene sequence.

Gene-Centered InfoGene Database

The *InfoGene* database database (http://www.softberry.com/inf/infodb.html) is created to collect and interactively work with information about known and predicted gene structures of human and other model genomes. Known genes are presented in 17 separate divisions (including human, mouse, *Drosophila,* nematode, *Arabidopsis,* rice, maize, and wheat), which contain records uniting available information about a particular gene from many GenBank (Release 119) entries. The human *InfoGene* division, for example, contains about 20,791 genes (including 16,141 partially sequenced genes), 54,558 coding regions, 83,488 exons, and about 58,000 donor and acceptor splice sites.

This information can be applied to create different sets of functional gene components for extraction of their significant characteristics as used in gene prediction systems. The interactive Java Viewer of Gene Structures has been designed by Igor Seledtsov and Victor Solovyev (1999) to visually inspect the gene structure of Infogene entries of known genes and predicted genes and to use for analysis of different gene prediction algorithms in annotating genomic sequences from genome sequencing projects. The viewer has four main panels (See figures 2.2, 2.12) : General View, Detail View, Locus Selection Panel, and the Output Message Panel. Both the General View Panel and the Detail View Panel have horizontal zoom scroll bars at the bottom of the windows. You can also zoom in and out by entering a scaling value from a pull-down menu: Action > Set Horizontal Scaling Factor.

General View Panel

The General View Panel shows all genes found in a given locus. If the locus contains overlapping genes, such as alternatively spliced ones, every such gene is displayed on its own line. As an example, let us look at the InfoGene locus HSAB001898. This locus should be automatically displayed in the General View Window if you have chosen the default setting from the Infogene page (the button "Show data" on this page). This example is also shown on the picture above. Genes are shown as red bars. When the mouse cursor points to one of the genes, the gene's name is displayed in the Output Message Panel. If you press and hold the right mouse button, detailed information about the gene is displayed in a separate temporary window. When the button is released, the temporary window disappears.

The same operation performed with the Shift key leaves the temporary window open after the release of the mouse button. The number of temporary information windows that can be opened at the same time is unlimited. You can mark genes or groups of genes in General View Panel by pressing and dragging the left mouse button. Marked gene(s) will then be displayed in the Detail View Window, replacing its previous content. If you press the Shift key at the same time as pressing and dragging the left mouse button, you can add new marked regions to the current ones. Then all marked regions will be displayed one under another in the Detail View Panel.

Detail View Panel

This panel offers detailed view of selected genes. Each selected gene or region is drawn on a separate line, the number of which is unlimited. Dark gray bars represent genes, yellow bars show exons, red bars show coding regions, and green bars show gene regions that are not included in a transcript. Symbols < and > at the end of exon or coding region mean that their exact boundaries are unknown. White separators that cut through genes separate the unsequenced regions. Colored triangles above genes represent functional signals: black—CAAT box, blue—TATA box, green—transcription start point, red—PolyA signal, pink—polyA site. When the mouse cursor points to any bar or triangle, information about this object is displayed in the Output Message Panel. If the mouse cursor goes across a gene, the gene name is displayed in the Output Message Panel.

Locus Selection Panel

This panel has five fields: Selection List, Locus Info Button, Back Button, Forward Button, and Search Input Window. Selection List shows the list of sequence identifiers that satisfy current criteria, set forth in the Divisions, Options, and Search Fields menus of the Viewer. The Divisions menu allows the user to choose the source of data: Genbank/Infogen known genes for several taxonomic group, or predicted genes for several organisms. The Options menu allows the user to chose a field to search in Infogen loci, GenBank Identifiers, GenBank Accession Codes, or Context. The latter option allows a search through all fields, performed by typing a search string (the wildcard * is allowed) into the Search Input Window and pressing Enter. Entries that satisfy search criteria are displayed in a Selection List. To display an entry from the list in the General View panel, double-click on an entry or select it and press the Enter key. The Forward and Back buttons display the next/previous thousand entries in selection list. Pressing the Locus Info button opens a separate window with detailed information on a given locus.

InfoGene exon-intron gene structures can be visualized by Dbscan program (http://www.softberry.com/scan.html), which searches for conserved regions in two sequences. This tool is useful to compare the localization of conservative regions with the localization of corresponding exon sequences or gene regulatory signals. In figure 2.11, we present the results of searching similar regions for mouse mRNA in a database of known human genes. We can see that all exons shown by red boxes (in the second window) have corresponding conserved regions (in the first window). In this way we can see exon boundaries in the mouse RNA. If we use predicted gene database and a given mRNA sequence, we can verify the corresponding predicted exons using this tool.

Predicted Genes in the *Drosophila* Genome

The Predicted Genes division includes an annotation of a draft of the *Drosophila* genome and a draft human genome sequence. The nucleotide sequence of nearly all euchromatic portion of the *Drosophila* genome (~120 MB) has been determined (Adams et. al. 2000) These sequences were annotated by predicting genes with the Fgenesh program and checking exon similarity with PfamA domains (Batemanet al. 2000). The results of this analysis are shown in table 2.13. In this table, in addition to computer predicted genes, also shows the results of removing (filtering out) most unreliable genes. Two criteria were used: (1) remove genes with the total length of protein coding region less than 30 amino acids; and (2) remove genes with total score of exons < 15. Such filtering proved useful to improve the accuracy of gene prediction (Salmov and Solovyer 2000). We should note that 20,622 genes include some pseudogenes and genes of mobile elements.

Table 2.13. Summary of predicted genes and proteins in *Drosophila* genome sequences.

	X	*2L*	*2R*	*3L*	*3R*	*4*	*Y*	*Unknown*	*Total*
Size (MB)	22.2	23.0	21.4	24.1	28.3	1.2	0.02	4.6	124.8
All genes	4071	4610	4573	4851	5954	133	1	691	24884
filtered	3349	3768	3915	4017	4962	105	1	504	20622
All exons	13036	15215	16310	16047	20382	679	10	1804	83483
filtered	11767	13713	15138	14561	18654	625	10	1467	75935
Exons-PfamA	1932	2148	2348	2130	2919	109	0	159	11745
filtered	1925	2141	2341	2126	2916	105	0	147	11701
United Pfam	1138	1193	1287	1216	1654	58	0	76	6622
Pfam types	431	475	499	460	546	43	0	40	1017

Table 2.14. PfamA domains found in predicted human genes

Number	*PfamA short name*	*Name*
467	Pkinase	Eukaryotic protein kinase domain
372	7tm_1	7 transmembrane receptor (rhodopsin family)
308	Myc_N_term	Myc amino-terminal region
256	Topoisomerase_I	Eukaryotic DNA topoisomerase I
224	Ig	Immunoglobulin domain
183	Rrm	RNA recognition motif
182	PH	PH domain
180	Myosin_tail	Myosin tail
166	EGF	EGF-like domain
159	Filament	Intermediate filament proteins
154	Syndecan	Syndecan domain
143	Ras	Ras family

Domain of the same type localized in neighboring exons were counted only once.

Predicted genes in the Human Genome

The nucleotide sequence of nearly 90 percent of the human genome (3 GB) has been determined by an international sequencing effort. Assembly of the current draft of the human genome was done by Prof. Haussler's Human Genome Project Team at UC Santa Cruz. Half of this sequence is occupied by repeat sequences and undefine nucleotides inserted during assembling. The Fgenesh program was used on this sequence (with masked repeats) to predict exons and assemble predicted genes. Annotation of similarity of each exon with the PfamA protein domain database (Bateman et. al. 2000) was produced by the Blast program (Altshul et. al. 1977) A total of 49,171 genes and 282,378 coding exons were predicted. On average, one gene was found per about 68,623 bp, and one exon per 11,949 bp. Complete summary of this analysis including the gene and exon numbers in different chromosomes, is presented at http:/ /www.softberry.com/inf/humd_an.html and can be viewed in the InfoGene database. Sequences of predicted exons and gene annotation data can also be copied from this site. One thousand one hundred and fifty-four types of PfamA different domains were found in the predicted proteins.

CHARACTERIZATION AND VERIFICATION OF PREDICTED GENES

Large-scale functional analysis of predicted, as well as known, genes might be done using expression micro array technology, which gives us the possibility of presenting all human genes on one or several Affymetrix type GeneChips. Traditionally genes are presented on the chips by unique oligonucleotides close to the 3′-end of the mRNA, but there are a lot of predicted new genes that have no known corresponding EST sequences. However, the expression of such genes could be studied using predicted exon sequences. We can present all predicted human exons (about 300,000) on a few chips and use expression profiling across many tissues to verify the predicted exons, observing if they are expressed in some of them.

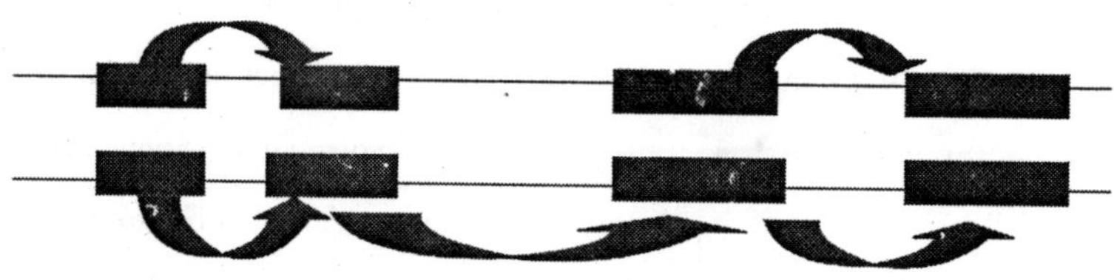

Fig. 2.10. Using expression data to verify gene boundaries and reality of exons predicted along genomic sequences. Real exons will have expression signals. Exons from one gene will have the same expression pattern in different tissues. Expression signal will increase toward 3′-exons.

Moreover, with this approach, we can verify the structure of genes (identify a subset of predicted exons that really belong to the same gene) based on the similar expression behavior of exons from the same gene in a set of tested tissues. Exons wrongly included in a predicted gene will have different expression patterns, and exons wrongly excluded by prediction will have similar expression patterns (Figure 2.11). It is interesting that such gene verification on a large scale can be done in parallel with identification of disease (tissue) specific drug target candidates. The recent chip designed by EOS Biotechnology included all predicted by Fgenesh and Genescan exons from chromosome 22, as well as predicted exons from human genomic sequences of phase 2 and 3. It was found that the predicted exon sequences present a good alternative to EST sequences, which opens a possibility of working with predicted genes on a large scale.

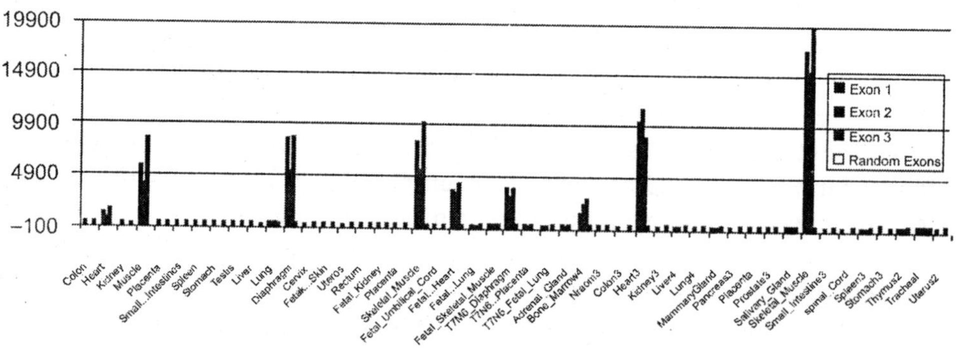

Fig. 2.11. Coordinative expression of three exons of human myoglobin gene from chromosome 22 (exons were predicted by Fgenesh program and used to design EOS Biotechnology Human genome chip). The high level of expression is observed only in several specific tissues.

An example of expression behavior of three exons of the myoglobin gene in different tissues is presented in figure 2.11 (expression data were received in EOS Biotechnology Inc.). Tissue-specific expression of this gene is clearly seen with the major peaks located in skeletal muscle, heart, and diaphragm tissues. The level of expression in these tissues is 10-100 times higher than the level of signals for other tissues, as well as the average level of expression signal for

Table 2.15. Web servers for eukaryotic gene and functional signal prediction

Program/task	WWW address
Fgenesh/HMM-based gene prediction (human, *Drosophila,* dicots, monocots, *C. elegans, S. pombe)*	http://genomic.sanger.ac.uk/gf/gf.shtml http://searchlauncher.bcm.tmc.edu:9331/ seq-search/gene-search.html http://www.softberry.com/nucleo.html
Genscan/HMM-based gene prediction (human, *Arabidipsis,* maize)	http://genes.mit.edu/GENSCAN.html
HMM-gene/HMM-based gene prediction (human, *C. elegans)*	http://www.cbs.dtu.dk/services/HMMgene/
Fgenes/Disciminative gene prediction (human)	http://genomic.sanger.ac.uk/gf/gf.shtml http://searchlauncher.bcm.tmc.edu:9331/ seq-search/gene-search. html
Fgenes-M/Prediction of alternative gene structures (human)	http://genomic.sanger.ac.uk/gf/gf.shtml http://www.softberry.com/nucleo.html
Fgenesh + /Fgenesh_c/gene prediction with the help of similar protein/EST	http://www.softberry.com/nucleo.html
Fgenesh-2/gene prediction using 2 sequences of close species	http://genomic.sanger.ac.uk/gf/gf.shtml http://www.softberry.com/nucleo.html
BESTORF/Finding best CDS/ORF in EST (human, plants, *Drosophila)*	http://genomic.sanger.ac.uk/gf/gf.shtml http://www.softberry.com/nucleo.html
Mzef/internal exon prediction (human, mouse, *Arabidopsis)*	http://argon.cshl.org/genefinder/
TSSW/TSSG/promoter prediction	http://searchlauncher.bcm.tmc.edu:9331/ seq-search/gene-search.html http://genomic.sanger.ac.uk/gf/gf.shtml
Promoter 2.0/promoter prediction	http://www.cbs.dtu.dk/services/Promoter/
CorePromoter/promoter prediction	http://argon.cshl.org/genefinder/CPROMOTER/ index.htm
SPL/splice site prediction (human, *Drosophila,* plants, yeast)	http://genomic.sanger.ac.uk/gf/gf.shtml http://www.softberry.com/nucleo.html
NetGene2/NetPGene/splice site prediction (human, *C. elegans,* plants)	http://www.cbs.dtu.dk/services/NetPGene/
Dbscan/searching for similarity in genomic sequences and its visualization altogether with known gene structure	http://www.softberry.com/nucleo.html

randomly chosen exons. We can observe that for specific tissues, all three exons demonstrate such a high level (with correlation coefficient 0.99; for random exons it is about 0.06). These exons were predicted correctly by the Fgenesh program and were used for selection of oligonucleotide probabilities. From this result we can conclude that the predicted exons can be used as a gene representatives. An additional application of expression data is the functional analysis and identification of alternatively spliced genes (exons), when in particular tissues some exons (or their parts) have very different expression intensities compared to the other exons from the same gene. Moreover, sometimes 3′-EST generated probabilities cannot be used for the identification of disease-specific gene variants in contrast with the using exon representation of a gene. (Fig. 2.12)

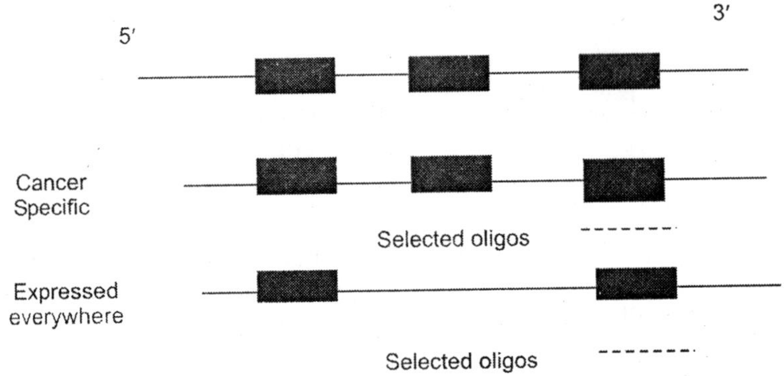

Fig. 2.12. Exon representation can be used to characterize alternatively spliced variants of genes. Oligonucleotldes selected in 3′-end of mRNA/EST sequences will not be selective for different gene variants. We will observe the sum of two genes' signal and we can miss cancer specificity of three exons' gene structure. Using oli-gonucleotides derived from exon sequences, we can detect different expressions of these two forms.

INTERNET RESOURCES FOR GENE FINDING AND FUNCTIONAL SITE PREDICTION

Prediction of genes, ORF, promoter, and splice sites finding by the methods described above is available on the World Wide Web. Table 9.15 presents just a few useful programs. It does not provide a comprehensive list.

ACKNOWLEDGEMENTS

I am grateful to Asaf Salamov and Igor Seledtsov for their collaboration in the development of gene-finding and other algorithms discussed here.

REFERENCES

Adams, M. D. et al. (2000). The genome sequence of Drosophila melanogaster. *Science* 287: 2185-2195. Aebi, M., and Weissmann, C. (1987). Precision and orderliness in splicing. *Trends Genet.* 3: 102-107.

Afifi, A. A., and Azen, S. P. (1979). *Statistical Analysis: A Computer Oriented Approach.* New York: Academic Press.

Altschul, S. F., Madden, T. L., Schaffer, A. A., Zhang, J., Zhang, Z., Miller W., and Lipman D. J. (1997). Gapped BLAST and PSI-BLAST: A new generation of protein database search programs. *Nucl. Acids Res.* 25(17): 3389-3402.

Ashburner, M., Misra, S., Roote, J., Lewis, S. E., Blazej, R., Davis, T., Doyle, C., Galle, R., George, R., Harris, N., Hartzell, G., Harvey, D., Hong, L., Houston, K., Hoskins, R., Johnson, G., Martin, C., Moshrefi, A., Palazzolo, M., Reese, M. G., Spradling, A., Tsang, G., Wan, K., Whitelaw, K., Celniker, S., and Rubin, G. M. (1999). An exploration of the sequence of a 2.9-mb region of the genome *of Drosophila melanogaster:* The Adh region. *Genetics* 153(1): 179-219.

Audic, S., and Claverie, J. (1997). Detection of eukaryotic promoters using Markov transition matrices. *Comput. Chem.* 21: 223-227.

Bateman, A., Birney, E., Durbin, R., Eddy, S., Howe, K., and Sonnhammer, E. (2000). The Pfam protein families database. *Nucl. Acids Res.* 28: 263-266.

Beck, T. W., Huleihel, M., Gunnell, M., Bonner, T. I., and Rapp, U. R. (1987). The complete coding sequence of the human A-raf-1 oncogene and transforming activity of a human A-raf carrying retrovirus. *Nucl. Acids Res.* 15(2): 595-609.

Benson, D. A., Boguski, M. S., Lipman, D. J., Ostell, J., Ouellette, B. F., Rapp, B. A., and Wheeler, D. L. (1999). GenBank. *Nucl. Acids Res.* 27(1): 12-17.

Berg, O. G., and von Hippel, P. H. (1987). Selection of DNA binding sites by regulatory proteins. *J. Mol Biol.* 193: 723-750.

Sequence Compression Program 3

With the imminent completion of the Human Genome project and the fast increase of many complete genomes of prokaryotes and eukaryotes, fundamental questions regarding the characteristics of these sequences arise the first of which is how to compare genomes. We introduce an effective general tool for such questions: compression programs for DNA sequences. We will first review some compression algorithms for biological sequences, and then switch our attention to the question of how to use such compression programs to compare genomes. Why are we interested in compressing DNA sequences? From a strictly mathematical point of view, compression implies understanding and comprehension from a more utilitarian point of view, as we will demonstrate soon, compression is a great tool for genome comparison and for studying various properties of genomes. Life represents order. It is neither chaotic nor random.

In other words, DNA sequences, which encode life, should be compressible. There is also strong biological evidence that supports this claim: it is well known that DNA sequences, especially in higher eukaryotes, contain many (approximate) tandem repeats; it is also well known that many essential genes (like rRNAs) have many copies; it is believed that there are only about a thousand basic protein folding patterns; it also has been conjectured that genes duplicate themselves sometimes for evolutionary or simply for "selfish" purposes. All these reasons give more concrete support that DNA sequences should be reasonably compressible. However, such regularities are often blurred by random mutation, translocation, cross-over, and reversal events, as well as sequencing errors. The compression of DNA sequences is a very difficult task. DNA sequences consist of only four nucleotide bases $\{a, c, g, t\}$; two bits are enough to store each base.

However, if one applies standard compression software such as the Unix "compress," "bzip2," and "gzip," or the MS-DOS archive programs "winzip" and "arj," they all *expand the* file with more than two bits per base, as shown in table 3.1 although all this compression software uses universal compression algorithms. These software tools are designed for English text compression, but the regularities in DNA sequences are subtler. One may treat compressibility study as the ultimate generalization of the simpler (and fruitful) biological studies such as G-C contents of various species.

More sophisticated studies on DNA sequences will give us a deeper understanding about the nature of these sequences. Different regions on a genome, different genes, different species may have different compression ratios. Such differences may imply, for example, different mutation rates in different genes. We will also discuss conditional compressibility where one compresses one sequence given another sequence as free information. Intuitively, conditional compressibility implies some relatedness between two sequences. However, directly using it results in an incorrect measure, which is not symmetric.

Table 3.1. Compression measured in bits per base.

Sequence	*Size (bases)*	*compress*	*arith-2*	*Biocom-press-2*	*GenCom-press-1*	*GenCom-press-2*
MTPACGA	100314	2.116	1.873	1.875	1.861	1.861
MPOMTCG	186608	2.202	1.966	1.938	1.898	1.898
CHNTXX	155844	2.187	1.934	1.617	1.614	1.614
CHMPXX	121024	2.075	1.837	1.685	1.669	1.670
HUMGHCSA	66495	2.194	1.938	1.307	1.092	1.097
HUMHBB	73323	2.195	1.918	1.877	1.813	1.814
HUMHDABCD	58864	2.230	1.943	1.877	1.800	1.809
HUMDYSTROP	38770	2.233	1.924	1.926	1.924	1.923
HUMHPRTB	56737	2.202	1.929	1.907	1.826	1.830
VACCG	191737	2.167	1.898	1.761	1.761	1.761
HEHCMVCG	229354	2.213	1.965	1.848	1.847	1.847

We define a proper distance to measure how much information two DNA sequences or two genomes share. As an example, we will show how our compression programs can be used to construct whole genome trees.

Unless otherwise mentioned, we will use the lower case letters *u, v, w, x, y* to denote finite strings over the alphabet {a, *c, g, t*}. $|u|$ denotes the length (*i.e.*, number of characters) of *u*. u_i is the *i*-th character of *u*. $u_i : j$ is the substring of *u* from position *i* to position *j*. The first character of *u* is u_0. Thus $u = u_{0\,:\,|u|-1}$. We use $\in$ to denote empty string and $|\in| = 0$.

GENCOMPRESS : A DNA SEQUENCE COMPRESSION PROGRAMME

Grumbach and Tahi (1994) proposed two lossless compression algorithms for DNA sequences, namely *Biocompress* and *Biocompress*-2 , in the spirit of the Ziv and Lempel (1977) data compression method. *Biocompress*-2 detects exact repeats and complementary palindromes located earlier in the target sequence, and then encodes them by repeat length and the position of a previous repeat occurrence. In addition, *Biocompress*-2 also uses arithmetic coding of order 2 if no significant repetition is found. In fact, the difference between *Biocompress* and *Biocompress*-2 is the addition of order-2 arithmetic coding. Rivals *et al.* (1995) give another compression algorithm, *Cfact,* which searches the longest exact matching repeat in an entire sequence using a suffix tree data structure.

The idea of *Cfact* is basically the same as *Biocompress*-2 except that *Cfact* is a two-pass algorithm. It builds the suffix tree in the first pass. In the encoding phase, the repetitions are coded with guaranteed gain; otherwise, two-bit per base encoding is used. This is similar to the codeword encoding condition in *Biocompress*-2 except that the order-2 arithmetic coding is not used in *Cfact*. *É*. Rivals *et al.* (1997) also designed a compression algorithm as a tool to detect the approximate tandem repeats in DNA sequences. Lempel and Ziv proposed two algorithms to compress universal data sequences. These are dictionary based compression algorithms that rely on exact repeats. The Lempel-Ziv algorithms can be viewed as having two components: the first component is to parse the input data sequence into variable-length strings based on the history of the dictionary. The second component is to replace the variable-length prefix by a proper binary codeword—concatenation of these codewords yields the encoder's output sequence in response to the input data sequence. We follow the same framework and generalize it to approximate matching for DNA sequences. *GenCompress* (Chenet.al. 2000) achieves significantly higher compression ratios than either *Biocompress*-2 or *Cfact*. Such improvement is key to its application in genome comparison. *GenCompress* is a one-pass algorithm. It proceeds as follows: For input w, assume that a part of it, say v, has already been compressed, and the remaining part is u, that is, $w = vu$. *GenCompress* finds an "optimal prefix" of u such that it approximately matches some substring in v; so that this prefix of u can be encoded economically. After outputting the code of this prefix, remove the prefix from u, and append it to the suffix of v. Continue the process till $u = \in$. We adopt the following constraint in *GenCompress* to limit the search.

If the number of edit operations (insert, delete, replace) located in any substring of length k in the prefix s of u for an edit operation sequence $\lambda(s, t)$ is not larger than a threshold value b, we say that $\lambda(s, t)$ satisfies the *condition* $C = (k, b)$ for compression. In *GenCompress,* we only search for approximate matches that satisfy condition C. This way we limit our search space. Experiments show that setting C to $(k,b) = (12, 3)$ gives good results. We defined a *compression gain function* G in order to evaluate if a particular approximate repeat provides profit in the encoding.

Table 3.2. Compression measured bits per base

Sequence	*Size (bases)*	*LZW 15*	*arith-2*	*Cfact*	*GenCompress-1*	*GenCompress-2*
atatsgs	9647	2.237	1.951	1.785	1.664	1.673
atefla23	6022	2.297	1.994	1.585	1.541	1.540
atrdnaf	10014	2.300	2.009	1.814	1.789	1.786
atrdnai	5287	2.239	1.994	1.468	1.419	1.410
hsg6pdgen	52173	2.168	1.937	1.928	1.785	1.800
xlxfg512	19338	2.084	1.923	1.490	1.376	1.385
mmzp3g	10833	2.244	1.953	1.911	1.854	1.857
celk07e12	58949	2.108	1.912	1.713	1.597	1.605

Let input $w = vu$, where v has already been processed. Given $G(s, t, \lambda)$ and C, the *optimal prefix* is a prefix s of u such that $G(s, t, \lambda)$ is maximized over all λ and t such that t is a substring of v and λ is an edit transcription from t to s satisfying condition C. The algorithm carefully finds the optimal prefix, and uses order-2 Arithmetic encoding whenever needed. *GenCompress* also detects the approximate complemented palindrome in DNA sequences. Some standard benchmark data has been used. These standard sequences (available at GeneBank 1999) come from a variety of sources and include the complete genomes of two mitochondria, *MPOMTCG, PANMTPACGA* (also called *MIPACGA*); two chloroplasts, *CHNTXX* and *CHMPXX* (also called *MPOCPCG*); five sequences from humans, *HUMGHCSA, HUMHBB, HUMHDABCD, HUMDYSTROP, HUMHPRTB;* and finally the complete genome from the two viruses, VACCG and *HEHCMVCG* (also called *HS5HCMVCG*). The compression ratios of *GenCompress,* as well as those of *Biocompress*-2 and some other compression algorithms, are presented in table 3.1.

The comparison of *GenCompress* with *Cfact* is presented in table 3.2 using the data from Rivals *et al.* (1995). Note that although *Cfact* looks for the best matches globally, whereas our *GenCompress* only searches for the best approximate match from the current prefix to the part of the text seen so far. *GenCompress* has a much better compression ratio than *Cfact.* From these experiments, it is clear that approximate matching plays a key role in finding similarities or regularities in DNA sequences. In conclusion, the compression results of *GenCompress* for DNA sequences indicate that our method based on approximate matching is more effective than others. *GenCompress* is able to detect more regularities and achieve better compression results.

GTAC : A DNA SEQUENCES ENTROPY ESTIMATOR

In many applications, we are more interested in estimating DNA sequence entropy than really achieving the final compression. The final compression size in bits usually overestimates the true entropy. There have been several attempts to characterize the entropy of DNA. One of the most common approaches is to estimate the probability of n-tuples for large n, and use this value to compute the block entropy (entropy of n-tuples). One problem with this approach is that it converges too slowly, requiring an exponentially large dataset. Even though genome databases are large and growing larger, the values that are obtained systematically overestimate the entropy due to the finite sample effect, and must be corrected. Several researchers address this problem and have developed methods to correct it, such as Lio et al. (1996), and Schmitt and Herzel (1997). Farach *et al.* (1994) developed a novel algorithm to estimate the entropy of DNA sequences called a match length entropy estimator.

This algorithm was used to test the differences between the entropy of introns and exons. Farach *et al.* also proved that their algorithm was universal, that is, that the entropy estimate will approach the true entropy as the size of the sequence increases, but only under the assumption,that the sequence is generated by a Markov process. Loewenstern and Yianilos (1999) developed CDNA, a program that estimates the entropy of DNA sequences. The motivation for CDNA comes from the observation that naturally occurring DNA sequences contain many more near repeats than would be expected by chance.

Two parameters that CDNA uses to capture the inexact matches are w, which represents the substring size, and h, which represents the Hamming distance. These parameters are used to create a panel of predictive experts p_w, h, each with differing values of w and h. CDNA then learns the weightings of these various experts, using Expectation Maximization, so that their predictive ability is maximized when combined into a single prediction. CDNA uses everything to the left of a nucleotide as the learning set to predict its value. The average over-all position is calculated and is the value recorded in table 3.3.

We have developed a new entropy estimator of DNA sequences, GTAC (Lanctot *et al.* 2000), which is based on the idea of Kieffer and Yang (1999) regarding the design and analysis of grammar based codes, and which recognizes the repeats and reverse complement property of DNA sequences. This entropy estimator is universal as it does not assume any source model and works for any individual sequence. This entropy estimator is proved to be universal. Table 3.4 compares GTAC with several entropy estimators. Kieffer and Yang (1999) recently put forth a new type of lossless source code called a grammar based code, and developed a new universal lossless source coding theory. In this theory, a grammar based code has the structure shown in figure 3.1.

Table 3.3. Comparison of entropy values in bits per symbol.

Sequence name	*Sequence length*	*UNIX compress*	*Biocompress-2*	*CDNA compress*	*GTAC*
PANMTPACGA	100314	2.12	1.88	1.85	1.74
MPOMTCG	186609	2.20	1.94	1.87	1.78
CHNTXX	155844	2.19	1.62	1.65	1.53
CHMPXX	121124	2.09	1.68	—	1.58
SCCHRIII	315339	2.18	1.92	1.94	1.82
HUMGHCSA	66495	2.19	1.31	0.95	1.10
HUMHBB	73308	2.20	1.88	1.77	1.73
HUMHDABCD	58864	2.21	1.88	1.67	1.70
HUMDYSTROP	38770	2.23	1.93	1.93	1.81
HUMHPRTB	56737	2.20	1.91	1.72	1.72
VACCG	191737	2.14	1.76	1.81	1.67
HEHCMVCG	229354	2.20	1.85	—	1.74

Note that UNIX-compress and Biocompress-2 are lossless compression algorithms. CDNA and GTAC are entropy estimators.

Table 3.4. Features of various entropy estimators

Algorithm	*Universal*	*Linear run time*	*Entropy estimate*
UNIX compress	yes	yes	worst
Match length	limited	yes	—
Biocompress-2	yes	yes	3rd best
CDNA	no	no	2nd best
GTAC	yes	yes	best

The idea is as follows. Given a sequence x, we generate a particular context-free grammar G_x such that one can recover x from G_x. For example, here is G_x for $x = aataaatgcaatatatatgc$.

$$\begin{aligned} S &\rightarrow BADBCCD \\ A &\rightarrow aa \\ B &\rightarrow At \\ C &\rightarrow at \\ D &\rightarrow Cgc \end{aligned}$$

Then we use an arithmetic coder to encode G_x.

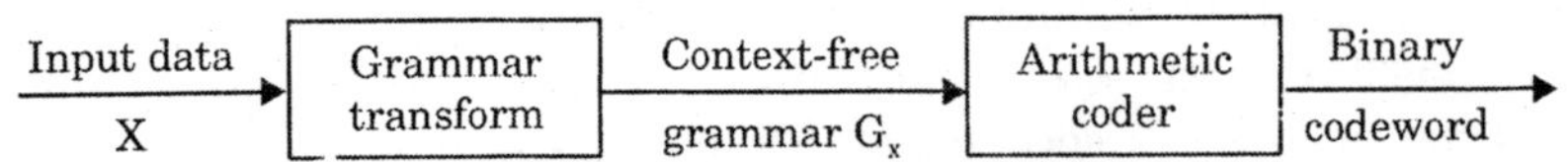

Fig. 3.1. Structure of a grammar-based code.

To characterize the resulting compression rate, let $\omega(G_x)$ be the sequence obtained by concatenating the righthand side of all production rules of G_x in some order and then deleting the first appearance of each variable (because they can be uniformly replaced by one special symbol and, during decoding, can be recovered by their order of appearance). Let

$$H(G_x) = \sum_s n(s) \log \frac{|\omega(G_x)|}{n(s)} \qquad ...(3.1)$$

where the summation is taken over all variables and terminal symbols in G_x, $n(s)$ denotes the number of times the variable (or terminal symbol) s appears in $\omega(G_x)$, and $|\omega(G_x)|$ denotes the length of the sequence $\omega(G_x)$. Also, in formula (3.1), the logarithm is relative to base 2, and the convention $0 \log \infty = 0$ is adopted. In terms of the terminology in Kieffer and Yang (1999), the quantity $H(G_x)$ is called the unnormalized entropy of the grammar G_x. For the CFG G_x shown in example 1, $\omega(G_x) = BCDaaAtatCgc$ and $H(G_x) = 34.26$. The following theorem, proved by Kieffer and Yang (1999), characterizes the resulting compression rate.

THEOREM 3.1

According to arithmetic coding or enumerative coding, one can assign a uniquely decodable binary codeword $B(G_x)$ to each admissible CFG G_x (or its equivalent form) such that

$$|B(G_x)| = f(G_x) + H(G_x) \qquad ...(3.2)$$

where $|B(G_x)|$ denotes the length of the binary codeword $B(G_x)$, and $f(G_x)$ represents the overhead paid to the universality of grammar based codes. In (3.2), $f(G_x)$ is negligible compared to $H(G_x)$ and is upper bounded, by

$$f(G_x) \leq 5|G_x| + 4$$

where $|G_x|$ denotes the total length of the righthand side of all production rules of G_x.

Because it can be shown that $f(G_x) \leq 5|G_x| \leq O(|x|/\log|x|)$, $H(G_x)/|x|$ goes to the actual per-bit-entropy of x (Kieffer and YanG 199). This justifies the use $H(G_x)$ as the entropy of x.

In order to estimate $H(G_x)$,we need to first construct G_x. Our task is to repeatedly find a longest match,. and then replace it by a nonterminal and create a new rule in the grammar G_x. Such a task, done trivially, easily costs $\Omega(n^3)$. Using a suffix tree data structure and by carefully maintaining changes on the suffix tree, a linear time algorithm is possible for computing G_x (Lanctot et. al. 2000). Note that the linear time algorithm is important here because the sequences we are compressing are often of many megabases. Table 3.3 compares the entropy estimates on the standard benchmark data using various programs. GTAC gives the lowest entropy estimates in most cases.

Together with Jonathan Badger, we have performed some preliminary experiments with interesting consequences. One such experiment is concerned with coding and noncoding regions in *E. coli*. Around 90 percent of the genome of higher eukaryotes is noncoding, whereas about 15 per cent of the genome of *E. coli* is noncoding. If noncoding regions have a definite role, they may be more regular than coding regions, which would support the conjecture that noncoding regions in prokaryotes are not junk. Our results confirmed this hypothesis. When comparing coding and noncoding regions of *E. coli*, we found the following entropy values:

- 1.85 bits/symbol for coding regions (4,090,525 bases)
- 1.80 bits/symbol for noncoding regions (640,039 bases).

A NEW DISTANCE MEASURE AND WHOLE GENOME COMPARISON

As we accumulate an enormous amount of nucleotide data (doubling every 18 months), one exciting challenge facing bioinformatics researchers is to provide tools to analyze such data. Entropy estimators provide a convenient tool for some aspects of such a task.

Given two sequences x and y, DNA or otherwise, we (Chen et. al. 2000; Liet al. 2000) have defined a distance measure between x and y as:

$$d(x, y) = 1 - \frac{K(x) - K(x \mid y)}{K(xy)}$$

where $K(x)$ is the Kolmogorov complexity of the string x, that is, the length in bits of the shortest program causing a standard universal computer to compute x as its unique output. $K(x \mid y)$ is the conditional Kolmogorov complexity, assuming the above program has y as extra free information. See Li and Vitanyi (1997) for a comprehensive introduction to the subject and its applications.

The numerator of the fraction is the amount of information y knows about x, which is equal to the amount of information x knows about y, by a theorem in Kolmogorov complexity to within a logarithmic error. This is called mutual algorithmic information, which is not itself a distance and does not satisfy the triangle inequality; however, $d(x, y)$ as defined above does (Liet al. 2001), ranging from a minimum of 0 when $x = y$, to a maximum of 1 when x and y are independent strings of equal or differing length. Distance function $d(x, y)$ is not only well-defined, it is also *universal* in the sense that if any distance measure uncovers some similarity between two sequences, so will d. Of course, it is well known that K is an uncomputable function; accordingly, we used our program *GenCompress,* discussed in section 3.2, to heuristically approximate it. Notice that *GTAC* is an unconditional entropy estimator, and at this point it cannot be used to do conditional entropy estimation. The corresponding conditional entropy theory and program is under development. Different approaches for comparing genomes or general sequences have been proposed. Grumback and Tahi (1994) proposed to use conditional compression.

Using ideas of Kolmogorov complexity, Varre, Delahaye, and Rivals defined "transformation distance." Essentially, this can be regarded as conditional compression using biologically related operations. Although very attractive, both of these measures are not symmetric, and hence cannot be used as distance in general. This situation can in fact be remedied by using Information Distance as defined by Bennett *et al.* (1998). However, information Distance, although symmetric, is also not a right measure in this case. The Information Distance is not suitable in genome comparison because it would over punish long deletions. Biologists proposed to use more involved and laborious methods like counting the number of shared genes in two genomes or comparing the ordering of the genes. These distances theoretically may be regarded as special cases of our $d(x, y)$, because each of these proposed measures may be regarded simply as one way of compression.

These distances, together with G+C content, edit distance, and reversal and rearrangement distances compare genomes using only partial genome information, and with a pre-assumed model of similarity, whereas our new distance uses all genome information and makes no assumption of evolutionary model, at least in theory. To demonstrate that theory is applicable, we have performed many experiments on genomes. One is described here.

It has been debated which two of the three main groups of placental mammals are more closely related: primates, ferungulates, and rodents. This is because by the maximum likelihood method, some proteins support the (ferungulates, (primates, rodents)) grouping, whereas other proteins support the (rodents, (ferungulates, primates)) grouping. Cao *et al.* (1998) aligned 12 concatenated mitochondrial proteins from the following species: rat (*Rattus norvegicus*), house mouse (*Mus musculus*), grey seal (*Halichoerus grypus*), harbor seal (*Phoca vitulina*), cat (*Felis catus*), white rhino (*Ceratotherium simum*), horse (*Equus caballus*), finback whale (*Balaenoptera physalus*), blue whale (*Balaenoptera musculus*), cow (*Bos taurus*), gibbon (*Hylobates lar*), gorilla (*Gorilla gorilla*), human (*Homo sapiens*), chimpanzee (*Pan troglodytes*), pygmy chimpanzee (*Pan paniscus*), orangutan (*Pongo pygmaeus*), and Sumatran orangutan (*Pongo pygmaeus abelii*), using opossum (*Didelphis virginiana*), wallaroo (*Macropus robustus*), and platypus (*Orni-thorhynchus anatinus*) as the outgroup, and built the maximum likelihood tree to confirm the grouping (rodents, (primates, ferungulates)).

Using the complete mitochondrial genomes of these species, we approximated our new distance $d(x, y)$ between each pair of species x and y. We then constructed a tree using the neighbour joining program in Adachi and Hasegawa's MOLPHY package. The tree is identical to the maximum likelihood tree of Cao et al. (1998). Because neighbour-joining is sometimes distrusted to further corroborate this grouping we applied our own hypercleaning program (Berry et. al. 2000) to the same distance matrix and obtained the same tree. The hypercleaning program constructs an evolutionary tree using the edges best supported by all possible four taxa subtrees (commonly called "quartets"). Thus using the new information-theoretic distances derived from the *complete* mtDNA genomes we have re-confirmed the hypothesis of (rodents, (primates, ferungulates)). The distance matrix can be found at the author's homepage: http://www.math.uwaterloo.ca/~mli/distance.html. The simple asymmetric measure $K(x \mid y)$ leads to a wrong tree using the same data and programs, as expected (data not shown).

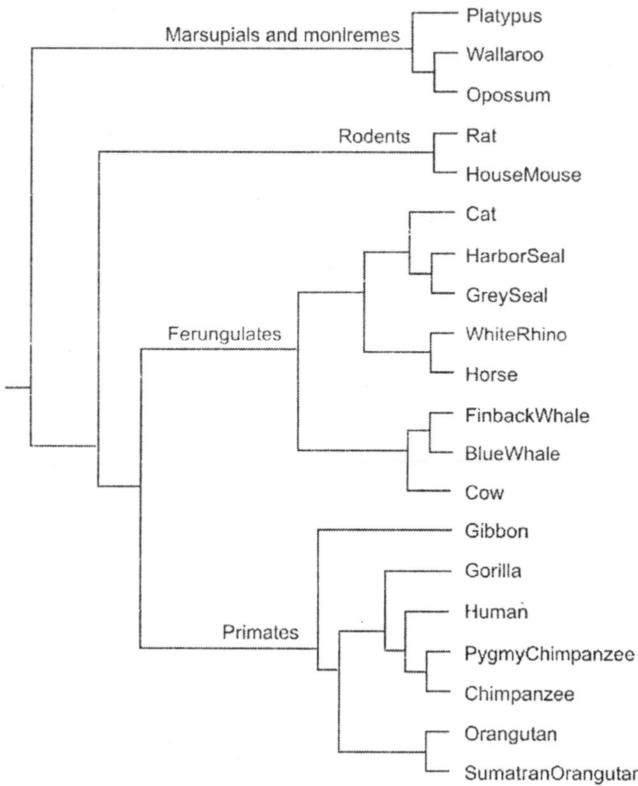

Fig. 3.2. The evolutionary tree built from complete mammalian mtDNA sequences.

The gene order and gene content approaches, although yielding symmetric distances, have the disadvantage of requiring laborious human analysis of the sequences and also are unlikely to provide enough information to distinguish closely related species such as the above data set. Note that this complete process is fully automatic and utilizes the information contained in noncoding regions in addition to the information contained in the genes.

Of course, the question is whether this method will in fact work on complete genomes. We have performed a small scale experiment with the following seven complete genomes.

- ***Archaea Bacteria:*** *Archaeoglobus fulgidus* (ul), *Pyrococcus abyssi* (u2), *Pyrococcus horikoshii* OT3 (u3).
- ***Bacteria:*** *Escherichia coli* K-12 MG1655 (u4), *Haemophilus influenzae Rd* (u5), *Helicobacter pylori* 26695 (u6); *Helicobacter pylori,* strain J99 (u7).

The resulting distance matrix, using 100 $(1 - d(x, y))$, is given in table 3.5 and the corresponding evolutionary tree is in figure 3.3.

Table 3.5. Distance $d(u, v)$ between all pairs u, v.

Sequence	*u1*	*u2*	*u3*	*u4*	*u5*	*u6*	*u7*
u1		4226821[b]	4226743	4228299	4228411	4228356	4228392
2178400[a]		0.018326[c]	0.019550	−0.000548	−0.002399	−0.001765	−0.002259
u2	3443540		3391432	3445299	3445242	3445257	3445264
1765118	0.023072		0.797546	0.00089	0.0000988	0.000812	0.000705
u3	3362288	3310372		3364086	3363996	3364031	3364045
1738505	0.023055	0.794383		−0.000391	0.000617	0.000109	−0.000109
u4	8920179	8920362	8920294		8914204	8919249	8919224
4639221	0.000373	−0.001084	−0.000537		0.048760	0.008160	0.008371
u5	3440205	3440216	3440229	3434165		3439033	3439068
1830138	0.000274	0.000145	−0.000044	0.049059		0.018303	0.017776
u6	3079174	3078992	3079021	3077924	3077935		1226333
1667867	−0.002217	0.000307	−0.000140	0.009068	0.016523		43.069863
u7	3075330	3075285	3075238	3074059	3073952	1219515	
1643831	−0.001314	−0.000782	−0.000062	0.009796	0.019680	43.171044	

[a]Number of bases in the input sequence.
[b]Number of bits of conditionally compressed file between u_i and u_j.
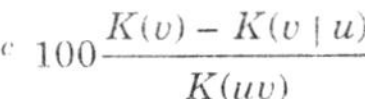
[c] $100\frac{K(v) - K(v \mid u)}{K(uv)}$.

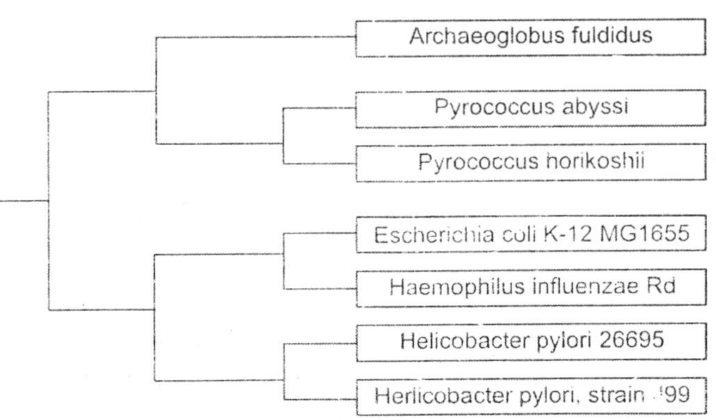

Fig. 3.3. The phylogeny for seven genomes derived from table 3.5

DISCUSSION

We hope this new methodology and the automatic tool we have developed could serve as an alternative approach to comparing genomes and constructing whole genome trees. It perhaps would serve as a quick and dirty initial way to compare genomes. This new method for whole genome comparison and phylogeny requires neither gene identification nor any human intervention; in fact, it is totally automatic, and mathematically well-founded. It works when there are no agreed upon evolutionary models, as further demonstrated by the successful construction of a chain letter phylogeny (Bennett *et al.* 2000) and when individual gene trees do not agree, as is the case for genomes.

However, there are many questions that remain to be answered and experiments to be performed. It is important to design a good conditional entropy estimator for DNA sequences. Only with a better and fast conditional entropy estimator can this method become effective. It is important to perform more experiments with this method to identify the range of data such a method works for. One potential objection to our method is that if noncoding regions are usually junk, then are we measuring junk? For prokaryotes, at least, this is not the case, as our experiments (performed by J.H. Badger, using *GTAC)* show that coding regions for *E. coli* have higher entropy than noncoding regions of *E. coli*. It is also interesting to extend our study here to other domains, such as program comparison. Currently we are working on this project.

Fragment Assembly of DNA

BIOLOGICAL BACKGROUND

The *sequence* a DNA molecule is to obtain the string of bases that it contains. In large-scale DNA sequencing we have a long target DNA molecule (thousand of hp) that we want to sequence. We may think of this problem as a puzzle in which we are given a double row of cards facing the table. We do not know which letter from the set (A, C, G, T) is written on each card, but we do know that cards in the same position of opposite strands from a complementary pair. Our goal is to obtain the letters using certain *hints*, which are (approximate) substrings of the rows. The long sequence to reconstruct is called the *target*. In the biological problem, we know the length of the target sequence approximately, within 10% or so.

It is impossible to sequence the whole molecule directly. However, we may instead get a piece of the molecule starting at a random position in one of the strands and sequence it in the canonical ($5' \rightarrow 3'$) direction for a certain length. Each such sequence is called a *fragment*. It corresponds to a substring of one of the strands of the target molecule, but we do not know which strand or its position relative to the begining of the strand in addition it may contain errors. By using the *shortgun method*, we obtain a large number of fragments and then we try to reconstruct the target molecule's sequence based on fragment overlap. Depending on experimental factors, fragment length can be as low as 200 or as high as 700. Typical problems involve target sequences 30,000 to 100,000 base-pairs long, and total number of fragments is in the range 500 to 2000.

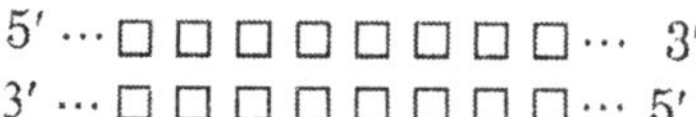

Fig. 4.1. Unknown DNA to be sequenced.

The problem is then to deduce the whole sequence of the target DNA molecule. Because we have a collection of fragments to put together, this task is known as *fragment assembly*. We note that it suffices to determine one of the strands of the original molecule, since the other can be readily obtained because of the complementary pair rule. In the remainder of this section we give additional details of a biological nature that are important in the design of fragment assembly algorithms.

THE IDEAL CASE

The best way of studying the issues involved is by looking at an example. be aware, however, that real instances are much larger than the examples we present. Suppose the input is composed of the four sequences.

ACCGT
CGTGC
TTAC
TACCGT

and we know that the answer has approximately 10 bases. One possible way to assemble this set is

```
--ACCGT--
----CGTGC
TTAC-----
-TACCGT--
---------
TTACCGTGC
```

Notice that we aligned the input set, ignoring spaces at the extremities. We try to align in the same column bases that are equal. The only guidance to assembly, apart from the approximate size of the target, are the *overlaps* between fragments. By overlap here we mean the fact that sometimes the end part of a fragment is similar to the beginning of another, as with the first and second sequences above. By positioning fragments so that they align well with each other we get a layout, which can be seen as a multiple alignment of the fragments.

The sequence below the line is the *consensus sequence*, or simply *consensus*, and is the answer to our problem. The consensus is obtained by taking a majority vote among all bases in each column. In this example, every column is unanimous, so computing the consensus is straightforward. This answer has nine bases, which is close to the given target length of 10, and contains each fragment as an exact substring. However, in practice fragments are seldom exact substrings of the consensus.

Complications

As mentioned above, real problem instances are very large. Apart from this fact, several other complications exist that make the problem much harder than the small example we saw. The main factors that add to the complexity of the problem are errors, unknown orientation, repeated regions, and lack of coverage. We describe each factor in the sequel.

Errors **:** The simplest erros are called *base call* errors and comprise base substitutions, insertions, and deletions in the fragments. Transpositions are also common but we can treat them as compositions of an insertion and a deletion or two substitutions.

Input	**Answer**
ACCGT	--ACCGT--
CGTGC	----CGTGC
TTAC	TTAC-----
TGCCGT	-TGCCGT--
	TTACCGTGC

Fig. 4.2. In this instance there was a substitution error in the second position of the last fragment, where A was replaced by G. The consensus is still correct because of majority voting.

Base call errors occurs in practice at rates varying from 1 to 5 errors every 100 characters. Their distribution along the sequence is not uniform, as they tend to concentrate towards the

3′ end of the fragment. As we can see from the examples, it is still possible to reconstruct the correct consensus even in the presence of errors, but the computer program must be prepared to deal with this possibility and this usually means algorithms that require more time and space.

For instance, it is possible to find the best alignments between two sequences in linear time if there are no errors. Apart from erroneous base calls, two other types of errors can affect assembly. One of them is the artifact of *chimeric* fragments, and the other is *contamination* by host or vector DNA. We explain each type in what follows.

Input	**Answer**
ACCGT	– – ACC – GT – –
CAGTGC	– – – – CAGTGC
TTAC	TTAC – – – – – –
TACCGT	– TACC – GT – –
	TTACC – GTGC

Fig. 4.3. In this instance there was an insertion error in the second position of the second fragment. Base appeared where there should be none. The consensus is still correct because of spaces introduced in the multiple alignment and majority voting. Notice that the space in the consensus will be discarded when reporting the answer.

Input	Answer
ACCGT	– – ACCGT – –
CGTGC	– – – – CGTGC
TTAC	TTAC – – – – –
TACGT	– TAC – GT – –
	TTACCGTGC

Fig. 4.4. In this instance there was a deletion in the third (or fourth) base in the last fragment. The consensus is still correct because of spaces in the alignment and majority voting.

Chimeric fragments, or *chimeras*, arise when two regular fragments from distinct parts of the target molecule join end-to-end to form a fragment that is *not* a contiguous part of the target. These misleading fragments must be recognized as such and removed from the fragment set in a preprocessing stage. Sometimes fragments or parts of fragments that do not have anything to do with the targer molecule are present in the input set. This is due to contamination from host or vector DNA.

The process of replicating a fragment consists of inserting it into the genome of a vector, which is an organism that will reproduce and carry along copies of our fragment. in the end, the fragment must be purified from the vector DNA, and here is where contamination occurs — if this purification is not complete. If the vector is virus, then the infected cell – generally a bacterial cell — can also contribute some genetic material to the fragment. Contamination is a rather common phenomenon in sequencing experiments.

Witness to this is the significant quantity of vector DNA that is present in the community databases. Scientists sometimes fail to screen against the vector sequence prior to assembly and submit a contaminated consensus to the database. As with chimeras, the remedy for this problem is to screen the data before starting assembly. The complete sequences of vectors commonly used in DNA sequencing are well known, and it is not difficult to screen all fragments against these known sequences to see whether a substantial part of them is present in any fragment. Chimeric fragment detection leads to interesting algorithmic problems, but we will not detail them any further in this book. Pointers to relevant references are given in the bibliographic notes.

Input	Answer
ACCGT	--ACCGT--
CGTGC	----CGTGC
TTAC	TTAC-----
TACCGT	-TACCGT--
TTATGC	TTACCGTGC
	TTA---TGC

Fig. 4.5. The last fragment in this input set is a chimera. The only way to deal with chimeras is to recognize and remove them from the input set before starting assembly proper.

Unknown Orientation : Fragments can come from any of the DNA strands and we generally do not know to which strand a particular fragment belongs to. We do know, however, that whatever the strand the sequence read goes from 5′ to 3′. Because of the complementarity and opposite orientation of strands, the fact that a fragment is a substring of one strand is equivalent to the fact that its reverse complement is a substring of the other. As a result, we can think of the input fragments as being all approximate substrings of the consensus sought either as given or in reverse complement.

As shown in the figure an assembly problem involving fragments in both orientations, initially unknown, but with no erros. In practice, we have to deal with both errors and unknown orientation at the same time. Because the orientations are unknown, in principle we should try all possible combinations, which are 2^n for a set with n fragments. Of course, this method is unacceptable and is not the way it is done in an assembly program, but it does hint at the complexity introduced by the orientation issue.

Input		Answer
CACGT	→	CACGT---------
ACGT	→	-ACGT---------
ACTACG	←	--CGTAGT------
GTACT	←	-----AGTAC---
ACTGA	→	---------ACTGA
CTGA	→	----------CTGA
		CACGTAGTACTGA

Fig. 4.6. Fragment assembly with unknown orientation. Initially we do not know the orientation of fragments. Each one can be used either in direct or reverse orientation. In the solution, we indicate by an arrow the chosen orientation: → means fragment as is, ← means its reverse complement.

Repeated Regions : Repeated regions or *repeats* are sequences that appear two or more times in the target molecule. Short repeats, that is, repeats that can be entirely covered by one fragments, do not pose difficulties. The worst problems are caused by longer repeats. Also the copies of a repeat do not have to be identical to upset assembly. If the level of similarity between two copies of a repeat is high enough, the differences can be mistaken for base call erros, Remember that the assembler must be prepared to deal with errors, so there is usually some degree of tolerance in overalp detection. The kinds of problems that repeats cause are twofold.

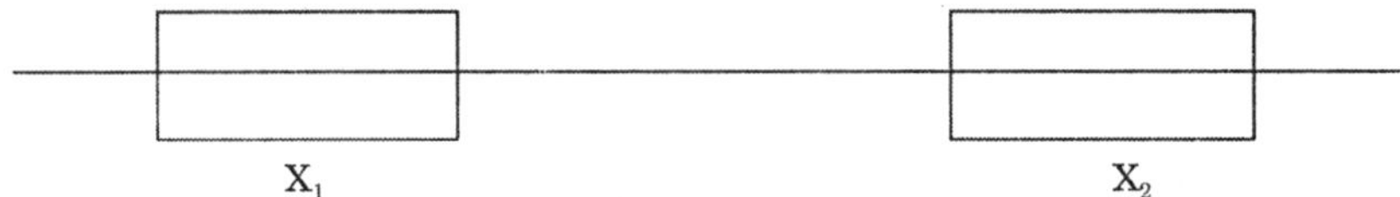

Fig. 4.7. Repeated regions. The blocks marked X_1 and X_2 are approximately the same sequence.

First, if a fragment is totally contained in a repeat, we may have several places to put it in the final alignment, as it may fit reasonably well in the several repeat copies. One could argue that it does not matter where we put it, since in any copy of the same repeat the consensus will be approximately the same. But the point is that, when the copies are not exactly equal, we may weaken the consensus by placing a fragment in the wrong copy. Second, repeats can be positioned in such a way as to render assembly inherently ambiguous; that is two or more layouts are compatible with the input fragments and approximate target length at equivalent levels of fitness. The first one features three copies of the same repeat, and the second has two interleaving copies of different repeats.

The common feature between these two cases is the presence of two different regions flanked by the same repeats. In the first example, both B and C are flanked by X and X. In the second, both B and D are flanked by X and Y. So far we discussed *direct repeats*, namely, repeated copies in the same strand. However, *inverted repeats*, which are repeated regions in opposite strands, can also occur and are potentially more dangerous. As few as two copies of a long, inverted repeat are enough to make the instance ambiguous. An example is given in Fig. 4.10.

Lack of Coverage : Another problem is lack of adequate coverage. We define the *coverage* at position i of the target as the number of fragments that cover this position. This concept is well defined, but it is impossible to compute it because we do not know the actual positions of the fragments in the target. Even fter assembly, all we have is our best guess about these positions. We can compute the *mean coverage*, though, by adding up all fragment lengths and dividing by the approximate target length.

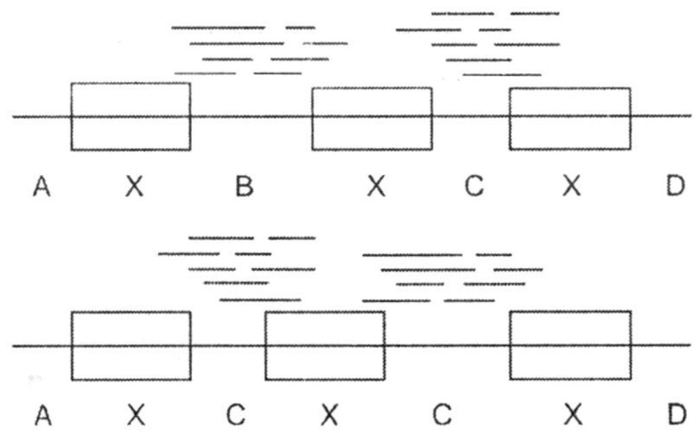

Fig. 4.8. Target sequence leading to ambiguous assembly because of repeats of the form XXX.

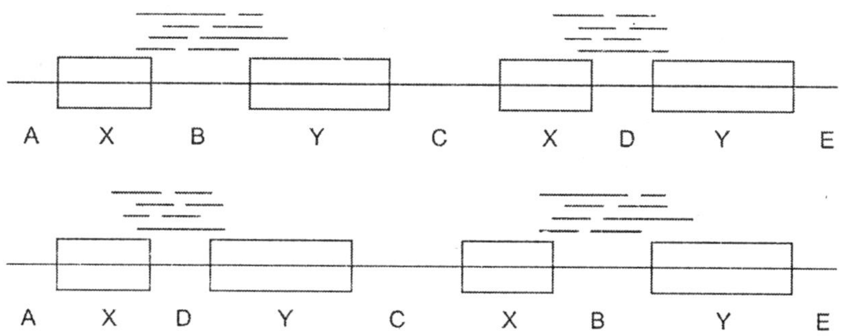

Fig. 4.9. Target sequence leading to ambiguous assembly because of repeats of the form XYXY.

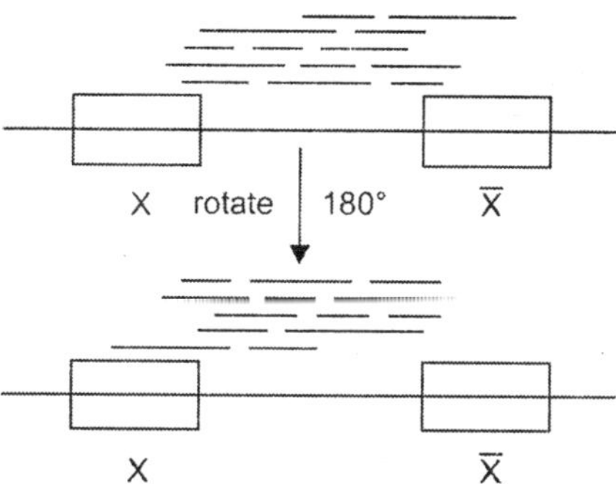

Fig. 4.10. Target sequence with inverted repeat. The region marked $\overline{X}$ is the reverse complement of the region marked X.

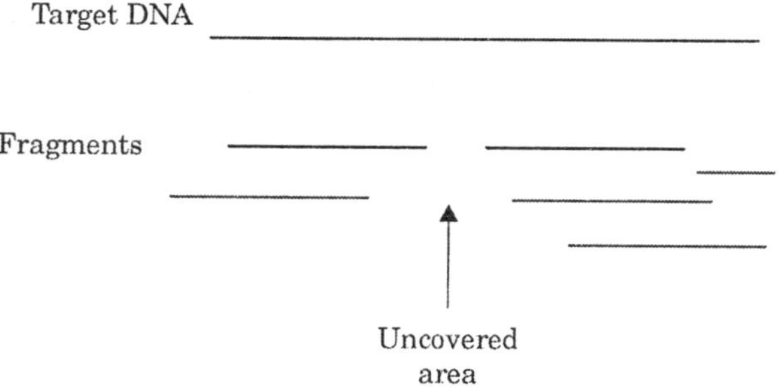

Fig. 4.11. Example of insufficient coverage.

If the coverage is equal to zero for one or more positions, then there is not enough information in the fragment set to reconstruct the target completely. In such cases, the best we can hope to achieve is a layout for every one of the contiguously covered regions, called *contigs*. Lack of coverage occurs because the sampling of fragments is essentially a random process; it is therefore possible that some parts end up not being well covered or not covered at all.

In general, we want not only one, but several different fragments covering any given point. The more fragments we have, the safer is our assessment of the consensus based on voting. It is also desirable to have fragments from both strands covering a given region, for it has been observed in practice that certain kinds of errors occur consistently in one strand only. Insufficient coverage can usually be dealt with by sampling more fragments, but we must be careful with this approach. If everything is well covered except for a small portion of the target, sampling at random can be a very inefficient way of covering the gaps. An approach, called *directed sequencing* or *walking* can be used in this case. To avoid the formation of these little gaps, some researchers advocate sampling at very high coverage rates, collecting fragments whose combined length is enough to cover the target molecule 8 times or more. It is important to know how many fragments we need to generate in order to achieve a given coverage.

The answer cannot be absolute, of course, since sampling is a random process, but reasonable bounds can be derived. We mention here an important result that is valid under certain simplifying assumptions. Let T denote the length of the target molecule. Assume that all fragments have about the same length l and that we can safely recognize overlaps of at least t bases. If we sample n fragments at random, the expected number p of *apparent contings* is given by the formula

$$p = ne^{-n(l-t)/T} \quad \text{...(4.1)}$$

The term *"apparent contig"* refers to the fact that we are assuming that we are able to recognize overlaps only if they have size at least t. Thus some of the contigs we have may in fact overlap by less that t bases, so they are really parts of a longer true contig but *"appear"* to be two separated contigs to us. Notice that p approaches zero as the number n of fragments grows, which seems strange because we know there will always be at least one apparent contig. This happens because formula is an approximation. It can be used also in the context of DNA mapping, where the number of contigs is much larger, so a difference of one or two is negligible. The formula is useful in the sense that it provides a reliable ballpark in a wide variety of situations. A formula for the fraction of the target molecule covered by the fragments is also available. With the same notation as above, the fraction covered by exactly k fragments is given by

$$r_k = \frac{e^{-c}c^k}{k!}, \quad \text{...(4.2)}$$

where $c = nl/T$ is the mean coverage.

ALTERNATIVE METHODS FOR DNA SEQUENCING

We mention here a few supplementary and alternative approaches to DNA sequencing. We start with *directed sequencing,* a method that can be used to cover small remaining gaps in a shotgun project. In direct sequencing a special primer is derived from the sequence near the end of a contig, so that fragments spanning the region including the end of this contig and the continuation of it in the target are generated. These new fragments are then sequenced and give the sequence adjacent to the contig, thereby augmenting it. Continuing in this fashion, we can cover the gap to the next contig.

The problem with this approach is that it is expensive to build special primers. Also, the next step can be accomplished only after the current one, so the process is essentially sequential rather than parallel (but can be done for all gaps in parallel).

Another technique that has become very popular is called *dual end sequencing.* In a shotgun experiment several copies of the target DNA molecule are broken randomly and short pieces are selected for cloning and sequencing. We recall from Section 1.5.2 that these pieces are called *inserts* because they are inserted into a vector for amplification. Inserts sizes range from 1 to 5 kbp, but only about 200 to 700 bases can be directly read from one extremity to yield a fragment.

However, it is also possible to read the other extremity if we have a suitable primer. The two fragments thus obtained belong to opposite strands, and they should be separated in the final alignment by roughly the insert size minus the fragment size. This extra information is extremely useful in closing gaps, for instance. Sometimes the dual end is sequenced only if it is necessary to close a gap. Notice that dual end sequencing exploits the fact that inserts are usually larger than the portion read from them. A radically different approach from shotgun methods has been proposed recently. Called *sequencing by hybridization* (SBH), it consists of assembling the target molecule based on many hybridization experiments with very short, fixed length sequences called *probes*.

A hybridization simply checks whether a probe binds (by complementarity) to a DNA molecule. The idea is to design a DNA *chip* that performs all the necessary hybridizations simultaneously and delivers a list of all strings of length w, or w-mers, that exist in the target. With current technology it is possible to construct such chips for probes of length up to eight bases. Larger probe sizes seem still prohibitive at this point. Important issues in SBH include the following. It is clear that not all target molecules can be reconstructed with a probe size as small as eight.

One important problem then is to characterize the molecules that can actually be reconstructed with a given probe size. Another problem is that the hyrbidization experiments do not provide the number of times a given w-mer appears in the target, just whether an w-mer does appear. Clearly, it would help to have the extra information on how many copies do appear. Finally, problems with errors in the experiments and orientation need to be tackled. One proposed strategy is to reduce the shotgun data to hybridization data by simply using all the w-mers of the fragments instead of the fragments themselves and to think of them as generated by hybridization experiments. However, this strategy throws away information, because it is generally impossible to recover the fragments from the w-mers, while it is possible to generate the w-mers from the fragments.

MODELS

We will now examine formalisms for fragment assembly. We present in this section three models for the problem: shortest common superstring (SCS), *RECONSTRUCTION*, and *MULTICONTIG*. Each throws light on the diverse computational aspects of the problem, although none of them completely addresses the biological issues. All three assume that the fragment collection is free of contamination and chimeras.

Shortest Common Superstring

One of the first attempts to formalize fragtment assembly was through a string problem in which we seek the shortest superstring of a collection of given strings. Accordingly, this is called the *Shortest Common Superstring* problem, or SCS. Although this model has serious shortcomings in representing the fragment assembly problem — it does not account for errors, for instance — the techniques used to tackle the resulting computational problem have application in other models as well. It is worthwhile, therefore, to study these techniques. The shortest common Superstring problem is defined as follows :

Problem : Shortest Common Superstring (SCS)

Input : A collection $\mathcal{F}$ of strings.

Output : A shortest possible string S such that for every $f \in \mathcal{F}$, S is a superstring of f.

Example 4.1

Let $\mathcal{F}$ = (ACT, CTA, AGT). The sequence S = ACTAGT is the shortest common superstring of $\mathcal{F}$. It obviously contains all fragments in $\mathcal{F}$ as substrings. To see that it is the shortest, notice that any string S' that has both u = ACT and v = AGT as substrings must have length at least 6. Moreover, with $|S'| = 6$ we have just the concatenations uv and vu. But CTA is a substring of $uv = S$ only.

In the context of fragment assembly, the collection $\mathcal{F}$ corresponds to the fragments, each one given by its sequence in the correct orientation, and S is the sequence of the target DNA molecule. Notice that the computational problem specifies that S should be a perfect superstring of each fragment, not an approximate superstring, so it does not allow for experimental errors in the fragments.

Furthermore, the orientation of each fragment must be known, which is seldom the case. Finally, even in a perfect assembly project in which all these factors could be somehow controlled, the shortest common superstring may not be the actual biological solution because of repeated sections in the target DNA sequence, as the following example shows.

Example 4.2

Suppose that the target molecule has two copies of an exact repeat and that fragments are sampled. Notice that the repeat copies are long and contain many fragments. In this case, even if the fragments are exact substrings of the consensus, and even if we know their correct orientation, finding the shortest common superstring may not be what we want.

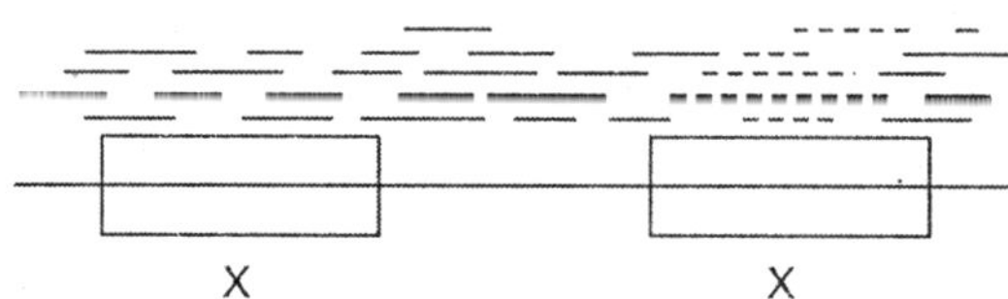

Fig. 4.12. Target sequence with long repeat that contains many fragments. This example shows that even with no errors and known orientation, the SCS formulation fails in the presence of repeats.

Indeed, Figure shows a different assembly, with a shorter consensus for the same fragment set. Observe that because the repeat copies are identical, a superstring may contain only one copy, which will absorb all fragments totally contained in any of the copies. The other copy can be shorter, as it must contain only fragments that cross the border between X and its flanking regions. The shortened version of X is denoted by X' in this figure.

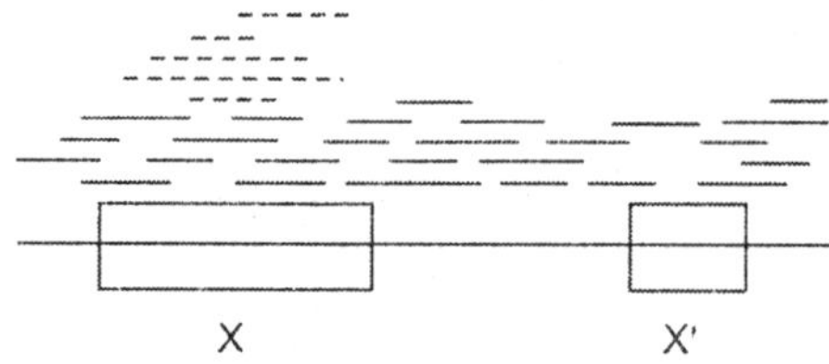

Fig. 4.13. Alternative assembly for the fragments in the previous figure. This assembly leads to a shorter consensus, because all fragments totally contained in the rightmost copy (the dashed fragments) were moved to the leftmost copy, causing a decrease in length of the rightmost copy.

Notice that although shorter, this alternative assembly is poorer in terms of both coverage and linkage. The coverage is uneven, with many fragments spanning X and much fewer spanning X'. The linkage is poor because no fragment links the leftmost part of X' to its rightmost part. In fact, this consensus is the concatenation of two unrelated parts u and v, with u going from the beginning to about the middle of X', and v going from the middle of X' to the end.

As far as shortest substrings are concerned, vu would be a perfectly equivalent consensus. The Shortest Common Superstring problem is NP-hard, but approximation algorithms exist. However, given all the shortcomings of this model with respect to the real biological problem, such algorithms are primarily of theoretical interest.

Reconstruction

This model takes into account both errors and unknown orientation. To deal with errors, we need a few preliminary definitions. We can configure the basic dynamic programming sequence comparison algorithm to suit many different needs. Here we will use distance rather than similarity and a version that charges for gaps in the extremities of the second sequence only. The scoring system is edit distance; that is, one unit of distance is charged for every insertion, deletion, or substitution, except for deletion in the extremities of the second sequence, which are free of charge.

The distance thus obtained will be called *substring edit distance*, to distinguish it from the classical edit distance that charges for end deletions in both strings. We denote it by d_s and define it formally by the expression

$$d_s(a, b) = \min_{s \in S} d(a, s),$$

where $S(b)$ denotes the set of all substrings of b and d is the classical edit distance. Notice that substring edit distance is asymmetric, that is, in general $d_s(a, b) \neq d_s(b, a)$.

```
– – – – – – GC – GATAG – – – –
CAGTCGCTGATCGTACG
```

Fig. 4.14. Optimal alignment for substring edit distance, which does not charge for end deletions in the second string.

Example 4.3

If a = CGCATAG and b = CAGTCGCTGATCGTACG, then the best alignment is indicated in Figure and the distance is $d_s(a, b) = 2$.

Let $\in$ be a real number between 0 and 1. A string f is an *approximate substring* of S at error level $\in$ when

$$d_s(f, S) \leq \in |f|,$$

where $|f|$ is the length of f. This means that we are allowed on average $\in$ errors for each base in f. For instance, if $\in = 0.05$ we are allowed five errors per hundred bases. We are now ready for the definition of fragment assembly according to the *Reconstruction* model.

Problem : *Reconstruction*

Input : A collection $\mathcal{F}$ of strings and an error tolerance $\in$ between 0 and 1.

Output : A shortest possible string S such that for every $f \in \mathcal{F}$ we have

$$\min(d_s(f, S), d_s(\bar{f}, S)) \leq \in |f|,$$

where $\bar{f}$ is the reverse complement of f.

The idea is to find a string S as short as possible such that either f or its reverse complement must be an approximate substring of S at error level $\in$. This formulation will assemble correctly all the examples except for the one involving a chimeric fragment. However, the general problem is still NP-hard. This is not surprising, though, as it contains the SCS as a particular case with $\in = 0$.

In summary, *reconstruction* models errors and orientation but does not model repeats, lack of coverage, and size of target.

Multicontig

The *multicontig* model adds a notion of good linkage to the answer. The previous models do not care about the internal linkage of the fragments in the layout — only the final answer matters. Thus, it is also necessary to accept answers fromed by several contigs, and for this reason it is called *"mutlicontig"*. We define first an error-free version of the *multicontig* model.

Given a collection $\mathcal{F}$ of fragments, we consider a multiple alignment, or layout. This layout must be such that every column contains only one kind of base. This is where the error-free hypothesis comes in. Also, to contemplate orientation, we require that either the fragment or its reverse complement be in the alignment, but not both. Let us call this layout $\mathcal{L}$. We number the columns from 1 to the length $|\mathcal{L}|$ of the layout. According to this numbering, each fragment f has a left endpoint $l(f)$ and a right endpoint $r(f)$, so that $|f| = r(f) - l(f) + 1$. We say that fragments f and g *overlap* in this layout if the integer intervals $[l(f)..r(f)]$ and $[l(g)..r(g)]$ intersect. The nonempty intersection $[l(f)..r(f)] \cap [l(g)..r(g)]$ is called the *overlap* between f and g.

The *size* of this overlap is just the size of the intersection. Because we are using integer intervals, all sets are finite. Among all the overlaps between fragments, we are interested in the most important ones, the ones that provide linkage. We say that an overlap $[x .. y]$ is *nonlink* if there is a fragment in $\mathcal{F}$ that properly contains the overlap on both sides, that is, if the fragment contains interval $[(x - 1) .. (y + 1)]$. If no fragment has this property, the overlap is a *link*. The *weakest link* in a layout is the smallest size of any link in it. Finally, we say that a layout is a *t-contig* if its weakest link is at least as large as t. If it is possible to construct a t-conting from the fragments of a collection $\mathcal{F}$, we say that $\mathcal{F}$ *admits* a t-contig. We can now formalize the fragment assembly problem according to the *MULTICOTIG* model. given a collection of fragments $\mathcal{F}$ and an integer t, we want to partition $\mathcal{F}$ in the minimum number of subcollections C_i, $l \leq i \leq k$, such that every C_i admits a t-contig.

Example 4.4

Let $\mathcal{F}$ = (GTAC, TAATG, TGTAA). We want to partition $\mathcal{F}$ in the minimum number of t-contigs. If $t = 3$, this minimum is two, as follows.

```
– – TAATG          GTAC
TGTAA – –
```

There is no way to make one single contig with all three fragments, so two contigs is the minimum possible in this case. Notice that GTAC is its own reverse complement.

If $t = 2$, the same solution is a partition in 2-contigs, since every 3-contig is a 2-contig. However, another solution exists now :

```
TAATG – – –        GTAC
– – – TGTAA
```

Here again it is impossible to assemble $\mathcal{F}$ into one t-contig.

Finally, if $t = 1$, we see that a solution with one t-contig exists :

```
TGTAA - - - - -
- - TAATG - - -
- - - - - - GTAC
```

A version contemplating errors can be defined as follows. The layout is not required to be error-free, but with each alignment we must associate a *consensus* sequence S of the same length, possibly containing space characters (–). The numbers $l(f)$ and $r(f)$ are still defined as the leftmost and rightmost columns of f in the alignment, but now we may have $|f| \neq r(f) - l(f) + 1$. The *image* of an aligned fragment f in the consensus is $S[l(f)..r(f)]$. Given an error tolerance degree $\in$, we say that S is an *$\in$-consensus* for this contig when the edit distance between each aligned fragment f and its image in the consensus is at most $\in|f|$. We are now ready for the formal definition.

Problem : *MULTICONTIG*

Input : A collection $\mathcal{F}$ of strings, an integer $t \geq 0$, and an error tolerance $\in$ between 0 and 1.

Output : A partition of $\mathcal{F}$ in the minimum number of subcollections C_i, $1 \leq i \leq k$, such that every C_i admits a t-contig with an $\in$-consensus.

Notice that in the *RECONSTRUCTION* model a fragment is required to be an approximate substring of S, but the particular place in S where it is aligned is not important. In contrast, in the *MULTICONTIG* formulation we need to specify explicitly where each fragment should go.

The *MULTICONTIG* formalization of fragment assembly is NP-hard, even in the simplest case of no errors and known orientation. It contains as a special case the problem of finding a Hamiltonian path in a restricted class of graphs.

This formulation models errors, orientation, and lack of coverage but has no provision to use information on the approximate size of the target molecule. In addition, it partially models repeats, in the sense that it can satisfactorily solve some instances with repeats, although not all of them.

ALGORITHMS

In this section we present two algorithms for the case of fragments with no errors and known orientation. One of them is known as the *greedy* algorithm, and many practical systems are based on the same idea, with additions that contemplate errors and unknown orientation. The other algorithm is based on the *multicontig* model and is useful when we can obtain an acyclic overlap graph by discarding edges with small weight. It should be mentioned that both algorithms are of little practical value in themselves, because of the restrictive hypothesis. Nevertheless, the ideas behind them can be useful in designing algorithms that deal with real instances of the problem.

Representing Overlaps

Common superstrings correspond to *paths* in a certain graph structure based on the collection $\mathcal{F}$. We can translate properties of these superstrings to properties of the paths. Because researchers have studied paths in graphs for a long time, many people feel more comfortable dealing with graphs, and relating a new problem to them is often a good idea. The *overlap multigraph* $\mathcal{OM}(\mathcal{F})$ of a collection $\mathcal{F}$ is the directed, weighted multigraph defined as follows. The set V of nodes of this structure is just $\mathcal{F}$ itself.

A directed edge from $a \in \mathcal{F}$ to a different fragment $b \in \mathcal{F}$ with weight $t \geq 0$ exists if the suffix of a with t characters is a prefix of b. In symbols, this is equivalent to

$$suffix(a, t) = prefix(b, t),$$

or

$$k^{|a|-t} a = b_k^{\ |b|-t},$$

or

$$(ak^t)b = a(k^t\, b),$$

where k is the killer agent. Notice that we must have $|a| \geq t$ and $|b| \geq t$ for this edge to exist. Observe also that there may be many edges from a to b, with different values of t. This is why the structure is called a multigraph instead of simply a graph. Note that we disallow self-loops, that is, edges going from a node to itself. On the other hand, edges of zero weight are allowed. A word must be said about edge weights. These wieghts represent overlaps between fragments, and we use the overlaps to join fragments into longer strings, sharing the overlapping part. This joining means that we believe that the two fragments come from the same region in the target DNA. Now this belief has to be based on solid evidence. In our case, this evidence is the size of the overlap.

A short overlap provides weak evidence, whereas a long overlap gives strong evidence. When dealing with hundreds of fragments, each one a few hundred characters long, it is foolish to assume that two fragments share ends based on an overlap of, say, three characters. For this reason, sometimes we impose a minimum acceptable amount of overlap, and throw away all edges with weight below this threshold. But, in principle, we will keep all the edges in the multigraph, including the zero weight edges. There are $n(n - 1)$ such edges, because any two strings share a common part of zero characters.

Paths Originating Superstrings

The overlap multigraph is important because directed paths in it give rise to a multiple alignment of the sequences that belong to this path. a consensus sequence can then be derived from this alignment, providing a common superstring of the involved sequences. Let us investigate this process more closely. Given a directed path P in $\mathcal{OM}(\mathcal{F})$, we can construct a multiple alignment with the sequences in P as follows. Each edge $e = (f,\ g)$ in the path has a certain weight t, which means that the last t bases of the tail f of e are a prefix of the head g of e. Thus, f and g can be aligned, with g starting t positions before f's end.

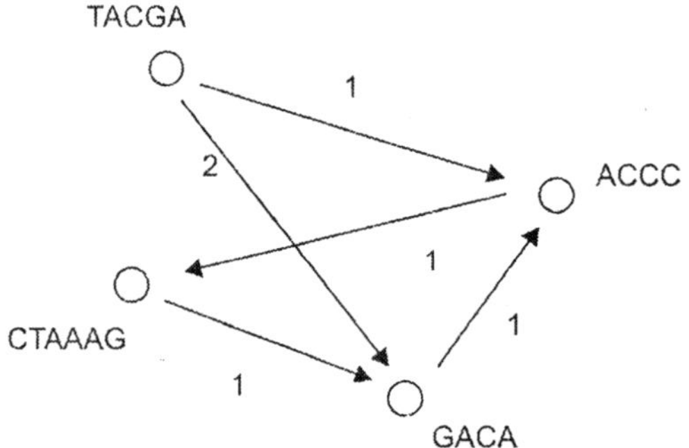

Fig. 4.15. Overlap multigraph with zero-weight edges omitted.

Only edges with strictly positive weight were drawn. The collection $\mathcal{F}$ is $\{a, b, c, d\}$, where

$$\begin{aligned} a &= \text{TACGA}, \\ b &= \text{ACCC}, \\ c &= \text{CTAAAG}, \\ d &= \text{GACA}. \end{aligned}$$

Usually, we indicate paths by a list of the form vertex, edge, vertex, edge, .., vertex, with each edge connecting its two neighbouring vertices in the list. However, in this example there are no parallel edges, so, for simplicity, we will indicate the paths by just a list of the vertices involved. The path $P_1 = dbc$ leads to the following alignment :

```
GACA - - - - - - - - -
- - - ACCC - - - - - -
- - - - - - CTAAAG
```

whereas path $P_2 = abcd$ originates the following alignment.

```
TACGA - - - - - - - - - - - -
- - - - ACCC - - - - - - - - -
- - - - - - - CTAAAG - - -
- - - - - - - - - - - - GACA
```

In general, let P be a path in $\mathcal{OM}(\mathcal{F})$, and let A be the set of fragments involved in P. Paths by definition cannot repeat vertices, so we have exactly $|A| - 1$ edges in P. The common superstring derived from P as above will be called $S(P)$.

The relationship between the total length of A, the path's weight, and the superstring length is given by the following equation.

$$||A|| = w(P) + |S(P)|, \qquad \text{...(4.3)}$$

where $||A|| = \Sigma_{a \in A} |a|$ is the sum of lengths of all sequences in A, and $w(P)$ is the weight of P. Is this really true? Let us check it for some simple cases.

Start with a path with just one fragment and no edges. Then $A = \{f\}$, $||A|| = |f|$, and $w(P) = 0$, since P has no edges. Furthermore, $S(P)$ is simply f. It is easy to see the Equation holds in this case. Let us try something larger. Suppose that $P = f_1 e_1 f_2 e_2 \dots f_l$ and that the formula holds for P. What happens if we consider an additional edge? We have $P' = Pe_{l+1} f_{l+1}$ and

$$w(P') = w(P) + w(e_{l+1}) = w(P) + k,$$

if $w(e_{l+1}) = k$.

On the other hand, $S(P') = S(P)\ \textit{suffix}\ (f_{l+1}, |f_{l+1}| - k)$, because the first k characters of f_{l+1} are already present in $S(P)$. We need to concatenate the remaining suffix of f_{l+1} only. Then,

$$\begin{aligned} |S(P')| &= |S(P)| + |f_{l+1}| - k \\ &= ||A|| - w(P) + |f_{l+1}| - k \\ &= ||A'|| - w(P') \end{aligned}$$

and the formula holds. We have just proved Equation (4.3) by induction.

Another way of seeing it is by using the killer character. If $P = f_1 e_1 f_2 e_2 \dots f_l$,

$$S(P) = f_1 k^{w(e_1)} f_2 k^{w(e_2)} \ldots k^{w(e_{l-1})} f_l.$$

The right-hand side is not ambiguous in this case because

$$(f_i k^{w(e_i)}) f_{i+1} = f_i (k^{w(e_i)} f_{i+1}),$$

by definition of edge weight.

Computing the length,

$$\begin{aligned} |S(P)| &= |f_1| - w(e_1) + |f_2| - w(e_2) + \ldots - w(e_{l-1}) + |f_l| \\ &= \sum_{i=1}^{l} |f_i| - \sum_{i=1}^{l-1} w(e_i) \\ &= \|A\| - w(P). \end{aligned}$$

So, every path originates a common superstring of the fragments involved. This is particularly important when we have a path that goes through every vertex. Such paths are called *Hamiltonian paths*. If we get a Hamiltonian path, we have a common superstring of $A = \mathcal{F}$. In this case Equation (4.3) becomes

$$|S(P)| = \|\mathcal{F}\| - w(P) \qquad \ldots(4.4)$$

and, since $\|\mathcal{F}\|$ is constant, that is, independent of the particular Hamiltonian path we take, we see that minimizing $|S(P)|$ is equivalent to maximizing $w(P)$.

Shortest Superstrings as Paths

We have seen that every path corresponds to a superstring. Is the converse true? Not always, as the following example shows. For the fragment set that originated the graph the superstring

GTATACGACCCAAACTAAAGACAGGG

does not corresond to any path, and it easy to see why. Superstrings may contain unnecessary characters not present in any fragment.

But shortest superstrings cannot waste any characters. So, do shortest superstrings always correspond to paths? The answer is affirmative, and we will prove this result in the sequel. We first observe that it is true for certain collections that are *substring-free*, defined below, and then generalize the result to any collection.

A collection $\mathcal{F}$ is said to be substring-free if there are no two distinct strings a and b in $\mathcal{F}$ such that a is a substring of b. The advantage of dealing with substring-free collections is that a partial converse of the path-to-superstring construction is easier to prove, as illustrated by the following result.

Theorem 4.1

Let $\mathcal{F}$ be a substring-free collection. Then for every common superstring S of $\mathcal{F}$ there is a Hamiltonian path P in $\mathcal{OM}(\mathcal{F})$ such that $S(P)$ is a subsequence of S.

Proof: The string S is a common superstring of $\mathcal{F}$, so for each $f \in \mathcal{F}$ there is an interval $[l(f) .. r(f)]$ of S such that $S[l(f) .. r(f)] = f$. For some fragments there may be many positions in which to anchor them. Never mind. Take one of them and fix it for the rest of this proof.

No interval of the form $[l(f) .. r(f)]$ is contained in another such interval, because $\mathcal{F}$ is substring-free. These intervals have the following important property: All left endpoints are distinct and all right endpoints are distinct, although there may be left endpoints equal to right endpoints Moreover, if we sort the fragments in increasing order of left endopoint, the right endpoints end up in increasing order as well. Indeed, if we had two strings f and g in $\mathcal{F}$

with $l(f) \leq l(g)$ and $r(f) \geq r(g)$, then the interval of g would be contained in the interval of f, which is impossible. So, we have the intervals sorted by increasing initial and final endpoints. Let $(f_i)_{i=1,\ldots,m}$ be this ordering. The fragments in this order make up the claimed path.

Consider the relationship between f_i and f_{i+1}, for each i such that $I \leq i \leq m-1$. If $l(f_{i+1}) \leq r(f_i) + 1$, then there is an edge in $OM(\mathcal{F})$ with weight $r(f_i) + 1 - l(f_{i+1})$. If $l(f_{i+1}) > r(f_i) + 1$, then take the edge of weight zero from f_i to f_{i+1}. The characters $S[j]$ for $r(f_i) + 1 \leq j \leq l(f_{i+1}) - 1$ will be discarded, because they do not belong to any interval and are therefore superfluous as far as making a common superstring is concerned. These characters are in S but not in $S(P)$, which is therefore a subsequence of S.

Corollary 4.1

Let $\mathcal{F}$ be a substring-free collection. If S is a shortest common superstring of $\mathcal{F}$, there is a Hamiltonian path P such that $S = S(P)$.

Proof: By the previous theorem, there is such a path with $S(P)$ a subsequence of S. But $S(P)$ is also a common suprstring, because P is Hamiltonian. Because S is a *shortest* common superstring, we have $|S(P)| \geq |S|$ and therefore $S = S(P)$.

We now generalize this result to any collection. First, a few definitions are in order. A collection of strings $\mathcal{F}$ *dominates* another collection $\mathcal{G}$ when every $g \in \mathcal{G}$ is a substring of some $f \in \mathcal{F}$. For instance, if $\mathcal{G} \subseteq \mathcal{F}$, then $\mathcal{F}$ dominates $\mathcal{G}$. Two collections $\mathcal{F}$ and $\mathcal{G}$ are said to be *equivalent*, denoted $\mathcal{F} \equiv \mathcal{G}$, when $\mathcal{F}$ dominates $\mathcal{G}$ and $\mathcal{G}$ also dominates $\mathcal{F}$. Equivalent collections have the same superstrings, so this notion is important for us.

Lemma 4.1

Two equivalent substring-free collections are identical.

Proof: If $\mathcal{F} \equiv \mathcal{G}$ and $\mathcal{F}$ is substring-free, then $\mathcal{F} \subseteq \mathcal{G}$. To see that, consider $f \in \mathcal{F}$. Since $\mathcal{F}$ is dominated by $\mathcal{G}$, there must be $g \in \mathcal{G}$ such that f is a substring of g. But $\mathcal{F}$ also dominates $\mathcal{G}$, so there is a superstring h of g in $\mathcal{F}$. By transitivity, f is a substring of h. But $\mathcal{F}$ is substring-free, so $f = h$, and, since g is "in between" them $f = g = h$. It follows that $f = g \in \mathcal{G}$. Since f is arbitrary, we conclude that $\mathcal{F} \subseteq \mathcal{G}$. The same argument with $\mathcal{F}$ and $\mathcal{G}$ interchanged shows that $\mathcal{G} \subseteq \mathcal{F}$. Hence, $\mathcal{F} = \mathcal{G}$.

The next theorem helps us understand how to obtain a substring-free collection from an arbitrary collection of fragments.

Theorem 4.2

Let $\mathcal{F}$ be a collection of strings. Then there is a unique substring-free collection $\mathcal{G}$ equivalent to $\mathcal{F}$.

Proof: Uniqueness is guaranteed by the previous lemma. Now let us turn to the existence. We want to show that for every $\mathcal{F}$ there is a substring-free collection $\mathcal{G}$ equivalent to $\mathcal{F}$. We do that by induction on the number of strings in $\mathcal{F}$. For $|\mathcal{F}| = 0$, $\mathcal{F}$ itself is substring-free, so we are done. For $|\mathcal{F}| \geq 1$, $\mathcal{F}$ may or may not be substring-free. If it is, we are again done. If it is not, then there are a and b in $\mathcal{F}$ such that a is a substring of b. Remove a from $\mathcal{F}$, obtaining a smaller collection $\mathcal{F}' = \mathcal{F} - [a]$. By the induction hypothesis, $\mathcal{F}'$ has an equivalent substring-free collection $\mathcal{G}$. But it is easy to verify that $\mathcal{F} \equiv \mathcal{F}'$. Obviously $\mathcal{F}$ dominates $\mathcal{F}'$, which is a subset of $\mathcal{F}$. On the other hand, the only string that is in $\mathcal{F}$ but not in $\mathcal{F}'$ is a, which is covered by $b \in \mathcal{F}'$. Hence, $\mathcal{G}$ is also equivalent to $\mathcal{F}$ and we are done.

This result tells us that if we are looking for common superstrings, we might as well restrict ourselves to substring-free collections, because given any collection there is a substring-free one equivalent to it, that is, having the same common superstrings. Incidentally, the proof of Theorem (also gives us a way of obtaining the unique substring-free collection equivalent to a given $\mathcal{F}$: Just remove from $\mathcal{F}$ all strings that are substrings of other elements of $\mathcal{F}$.

THE GREEDY ALGORITHM

We now know that looking for shortest common superstrings is the same as looking for Hamiltonian paths of maximum weight in a directed multigraph. Moreover, because our goal is to maximize the weight, we can simplify the multigraph and consider only the heaviest edge between every pair of nodes, discarding other, parallel edges of smaller weight. Any path that does not use the heaviest edge between a pair of nodes can be improved, so it is not a maximum weight path. Let us call this new graph the *overlap graph* of $\mathcal{F}$, denoted by $\mathcal{OG}(\mathcal{F})$. In an actual implementation we would not construct these edges one by one.

Using suffix trees, we can build the edges simultaneously, saving both space and time in the computer (see the bibliographic notes). The same structures can help in deciding which fragments are substrings of other fragments and may therefore be left out of the graph. The following algorithm is a *"greedy"* attempt at computing the heaviest path. The basic idea employed in it is to continuously add the heaviest available edge, which is one that does not upset the construction of a Hamiltonian path given the previously chosen edges. Because the graph is complete, that is, there are edges between every pair of nodes, this process stops only when a path containing all vertices is formed. In a Hamiltonian path, or in any path for that matter, we cannot have two edges leaving from the same node, or two edges leading to the same node.

In addition, we have to prevent the formation of cycles. So these are the three conditions we have to test before accepting an edge in our Hamiltonian path. Edges are processed in nonicreasing order by weight, and the procedure ends when we have exactly $n - 1$ edges, or, equivalently, when the accepted edges induce a connected subgraph. For each node, we keep its current indegree and outdegree with respect to the accepted edges. This information is used to check whether a previously chosen edge has the same tail or head as the currently examined edge. In addition, we keep the disjooint sets that form the connected components of the graph induced by the accepted edges. It is easy to see that a new edge will form a cycle if and only if its tail and head are in the same component. Therefore, this structure is needed to check for cycles. The disjoint sets are maintained by a disjoint-set forest data structure. We described the greedy algorithm in terms of graphs, but it can be implemented using the fragments directly.

The algorithm corresponds to the following procedure being continuously applied to the collection of fragments, until only one fragment remains. Take the pair (f, g) of fragments with largest overlap k, remove the two fragments from $\mathcal{F}$, and add $f\mathcal{K}^k g$ to $\mathcal{F}$. We assume that the collection $\mathcal{F}$ is substring-free.

This greedy strategy does not always produce the best result, as shown by the following example.

```
Algorithm Greedy
Input : weighted directed graph OG(F) with n vetices
Output : Hamiltonian path in OG(F)
        // Initialize
        for i ← 1 to n do
             in[i] ← 0 // how many selected edges enter i
             out[i] ← 0 // how many selected edges exist i
             Make Set(i)
// Process
Sort edges by weight, heaviest first
for each edge (f, g) in this order do
        // test edge for acceptance
        if in[g] = 0 and out[f] = 0 and FindSet(f) ≠ FindSet(g)
             select (f, g)
             in[g] ← 1
             out[f] ← 1
             Union(FindSet(f), FindSet(g))
        if there is only one component
             break
        return selected edges
```

Fig. 4.16. Greedy algorithm to find Hamiltonian path.

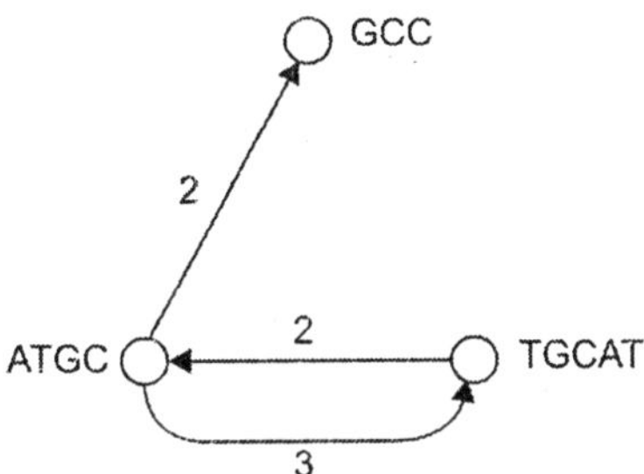

Fig. 4.17. A grahp where the greedy algorithm fails.

Example 4.5

Suppose we have

$$\mathcal{F} = [\text{GCC, ATGC, TGCAT}].$$

The overlap graph looks like Figure, where we omitted the zero edges to avoid cluttering the drawing. In this graph, the edge of weight 3 is the first one to be examined and is obviously accepted because the first edge is always accepted. However, it invalidates the two edges of weight 2, so the algorithm is forced to select an edge of weight zero to complete the Hamiltonian path.

The two rejected edges form a path of total weight 4, which is the heaviest path in this example. This example shows that the greedy algorithm will not always return the shortest superstring. Is there an algorithm that works in all cases? Apparently not. We are trying to solve the SCS problem through the Hamiltonian path (HP) problem; the HP problem is NP-complete. Perhaps we could have better luck trying another approach to solve the SCS problem. Unfortunately this is probably not the case either, since, as already mentioned, it can be shown that the SCS problem is NP-hard.

ACYCLIC SUBGRAPHS

The hardness results we mentioned in the previous section apply to an arbitrary collection of fragments. In this section we consider the problem of assembling fragments without errors and known orietation assuming that the fragments have been obtained from a *"good sampling"* of the target DNA. As shown in the sequel, this assumption enables us to derive algorithms that deliver the correct solution and are particularly efficient. What do we mean by a "good sampling"? Basically, we want the fragments to cover the entire target molecule, and the collection as a whole to exhibit enough linkage to guarantee a safe assembly.

The definition of t-contig. Our definitions here are based on similar concepts.

Suppose that S is a string over the alphabet {A, C, G, T}.

An *interval* of S is an integer interval $[i .. j]$ such that $1 \leq i \leq j + 1 \leq |S| + 1$. A *sampling of* S is a collection A of intervals of S. The sampling A *covers* S if for any i such that $1 \leq i \leq |S|$ we have at least one interval $[j .. k] \in A$ with $i \in [j .. k]$. We say that two intervals α and β are *linked at level t* if $|\alpha \cap \beta| \geq t$. The entire sampling A is said to be *connected at level t* if for every two intervals α and β in A there is a series of intervals α_i for $0 \leq i \leq l$ such that $\alpha = \alpha_0$, $\beta = \alpha_l$ and α_i is linked to α_{i+1} at level t for $0 \leq i \leq l - l$. Note that we are using parameter t to measure how *strong* a link is.

A sampling A Will be considered good if it is connected at level t. Our goal is to study fragment collections coming from connected at level t samplings that cover a certain string S. The value of t is around 10 in typical, real instances, but in our study we will not fix it. We do assume that t is a nonnegative integer, though.

The fragment collection *generated* by a sampling A is

$$S[A] = \{S[\alpha] \mid \alpha \in A\}.$$

We say that a sampling A is *subinterval-free* if there are no two intervals $[i .. j]$ and $[k .. l]$ in A with $[i .. j] \subseteq [k .. l]$. The following observation about subinterval-free samplings will be coming into play again and again. In a subinterval free sampling, no two intervals can have the same left endpoint. Otherwise, whichever has the larger right endpoint would contain the other. An analogous argument holds for right endpoints.

We consider also modified forms of our overlap graphs. Given a collection $\mathcal{F}$ and a nonnegative integer t define $\mathcal{OM}(\mathcal{F}, t)$ as the graph obtained from $\mathcal{OM}(\mathcal{F})$ by keeping only the edges of weight at least t. A similar construction defines $\mathcal{OG}(\mathcal{F}, t)$. This reduces the overall number of edges, keeping only the stronger ones.

We will use the symbol $\alpha \xrightarrow{w} \beta$ to mean that $|\alpha| \geq w$, $|\beta| \geq w$, and $l(\beta) + w - 1 = r(\alpha)$. similarly, $f \xrightarrow{w} g$ means that there is an edge from f to g in $\mathcal{OM}(\mathcal{F})$ of weight w.

Suppose now that we got such a collection $\mathcal{F}$ and would like to recover S. How can we do that? What are the good algorithms for it? It turns out that the answer depends strongly on the repeat structure of S. If S does not have any repeats of size t or more, then it is easy to reconstruct it. But if S has such repeats, the problem gets more complex, and in some cases S cannot be recovered without ambiguity.

One thing that is independent of the repeat structure of S is the existence of a Hamiltonian path in the graph $\mathcal{OM}(\mathcal{F}, t)$. We state and prove this basic result here, before embarking on an analysis of the repeat structure of S. First, an auxiliary lemma.

Lemma 4.2

Let S be a string over (A, C, G, T) and A be a subinterval-free, connected at level t sampling of S, for some $t \geq 0$. If α is an interval in A such that there is another interval β in A with $l(\alpha) < l(\beta)$, then β can be chosen so that

$$r(\alpha) + 1 - l(\beta) \geq t.$$

Proof: We know A is connected at level t, therefore there is a sequence $(\alpha_i)_{0 \leq i \leq l}$ of intervals in A such that $\alpha = \alpha_0$, $\beta = \alpha_l$, and $|\alpha_i \cap \alpha_{i+1}| \geq t$ for $0 \leq i \leq l - 1$.

Let i be the smallest index such that $l(\alpha) < l(\alpha_l)$. Because $\beta = \alpha_l$ satisfies this property, we know that $i \leq l$. On the other hand, $i > 0$ because $\alpha = \alpha_0$ does not satisfy the property. Because i is the smallest index with the property, we have $l(\alpha_{i-l}) < l(\alpha) < l(\alpha_i)$. If follows that $r(\alpha_{i-l}) < r(\alpha) < r(\alpha_i)$, since A is subinterval-free.

Fig. 4.18. Three mutually overlapping intervals.

We also know that $|\alpha_{i-1} \cap \alpha_i| \geq t$. But $|\alpha \cap \alpha_i| \geq |\alpha_{i-1} \cap \alpha_i| \geq t$, so α_i is an interval in A such that

$$r(\alpha) + 1 - l(\alpha_i) \geq t.$$

Now, the Hamiltonian path existence result.

Theorem 4.3

Let S be a string over [A, C, G, T] and A be a subinterval-free, connected at level t sampling of S, for some $t \geq 0$. Then the multigraph $\mathcal{OM}(\mathcal{F}, t)$, where $\mathcal{F}$ is the fragment collection generated by A, admits a Hamiltonian path P. Furthermore, if A cover S, then P can be chosen so that $S(P) = S$.

Proof : Let $(\alpha_i)_{l \leq i \leq n}$ be all the intervals in A sorted by left endpoint. Because A is subinterval-free, this sequence is also sorted by right endpoint. Let $f_i = S[\alpha_i]$ for $1 \leq i \leq n$. We claim that $f_i \xrightarrow{k_i} f_{i+l}$ for some $k_i \geq t$, so that the f_i in the above order form a Hamiltonian path.

Let α_i be any interval except the last one. By the previous lemma, there is another interval $\beta \in A$ such that $l(\alpha_i) < l(\beta)$ and $r(\alpha_i) + 1 - l(\beta) \geq t$. The sequence $(\alpha_i)_{l \leq i \leq n}$ is sorted by left endpoint, so $l(\alpha_i) < l(\alpha_{i+1}) \leq l(\beta)$. (The intervals α_{i+1} and β may be the same). Hence,

$$k_i = |\alpha_i \cap \alpha_{i+1}| = |[(\alpha_{i+1}) \,..\, r(\alpha_i)]| = r(\alpha_i) - l(\alpha_{i+1}) + 1 \geq r(\alpha_i) + 1 - l(\beta) \geq t.$$

This shows that an edge $f_i \rightarrow f_{i+l}$ really exists in $\mathcal{OM}(\mathcal{F}, t)$ and concludes the proof of the first part.

The second part, which says that $S(P) = S$, is a consequence of the following fact. Every time we have $\alpha \xrightarrow{w} \beta$, we also have $S[\alpha] \xrightarrow{w} S[\beta]$ and $S[\alpha]\ k^w\ S[\beta] = S[\alpha \cup \beta]$.

The presence of repeated regions, or repeated elements, in the target string S is related to the existence of cycles in the overlap graph. In this and following sections we make this relationship more explicit.

A repeated element, or just a *repeat* in S, occurs when two different intervals of S give rise to the same substring. In other words, there are two intervals $\alpha \neq \beta$ such that $S[\alpha] = S[\beta]$. The *size* of this repeat is $|S[\alpha]|$, which is the same as $|S[\beta]|$. The repeat is *self-overlapping* when $\alpha \cap \beta \neq \theta$, otherwise we have *disjoint occurrences* of the repeat.

Our next result shows that cycles in an overlap graph are necessarily due to repeats in S. The converse is not necessarily true; that is, we may have repeats but still an acyclic overlap graph. Again, a couple of intermediate results are needed before we can prove an important fact.

Lemma 4.3

If $\alpha \xrightarrow{w} \beta$ then $l(\alpha) \leq l(\beta)$ and $r(\alpha) \leq r(\beta)$.

Proof: From $|\alpha| \geq w$ we have $l(\alpha) + w - 1 \leq r(\alpha)$. From $\alpha \xrightarrow{w} \beta$ we get $l(\beta) + w - 1 = r(\alpha)$. Therefore $l(\alpha)$. Therefore $l(\alpha) + w - 1 \leq l(\beta) + w - 1$, that is, $l(\alpha) \leq l(\beta)$. Similarly, $|\beta| \geq w$ implies $l(\beta) + w - 1 \leq r(\beta)$, or $r(\alpha) \leq r(\beta)$, in view of previous equations.

We need a few more definitions before proceeding. A *false positive* at level t is a pair of intervals α and β such that there is a $w \geq t$ with $S[\alpha] \xrightarrow{w} S[\beta]$ but $\alpha \xrightarrow{w} \beta$ does not hold. The name refers to the fact that we assumed that $S[\alpha]$ and $S[\beta]$ came from intersecting regions in the target molecule because their sequences overlap, but this is not the case. They *look* overlapping but they are not, so they constitute a false positive.

Lemma 4.4

The existence of a false positive at level t implies the existence of a repeat of size at least t.

Proof: Because $S[\alpha] \xrightarrow{w} S[\beta]$, we have $|\alpha| \geq w$ and $|\beta| \geq w$. If $l(\beta) = r(\alpha) - w + 1$, then we would have $\alpha \xrightarrow{w} \beta$, since $\alpha \neq \beta$ by hypothesis. Hence, $l(\beta) \neq r(\alpha) - w + 1$.

The relation $S[\alpha] \xrightarrow{w} S[\beta]$ also implies that $S[\alpha'] = S[\beta']$, where α' is the last w elements of α and β' is the first w elements of β. Because $l(\beta) \neq r(\alpha) - w + 1$, α' and β' are different intervals giving rise to the same substring; that is, we have a repeat of size $w \geq t$.

Theorem 4.4

Let $\mathcal{F}$ be a collection generated by a sampling A of S. If $\mathcal{OG}(\mathcal{F}, t)$ has a directed cycle, then there is a repeat in S of size at least t.

Proof: Let $f_1 \to f_2 \to \dots \to f_1 \to f_1$ be a directed cycle in $\mathcal{OG}(\mathcal{F}, t)$. Consider intervals $\alpha_1, .., \alpha_l$ such that $S[\alpha_i] = f_i$, for $1 \leq i \leq l$. We will show that at least one of the pairs (α_i, α_{i+1}), including (α_l, α_1), is a false positive, which implies the existence of a repeat in S by Lemma 4.4.

Suppose they are all true positives. Then $\alpha_i \to \alpha_{i+1}$ for $1 \leq i < l$ and $\alpha_i \to \alpha_1$. It follows that $l(\alpha_1) \leq l(\alpha_2) \leq .. \leq l(\alpha_l) \leq l(\alpha_1)$. So, all the $l(\alpha_i)$ are equal. An analogous observation holds for the $r(\alpha_i)$. But then $\alpha_1 = \alpha_2 = \dots = \alpha_l$, which is impossible, since the f_i are distinct by hypothesis. Hence, there is at least one false positive and the result is proved.

All these considerations lead to a satisfactory solution for fragment assembly when the target string S does not have repeats.

Theorem 4.5

Let S be a string over {A, C, G, T}, A a subinterval-free, connected at level t sampling of S that covers S, and $\mathcal{F} = S[A]$. If S has no repeats of size t or larger, then the graph $\mathcal{OG}(\mathcal{F}, t)$ has a unique Hamiltonian path P and $S(P) = S$.

Proof : We have ssen that $\mathcal{OG}(\mathcal{F}, t)$ is acyclic (Theorem 4.4) and that $\mathcal{OM}(\mathcal{F}, t)$ has a Hamiltonian path P with $S(P) = S$.

To begin with, notice that $\mathcal{OM}(\mathcal{F}, t) = \mathcal{OG}(\mathcal{F}, t)$ if S has no repeats of size t or larger. This follows from the absence of false positives. It is clear that the vertex sets of $\mathcal{OM}(\mathcal{F}, t)$ and $\mathcal{OG}(\mathcal{F}, t)$ are the same. It is also clear that every edge of $\mathcal{OG}(\mathcal{F}, t)$ is also in $\mathcal{OM}(\mathcal{F}, t)$. What we need is to convince ourselves that all edges in $\mathcal{OM}(\mathcal{F}, t)$ are in $\mathcal{OG}(\mathcal{F}, t)$.

Let $f \xrightarrow{w} g$ be an edge in $\mathcal{OM}(\mathcal{F}, t)$. Then $w \geq t$. If this edge is not in $\mathcal{OG}(\mathcal{F}, t)$, this is due to the existence of another edge $f \xrightarrow{x} g$ with $x > w$. Let α and β be intervals in A with $S[\alpha] = f$ and $S[\beta] = g$. Since there are no false positives, we must have at the same time $\alpha \xrightarrow{w} \beta$ and $\alpha \xrightarrow{x} \beta$, which is impossible. The weight in this edge is uniquely determined by $w = r(\alpha) - l(\beta) + 1$.

So, $\mathcal{OG}(\mathcal{F}, t)$ is an acyclic graph with a Hamiltonian path. It turns out that every time an acyclic graph has a Hamiltonian path this path is unique. To see why this is true, consider the nodes in the graph that have indegree zero (the *sources*). Because the graph is acyclic, it must have at least one source. But, having a Hamiltonian path, it can have at most one source: the starting point of this path. All other vertices have at least one incoming edge. We conclude that our graph must have exactly one source, and it must be the starting point of the Hamiltonian path.

Now, if we remove this source from the graph, the remaining graph is still acyclic, and has a Hamiltonian path, because the vertex we removed is an extremity of the old Hamiltonian path.

Proceeding in this way, we see that the Hamiltonian path is unique, and we know how to construct it: just keep removing sources from the graph. The algorithm just described is well known in computer science. It is called *topological sorting* and, in its more general context, is used to find an ordering of nodes that is consistent with an acyclic set of edges in the following sense: For all edges $f \to g$ the node f appears before g in the ordering. If the graph has a Hamiltonian path besides being acyclic, then this ordering is unique, as we just saw. We end this section with an example where *Greedy* does not work, but there is an acyclic subgraph for a suitable threshold t, so that topological sorting will find the correct path.

When we say that *Greedy* does not work, we mean that if does not do a good job regarding assembly, although it does do a good job finding a shortest common superstring. In this example the fragments come from a target sequence that does have repeats longer than t. This goes to show that the absence of long repeats is not necessary to yield an acyclic graph. It is merely sufficient, in mathematical jargon.

Example 4.6

Consider the following target string S and fragments w, z, u, x and y:

S = AGTATTGGCAATCGATGCAAACCTTTTGGCAATCACT
w = AGTATTGGCAATC
z = AATCGATG

u = ATGCAAACCT

x = CCTTTTGG

y = TTGGCAATCACT

Choosing a threshold value of $t = 3$, we get the overlap graph. A greedy assembly will result in a shortest common superstring, formed by concatenation of the contig wk^9y with the contig $zk^3\ uk^3\ x$ in any order. This solution has length 36, one character shorter than the target sequence, which is also a superstring. However, the solution given by the Hamiltonian path has better linkage.

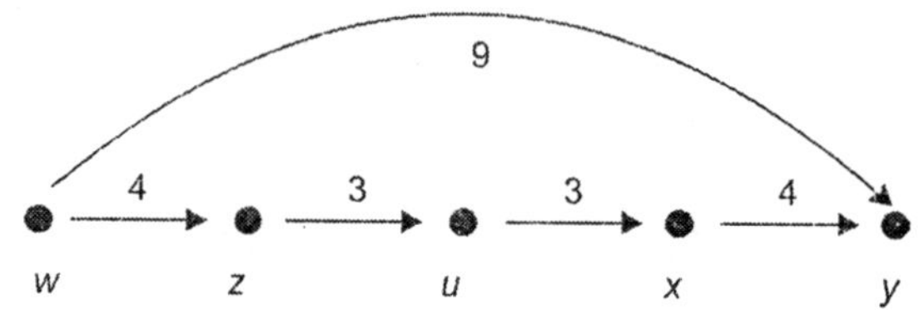

Fig. 4.19. Overlap graph for a given set of five fragments.

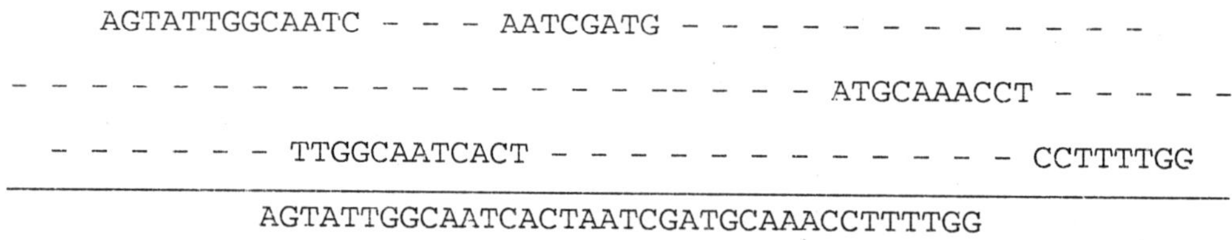

Fig. 4.20. A bad solution for an assembly problem, with a multiple alignment whose consensus is a shortest common superstring. This solution has length 36 and is generated by the Greedy algorithm. However, its weakest link is zero.

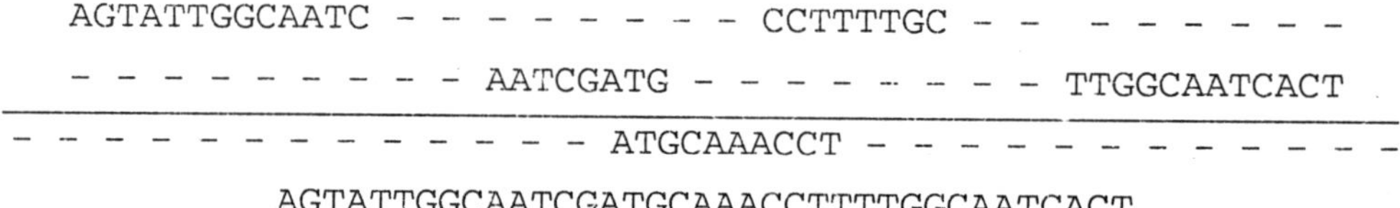

Fig. 4.21. Solution according to the unique Hamiltonian path. This solution has length 37, but exhibits better linkage. Its weakest link is 3.

HEURISTICS

We have seen that none of the formalisms proposed for fragment assembly are entirely adequate, so in practice we must often resort to heuristic methods. We review some of these in this section, starting by pointing out relevant properties of a good assembly layout. Fragment assembly can be viewed as a multiple alignment problem with some additional features.

The most striking such feature is perhaps the fact that each fragment can participate with either the direct or the reverse-complemented sequence. A second special feature is that the sequences themselves are usually much shorter than the alignment itself. This leads us to charge internal gaps and external gaps in the fragments differently, with internal gaps being much more penalized. If we charge both kinds of gaps the same or nearly the same penalty, then there would be no incentive to keep fragment characters together, potentially causing them to scatter along the entire length of the alignment.

Moreover, the fact that fragments are short compared to the alignment forces us to look at other criteria, besides the score, to assess the quality of an alignment. These extra criteria include coverage and linkage. The latter measures how strongly the sequences are linked in terms of their overlap. We now discuss these criteria in more detail.

Scoring

In each column of the multiple alignment we have a certain coverage. The ideal situation is when all sequences participating in a column have the same character there. In general uniformity is good and variability is bad in a column.

This suggests the use of a concept called *entropy* to measure the quality of a column. Entropy is a quantity that is defined on a group of relative frequencies, and it is low when one of these frequencies stands out from the others, and high when they are all more or less equal. In our context, the lower the entropy, the better. The frequencies we use are the frequencies of each character, where a character here can be an ordinary base A, C, G, T or an internal space.

Suppose that we have n sequences covering a certain column, and relative frequencies PA, PC, PG, PT, and p space for each character. The entropy of this distribution is defined as

$$E = -\sum_c Pc \log Pc.$$

If $Pc = 1$ for a certain character c and all other frequencies are zero, then we have (assume $0 \log 0 = 0$).

$$E = 0.$$

If, on the other hand, all frequencies are equal, we have $Pc = 1/5$ for each c and then

$$E = -5 \times \frac{1}{5} \log \frac{1}{5} = \log 5,$$

which is the maximum value possible for five frequencies.

We used the sum-of-pairs scoring scheme when we discussed multiple alignments. This scheme combines in a single number both the coverage and the uniformity of a column. The larger the coverage, the larger the sum-of-pairs value, provided the uniformity is high. Also, for constant coverage, the sum-of-pairs value increases with uniformity. In constrast, the entropy is independent of coverage. It measures uniformity alone. Depending on the situation, a scientist may want to use sum-of-pairs, a combined measure, or the coverage and entropy separately.

Coverage

We defined the coverage of a set of fragments with respect to the target from which they were obtained. Here we define an analogous concept for a layout. A fragment covers a column i if it participates in this column either with a character or with an internal space. Let $l(f)$ and $r(f)$ be the leftmost position and the rightmost position, respectively, of a character of f(or of its reversed complement) in the layout. The previous definition means that fragment f covers column i if $(f) \leq i \leq r(f)$. Having defined what the coverage is in each column, we may speak of *minimum*, *maximum*, and *mean* coverage of a layout.

The meanings are as expected. If the coverage ever reaches value zero for a column i, then we do not really have a connected layout. Indeed, no fragment image spans column i, so the parts of the layout to the left and to the right of i are in some sense independent. For instance, the right part could be placed to the left of the other, giving an alternative layout as good as the original one. This means that if we have more columns with zero coverage, any permutation of the intervening regions is acceptable.

Each one of these regions corresponds to what we have called a *contig*. Coverage is also important to give confidence to the consensus sequence. if there are several fragments covering a column, we have more information on the particular base call for this column. It is also important to have enough fragments from both strands in the layout.

Linkage

Apart from the score and the coverage, the way individual fragments are linked in the layout is another determinant of layout quality. The coverage is high, but there are no real links between the fragments. Each block is composed of many alomost exact copies of a certain sequence, but the blocks themselves do not seem related at all. Fragment images should have overlapping ends to show some evidence of linkage.

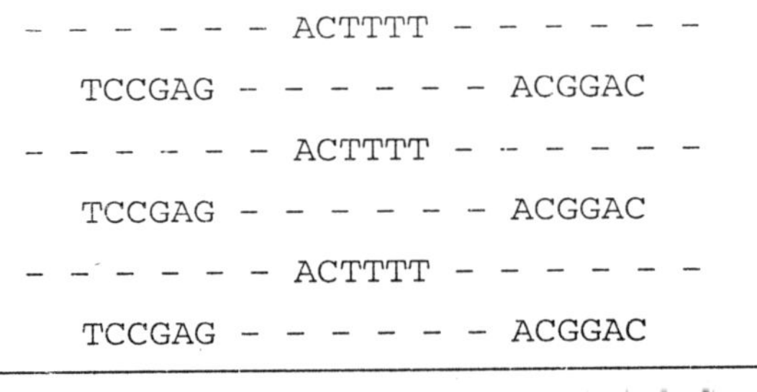

Fig. 4.22. Good coverage but bad linkage.

Assembly in Practice : A good assembly program must balance these requirement and find all solutions that satisfy them reasonably well. This is no easy task, and, coupled with the fact that no perfect formalisms exist, it makes algorithm design very difficult. As a result, practical implementations often divide the whole problem in three phases, each one using results produced by the previous phases. A typical division encompasses the following phases :

- Finding overlaps
- Building a layout
- Computing are consensus

Such partitioning makes the problem more manageable, because each phase can be treated more or less independently as a different problem. however, it becomes more difficult to understand the relationship between the initial input and the final output. Often, the methods used in these phases are heuristic, and no clear guarantees can be given with respect to solution quality. Nevertheless, many existing implementations perform well in practice and it is worthwhile to know some of the techniques used.

FINDING OVERLAPS

The first step in any assembly program is fragment overlap detection. We need to try all pairs of fragments and their reverse complements to determine whether there is a prefix of a fragment that matches well with the suffix of another. In general, we will need all such pairs. Also, we are interested in fragments entirely contained in other fragments. All these comparisons are made within an error threshold.

We mention here a way of comparing two fragments with the goal of finding such overlaps. In a real problem we could apply this method to all pairs, including reverse complements, but this is seldom done, because it would take too long due to its being based on a quadratic time algorithm. However, it is one of the most reliable methods of assessing overlap.

The comparision method consists in applying the dynamic programming algorithm, which is a variant of the basic alogrithm. This variant does not charge for gaps in any extremity. Figure presents the result of using such an algorithm for two fragments. The score system used in this comparison is 1 for matches, –1 for mismatches, and –2 for gaps. Notice that the algorithm does not charge for the gaps after the first sequence or before the second one, so it was able to find a prefix-suffix similarity. The same algorithm would find an approximate containment.

ORDERING FRAGMENTS

In this section we briefly discuss one aspect of layout construction in a real assembly problem, namely, the issue of finding a good ordering of the fragments in a contig. Given such an ordering, and knowing that each fragment overlaps with the next one, a layout can be built using the methods of the next section. Unfortunately, there is no algorithm that is simple and general enough, even resorting to heuristics. Therefore, rather than presenting an algorithm, we point out important considerations that should be taken into account when trying to construct good fragment orderings.

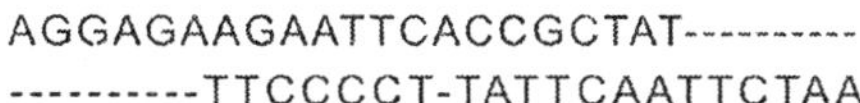

T T C C C C T T A T T C A A T T C T A A

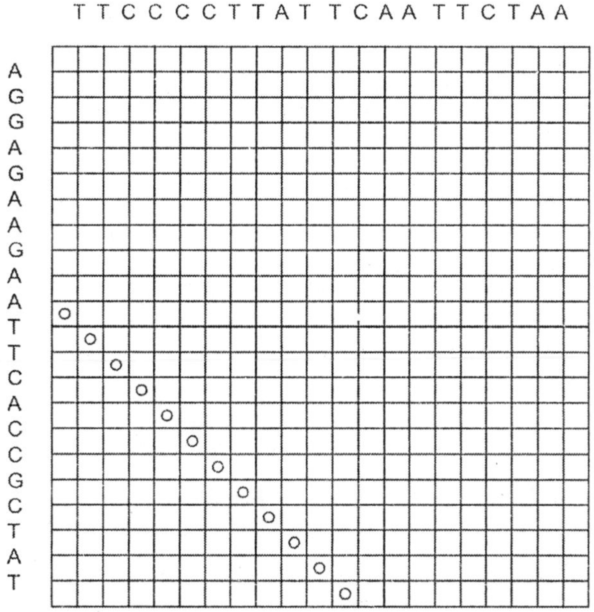

A G G A G A A G A A T T C A C C G C T A T

Fig. 4.23. An optimal semiglobal alignment and its corresponding path in the dynamic programming matrix. The algorithm is configured to find prefix-suffix or containment alignments, which are appropriate for fragment assembly.

In a real instance we have to deal with reverse complements, so it is necessary to extent the input collection $\mathcal{F}$ with the reverse complements of the fragments. Define

$$\bar{F} = [\bar{f} \mid f \in \mathcal{F}]$$

and $$\mathcal{DF} = \mathcal{F} \cup \bar{\mathcal{F}}.$$

In a real problem we have to worry about errors as well, and this means considering approximate matching. We will use the notation $f \rightarrow g$ in a slightly different sense from previously: It now indicates that the end of fragment f is approximately the same as the beginning of fragment g in some sense. We also assume that whenever $f \rightarrow g$ we also have $\bar{g} \rightarrow \bar{f}$, because it is resonable to expect that whatever criterion is used to assess the similarity between two sequences, the same criterion would apply to their reverse complements. This can be shown as follows: IF $f \rightarrow g$, we have that $f = uv$, $g = wx$, and v and w are similar. This implies that $\bar{g} = \bar{x}\bar{w}, \bar{f} = \bar{v}\bar{u}$, with $\bar{w}$ and $\bar{v}$ similar, which is equivalent to $\bar{g} \rightarrow \bar{f}$. We will use the notation $f \subseteq g$ meaning that f is an approximate substring of g. An argument analogous to the one just presented shows that $f \subseteq g \Rightarrow \bar{f} \subseteq \bar{g}$.

The graph we will be interested in is then $\mathcal{OG}(\mathcal{DF})$, with edges indicating approximate overlap. In addition it is important to know which fragments are approximate substrings of others.

Finding a good ordering of overlapping fragments is essentially equivalent to finding directed paths in $\mathcal{OG}(\mathcal{DF})$. These paths originate contigs. Notice that for any path

$$f_1 \rightarrow f_2 \rightarrow \ldots \rightarrow f_k$$

in this graph there is a corresponding "complementary" path

$$\overline{f_k} \rightarrow \overline{f_{k-1}} \rightarrow \ldots \rightarrow \overline{f_1},$$

which gives rise to a contig with a consensus that is the reverse complement of the first path. This means that both strands of the target DNA are being constructed simultaneously. Ideally we would like to get a pair of complementary paths such that every non-contained fragment belongs to exactly one path in the pair. Contained fragments can be positioned afterward so their presence in the paths is not essential. Actually, forcing their presence may prevent the formation of a path.

In rare occasions contained fragments may provide an important link, so we consider them optional fragments when building a path. Two obstacles may prevent the ideal goal of a pair of paths covering all noncontained fragments: lack of coverage and repeats. A disconnected overlap graph is usually an indication of lack of coverage. Repeats cause the appearance of cycles in the overlap graph, if coverage is good, or may show up as edges $f \rightarrow g$ and $f \rightarrow h$ with g and h unrelated (g and h are unrelated if $g \nrightarrow h, h \nrightarrow g, g \not\subseteq h$, and $h \not\subseteq g$). This happens when f is a fragment near the end of a repeat X and g and h are fragments leading from X to distinct flanking regions in the target. A cycle involving both f and $\bar{f}$ for a certain fragment f indicates the presence of an inverted repeat. Another indication of repeated regions is unusually high coverage (See figure 4.13).

All fragments from all copies of a given repeat are piled together, causing coverage to increase to about the average value times the number of repeat copies (assuming an approximately uniform distribution of fragments along the target). In conclusion, there are four issues to keep in mind when building paths: (1) Every path has a corresponding complementary path; (2) it is not necessary to include contained fragments; (3) cycles usually indicate the presence of repeats; and (4) unbalanced coverage may be related to repeats as well.

ALIGNMENT AND CONSENSUS

We briefly discuss in this section the problem of building a layout from a path in an overlap graph. This is straightforward if the overlaps are exact, but approximate overlaps require extra care. We present here two techniques related to alignment construction. The first one helps in building a good layout from a path in the presence of errors. The second one focuses on locally improving an already constructed layout. To introduce the first technique, we use the following example.

Example 4.7

Suppose we have a path $f \rightarrow g \rightarrow h$ with

f = CATAGTC

g = TAACTAT

h = AGACTATCC

The two optimal semiglobal alignments between f and g are

```
CATAGTC - - -            CATAGTC - - -
- - TAA - CTAT           - - TA - ACTAT
```

We do not have any reason to prefer one of them over the other, since their score is the same. However, when sequence h enters the scene, it becomes clear that the second optimal alignment (where the second A of g is matched to a T in f) is better.

```
CATAGTC - - - - -
- - TA - ACTAT - -
 - - - AGACTATCC
  CATAGACTATCC
```

The consensus S satisfies $d_s(f, S) = l$, $d_s(g, S) = 1$, and $d_s(h, S) = 0$. Had we aligned A with G, the voting in column 6 would have been ambiguous, with T, A, and – all tied with one occurrence each.

In the consensus any choice other than A in this column would have led to a worse result, in the sense that the sum $d_s(f, S) + d_s(g, S) + d_s(h, S)$ would be greater than with the first solution. To avoid such problems, we use a structure that represents alignments distinguishing between bases that are actually matched to some other base, and bases that are *"loose"*, as is the case of the second base A of sequence g in the above example.

We can implement this idea as follows. Each sequence is represented by a linked list of bases. When an alignment is formed, matched bases are unified and become the same node in the overall structure. If unsure about matching a particular base, we can leave it unmatched, for the sequences f and g of Example 4.7. Then, as new sequences are added, some of their bases will be unified according to suitable alignment. The structure can be seen as an acyclic graph, and a traversal in topological order yields a layout. Applying this technique, we get the best possible layout. When h is added, an arrow from G to the A node just below it will be added, as well as additional bases that belong to h alone.

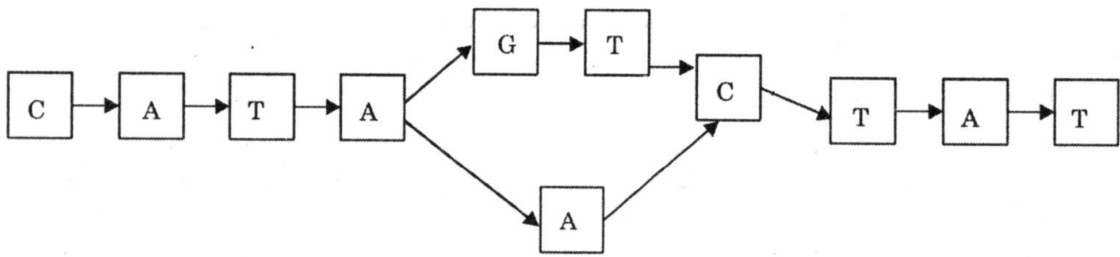

Fig. 4.24. Linked structure for approximate overlaps.

ACT - GG	ACT - GG
ACTTGG	ACTTGG
AC - TGG	ACT - GG
ACT - GG	ACT - GG
AC - TGG	ACT - GG
ACTTGG	ACT - GG

Fig. 4.25. Two layouts for the same sequences. The layout to the right is better, but an assembly program might produce the left one due to ambiguities in pairwise alignments.

Our second technique concerns improving a given layout. One common problem in practice is illustrated by the following example.

Example 4.8

Figure shows two layouts for the same set of sequences. In the left-hand layout, the consensus ends up with two Ts by majority voting, but four of the five sequences have just one T. The layout to the right is better, for its consensus agrees with the majority of the sequences.

One idea to solve such problems is to perform a multiple alignment locally. We select a short range of consecutive columns that need doctoring, extract from each fragment its portion contained in these columns, and align the portions. For moderate values of the local coverage and width of the range the multiple alignment algorithm will run in a reasonable amount of time, despite its exponential complexity. A sum-of-pairs scorling scheme seems adequate here. In Example, the right-hand layout maximizes the sum-of pairs score, assuming $p(a, b) = 1$ if $a = b$, $p(a, b) = 0$ fi $a \neq b$, $p(a, -) = -1$, and $p(-, -) = 0$.

SUMMARY

In this chapter we studied the fragment assembly problem, which appears in large-scale DNA sequencing. Basically, it consists of assembling in the correct order and orientation a collection of fragments coming from a long, unknown DNA sequence. We began by reviewing the biological setting of this problem, and how the biolgoical factors influence the input. Then we presented attempts that have been made to capture the problem into a mathematical framework.

The proposed models fail to address all the issues involved. In particualr, it is hard to deal with repeated regions and to take into account information about the *size* of the target sequence. Also, all models assume that the collection of fragments is free of chimeras and that there is no vector contamination. Algorithms based on the models were also presented. Fragment assembly requires a particular kind of multiple aligment, also called layout, in which we have to worry about fragment orientation, coverage, and linkage. We do not charge for end spaces when scoring this alignment. Because of the difficulties in finding a precise formulation for the problem, heuristic methods are often used in practice. The problem is usually broken into three phases: finding overlaps between fragments, building a layout, and computing the consensus. Programs based on this approach perform well in real instances without repeats. We presented an overview of such heuristics. New techniques, such as sequencing by hybridization, may radically change the way large-scale sequencing is done. if this happens, new algorithms will be required.

5 Biological Background

A *physical map* of a piece of DNA tells us the location of certain *markers* along the molecule. These markers are generally small but precisely defined sequences. Such a map helps molecular biologists further explore a genome. For example, suppose a certain stretch of DNA has been completely sequenced, giving us sequences. If we know which chromosome S came from, and if we have a physical map of this chromosome, we could try to find one of the map's markers in S. If we succeed, we have located S in the chromosome. So the question now is. How do we create such maps? The first task is to obtain several copies of the DNA molecule we want to map (known as the target DNA). Each copy must then be broken up into several fragments, using restriction enzymes.

Mapping is done by carefully comparing the fragments obtained, in particular by observing overlaps. Note that in general each fragment is a DNA piece still too long to be sequenced, so we cannot determine overlap by sequencing and comparing fragments, as is done in fragment assembly. Instead, we obtain overlap information by generating *fingerprints* of the fragments. A fingerprint should describe part of the information contained in a fragment in some unique way, just like our fingerprints uniquely describe some part of ourselves. Two popular ways of setting fingerprints are *restriction site analysis* and *hybridization*.

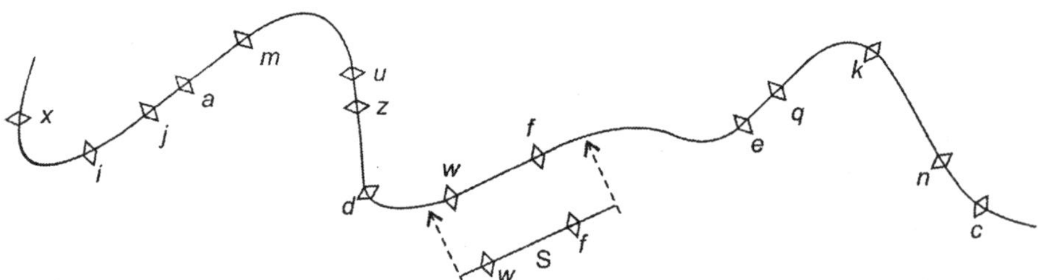

Fig. 5.1. Sketch of a physical map. In this figure a very long DNA molecule (*e.g.*, a chromosome) is represented by the curved solid line. Markers along the molecule are represented by the diamond symbols. Each marker is distinct, and this is shown by the letters associated with each. Segment S represents a DNA fragment that has been sequenced, containing, for example, a gene. Thanks to the map, S can be located on the chromosome.

In restriction site mapping the aim is to locate the restriction sites of a given enzyme on the target DNA. The restriction sites are thus the markers previously mentioned. The technique used for location is baded on fragment length measurement; a fragment's length is thus its fingerprint. It hybridization mapping we check whether certain small sequences bind to fragments; the subset of such small sequences that do bind to the fragment become its fingerprint.

The fingerprint of a fragment can also be given by hybridization information and fragment lengths, but in this chapter we consider each technique separately for clarity of exposition. In any case, by comparing fingerprints we try to determine whether the fragments overlap and thus determine their relative order (which was lost during the break-up process). Just as in other problems from molecular biolgoy, possible lack of information and the presence of numerous experimental errors make the physical mapping problem especially hard. In particular, it may not be possible with a given collection of fragments to obtain one contiguous physical map. This may happen simply because the fragmentation process did not produce fragments covering certain sections of the target DNA. When this happens, the physical map pieces are called contigs (the same name used in fragment assembly). This is a feature common to all physical mapping processes. We now discuss each of the techniques mentioned above in more detail.

RESTRICTION SITE MAPPING

In this section we briefly present two possible techniques for mapping based on measuring lengths of fragments between restriction sites. One way to obtain fragments of many different sizes is to apply not one, but *two* restriction enzymes to the target DNA. We will explain this technique by an example. Suppose we have two restriction enzymes, A and B, each recognizing a different sequence (restriction site).

Applying A to one copy of the target DNA, we obtain fragments measuring 3, 6, 8 and 10 length units (for example, thousands of base pairs). Applying B to another copy of the target DNA, we obtain fragments measuring 4, 5, 7 and 11 units. Applying now *both* enzymes to a third copy of the target DNA we obtain fragments measuring 1, 2, 3, 3, 5, 6 and 7 units. The problem now is, How can we order these fragments in such a way that the order we determine is consistent with experimental results? This is known as the *double digest problem*, because the process of breaking a molecule by an enzyme is known as a digest. A variant of the double

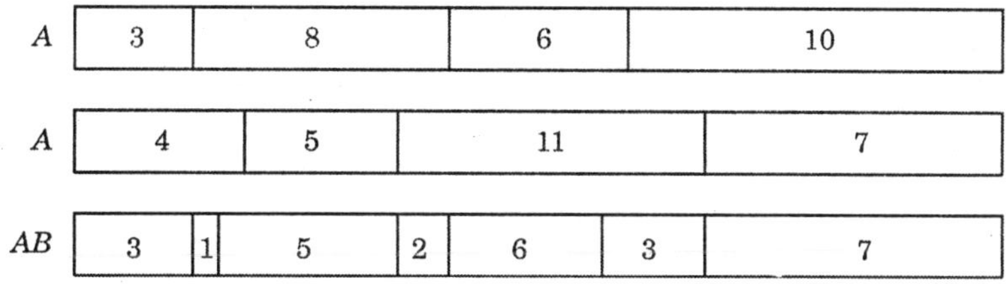

A: 3 | 8 | 6 | 10

A: 4 | 5 | 11 | 7

AB: 3 | 1 | 5 | 2 | 6 | 3 | 7

Fig. 5.2. Solution to a double digest problem.

digest approach is *partial digest*. Here we subject the target DNA to one enzyme only but perform many experiments on copies of the DNA, varying the time during which the enzyme acts on each copy.

By giving more or less time to the enzyme, more or less restriction sites will be recognized, thus yiedling fragments of different lengths. Ideally the experiments should provide us with at least one fragment for every pair of restriction sites.We then try to pinpoint the location of the restriction sites by analyzing fragment lengths. For example, enzyme A recognizes three restriction sites. A partial digest experiment based on this enzyme would result in the following fragment sizes: 3, 11, 17 and 27 (fragments between the left endpoint and all other sites); 8, 14 and 24 (fragments between the first restriction site and those to its right); 6 and 16 (fragments between the second restriction site and those to its right); and 10 (last fragment). Several kinds of experimental errors may occur with digestion data.

First, there is uncertainty in length measurement. This is done by gel electrophoresis where there coul be an error of up to 5%. Second, if fragments are too small, it may not be possible to measure their lengths at all. Third, some fragments may be lost in the digestion process, leading to gaps in the DNA coverage. As can be expected, all these errors bring complications to the mapping process. We will see later on that even when there are no errors the associated computational problems seem very hard.

HYBRIDIZATION MAPPING

Recall that, in hybridization mapping, overlap information between fragments is based on partial information about each fragment's content. Before explaining what this partial information is, we note that in this context fragments are called clones, replicated using a technique called *cloning*. After each copy of the target DNA has been broken up (in a different way), and after each resulting fragment has been cloned, we obtain a collection of many thousands of clones (a *clone library*), each clone being typically several thousands of base pairs long. Partial information about a clone is obtained through *hybridization experiments*. In these experiments we try to verify whether a small sequence known as a *probe* binds (or *hybridizes*) to the clone. If the binding does occur, this means that the clone contains the sequence complementary to the probe sequence.

We can do this experiment on a clone using several different probes. The clone's fingerprint will be the set of probes that successfully hybridized to the clone. Two clones that share part of their respective fingerprints are likely to have come from overlapping regions of the target DNA. For example, if we know that probes *x, y* and *z* bound to clone *A*, whereas probes *x, w* and *z* bound to clone *B*, we have good reason to believe that clones *A* and *B* overlap each other, for example (unless there are *repeats*; more on this later). Notice that with this kind of information we will not in general be able to tell the *location* of the probes along the target DNA, but only their relative order. We return to this important point further on.

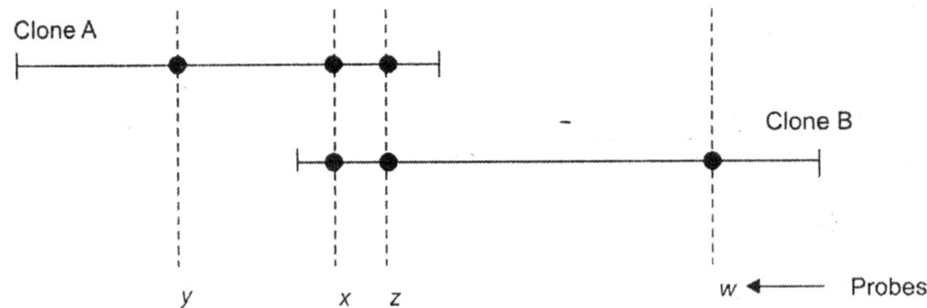

Fig. 5.3. An example of how four probes can hybridize to two clones.

Many kinds of errors may happen in hybridization experiments. First, a probe may fail to bind where it should; this creates a *false negative*. Second, a probe may bind to a site where it should not; this creates a *false positive*. Sometimes there is human misreading of experimental results, also resulting in false negatives or false positives. Errors may have appeared even before the hybridization itself. During the cloning process, two separate pieces of the target

DNA may join and be replicated as if they were one single clone. This is known as a *chimeric clone*, and from it false inference about relative probe order can be made.

Chimerism occurs frequently and is thus one of the most severe problems in hybridization mapping. Some estimates say that in many clone libraries between 40% and 60% of all clones are in fact chimeric. Another type of error in the cloning process is the *deletion*, in which an internal piece of a clone is lost, thus again joining two fragments that are not contiguous in the target DNA. Two other situations can cause problems to hybridization mapping.

One happens when probes are not unique, meaning that there may be more than one site along the target DNA to which the probe can bind; these are the *repeats*. A technique called *Sequence Tagged Site* (STS) avoids this problem by generating probes that hybridize to unique sites along the target DNA. Another problematic situation is simply lack of data, as it may not be feasible to perform all hybridization experiments required. Pooling techniques have been developed to deal with this problem.

MODELS

Having looked at the various mapping strategies, we now review some of the corresponding mathematical models. Our aim is to explore the computational complexity of the mapping problems derived from these models. As we will see, mapping appears to be algorithmically hard, but it also brings up several interesting issues in problem solving that we discuss at the end of this section.

Restriction Site Models

Initially we will look at a model for the double digest problem, which we shorten to DDP. In this model we consider each fragment to be represented by its length. We will assume that the digestion process is perfect, meaning that lengths have no measurement errors, and all fragments are present. This is clearly unrealistic, but as we will see shortly even with these oversimplifications the resulting computational problem appears to be very hard. Placing the experimental results in the model framework, we get a collection of lengths (positive integers) for each experiment performed. Digesting the target DNA by the first enzyme gives us collection $A = \{a_1, a_2, ..., a_n\}$. Some of these fragments may have the same length, which means that this collection is actually a multiset.

Similarly, from the second enzyme we get $B = [b_1, b_2, ..., b_m]$. Finally, from the digestion with both A and B we get the multiset $O = [o_1, o_2, ..., o_k]$. Given these multisets, what is it that we want to find? We want a permutation π_A of the elements from A and a permutation π_B of the elements from B, such that the following can be achieved. We plot the lengths from A on a line according to the order given by π_A.

Similarly, we plot the lengths from B following π_B, but we do it *on top* of the previous plot. Due to overlap, we may now have several new subintervals. It should now be possible to establish a one-to-one correspondence between each resulting subinterval and each element from O. If this can be done, we will say that π_A and π_B are solutions to the DDP. Finding these permutations is an NP-complete problem, and here is the proof. First, it is not difficult to see that given a solution to the DDP it is easy to check whether it is a true solution. All we have to do is sort O and the plotted subintervals to determine whether we get identical results. To complete the proof, we note that, as stated, the DDP is simply a generalization of the *set partition problem*, a well-known NP-complete problem.

In the set partition problem we are given a set of integers $X = [x_1, x_2, ..., x_t]$ and want to know whether we can partition X into sets X_1 and X_2 such that the sum of all elements in X_1 is the same as those in X_2. This corresponds to a double digest problem in which $A = X$, $B = [K/2, K/2]$, and $O = A$, where $K = \Sigma_{x \in X} x_i$. In other words, the set partition problem is a DDP in which one of the enzymes produced only two fragments, and both of the same length. Considering that we are always seeking the *true* solution to the DDP, this problem has one other facet that makes this search a great deal harder.

The number of solutions may be exponential. To see this, consider the case where between two sites cut by enzyme A there are three site b_1, b_2, b_3 cut only by enzyme B. In this case we will not be able to tell the order of fragments $[b_1, b_2]$ and $[b_2, b_3]$. In general, with a run of k restriction sites from one enzyme we will have $(k - 1)!$ solution. In fact, the situation is even worse than that because of the so-called *coincidences*. A coincidence happens when a restriction site for enzyme A is very close to a restriction site for enzyme B (they cannot exactly coincide; why?). When we have consecutive coincidences, the corresponding fragments for the A digestion and the B digestion will appear to have the same length (due to measurement inaccuracy). Therefore a run of n consecutive coincidences will also result in $n!$ possible solutions.

Taking this argument still further, it is possible to show that the number of solutions increases exponentially with the target DNA length (see the bibliographic notes). Thus we see that even with a very simple model double digestion poses serious computational problems. The NP-completeness of partial digest has not been proved, and in addition the problem seems to be more tractable for the following reason. In mapping by partial digest there are n restriction sites (including endpoints), and by experiment we obtain at least one fragment for every pair of sites, or $\binom{n}{2}$ fragments. When measured, these fragments yield a multiset of distances. The problem is to determine the location of the sites on the line such that the set of distances between these locations is the same as the lengths obtained by the digests. What makes the problem easier is that researchers have shown that the maximum number of solutions is small, thus making partial digest, from a combinatorial point of view, a good alternative to double digest.

Interval Graph Models

We can have a first look at the computational complexity of hybridization mapping (or more generally, *fingerprint mapping*) using simple models based on *interval graphs*. These models abstract away many of the complications of mapping; yet, simple as they are, they also give rise to problems that are NP-hard, as was the case with the double digest problem seen earlier. In these models we create graphs whose vertices represent clones and whose edges represent overlap information between clones.

If we had complete and correct information about clone overlapping, the resulting graph would be an interval graph, and this is a key observation in these models. In the first model, we create two graphs. The first graph is $G_r = (V, E_r)$. If $(i, j) \in E_r$, this means that we know for sure that clones i and j overlap. The second graph is $G_t = (V, E_t)$, where E_r represents known *plus* unknown overlap information (thus $E_r \subseteq E_t$). G_t is not necessarily complete, because we may know for sure that certain pairs of clones do not overlap and hence the corresponding edge would be left out of E_t. The problem we want to solve is this: Does there exist a graph $G_s = (V, E_s)$ such that $E_r \subseteq E_s \subseteq E_r$ and such that G_s is an interval graph? Notice that it is not necessarily the case that G_r or G_t are interval graphs (in which case a solution would be trivial).

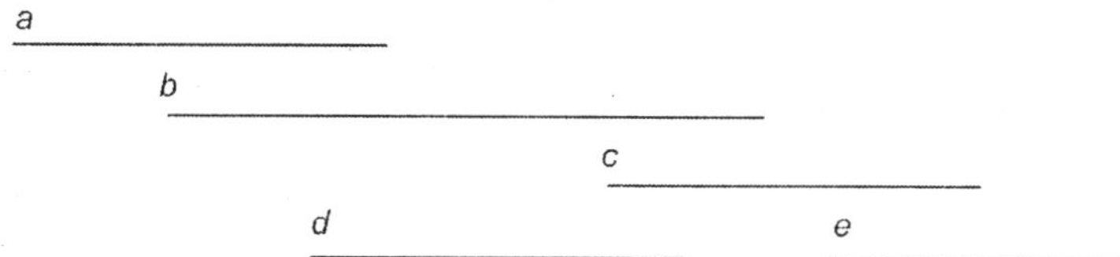

Fig. 5.4. Example of clone overlapping patterns.

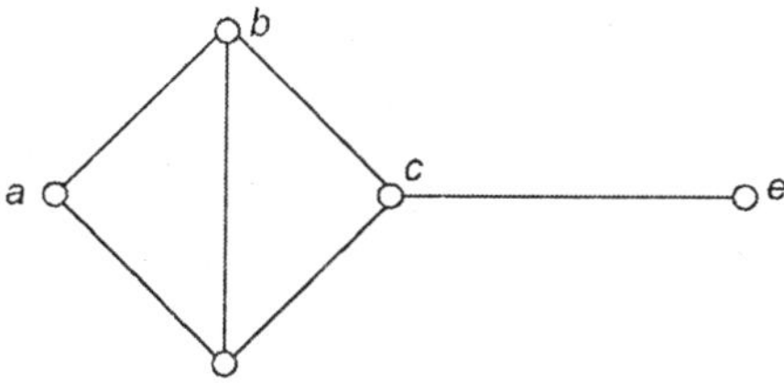

Fig. 5.5. The interval graph corresponding to Figure 5.4.

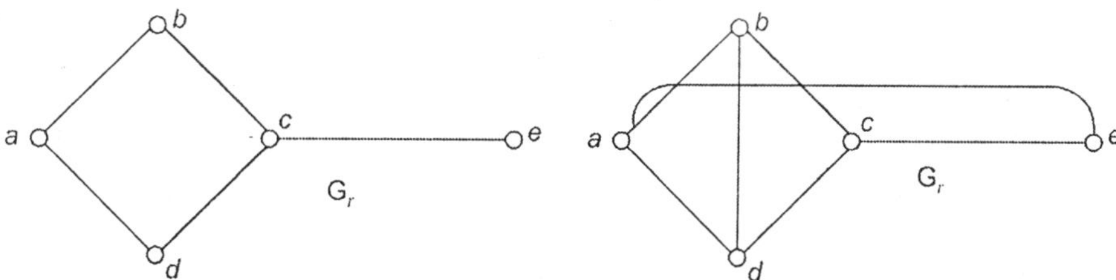

Fig. 5.6. Possible graphs G_r and G_t for the example from Figure 5.4. None of them is an interval graph.

In the second model, we do not assume that the known overlap information is reliable. We construct a graph $G = (V, E)$ using that information anyway. The problem we want to solve is this: Does there exist graph $G' = (V, E')$ such that $E' \subseteq E$ G' is an interval graph and $|E'|$ is maximum? The requirement that $E' \subseteq E$ means that, to get a solution for this problem, we may have to discard some edges of G, which would be interpreted as false positives. The solution is the interpretation that has the minimum number of false positives. In the third model, we use overlap information together with information about the source of each clone as follows. Because clones come from different copies of the same molecule, a reasonable idea is to label each clone with the identification of the molecule copy it came from. We assume we have k copies of the target DNA, and different restriction enzymes were used to break up each copy.

We now build a graph $G = (V, E)$, with known overlap information between clones. Because clones are labeled, we can think that the corresponding vertices are *colored*, using k colors (one color for each molecule copy). The graph constructed will not have an edge between vertices of the same color, because the correspond to clones that came from the same molecule copy and hence cannot overlap. We thus say that the graph has a *valid* coloring. The problem we want to solve is this: Does there exist graph $G' = (V, E')$ such that $E \subseteq E'$, G' is an interval graph, and the coloring of G is valid for G'? In other words, can we add edges to G transforming it into an interval graph without violating the coloring? We have just described three different models for hybridization mapping, and all are relatively simple. However, it turns out that all three problems described are NP-hard. We will not give the proofs here; they can be found in the references cited at the end of the chapter. We now proceed to describe more specialized models.

THE CONSECUTIVE ONES PROPERTY

The models described in the previous section apply to hybridization mapping, but in principle they can be used in any situation where we can obtain some kind of fingerprint for each fragment. We now present a model that uses as a clones fingerprint the set of probes that bind to it. In this model we make the following assumptions:

- The reverse complement of each probe's sequence occurs only one along the target DNA ("probes are unique").
- There are no errors.
- All "clones x probes" hybridization experiments have been done.

The reader might think that another NP-hard problem is coming up. For once, this is not the case. As we will see the resulting problem is polynomial-time solvable, although some of its generalizations are indeed NP-hard. Assuming there are n clones and m probes, the experimental data will enable us to build an $n \times m$ binary matrix M, where entry M_{ij} tells us whether probe j hybridized to clone $i(M_{ij} = 1)$ or not $(M_{ij} = 0)$. Obtaining a physical map from this matrix becomes the problem of finding a permutation of the columns (probes) such that all is in each row (clone) are consecutive.

A binary matrix for which such a permutation can be found is said to have the *consecutive is property* for rows, or CIP for short. Below we also use the terms CIP permutation and CIP problem, with the obvious meanings. Verifying whether a matrix has this property and then finding a valid permutation is a well-known problem for which polynomial algorithms exist. However, the model is clearly too simple. If errors are present, even in the true permutation some rows may not have all is in consecutive positions. So why bother? The answer is that looking at the error-free case helps us to gain insights that are useful when designing algorithms for data with errors. In particular, we can use the following important assumption.

Laboratory reserchers try to minimize the number of errors in thier experiments. If their experiments were *perfect*, the resulting hybridization matrix *would have* the CIP. This means that any algorithm that tries to find or approximate the true column permutation should give solutions that minimize the number of situations that must be explained by experimental error. Hence, such algorithms should be able to find a CIP permutation if one exists. Note that even if a CIP permutation exists, we cannot claim that it is the true permutation. For one thing, there may be several such permutations for a given input matrix. How can we say which is the true one? The answer is that we cannot, in general. We can only say that all permutations that make all is consecutive are likely candidates for the true permutation.

For this reason an algorithm for the CIP problem should give us not one possible solution, but all of them. In addition, given that errors do exist, it may even be the case, as noted above, that the true permutation is not a CIP permutation, even if there is one. On the other hand, how can we modify the model to account for errors? There are many possibilities. Instead of requiring exactly one block of consecutive is per row, we may try to find a column permutation such that in each row there are at most k blocks of consecutive is, where k's value could be 2 or 3. Or we may try to minimize the total number of blocks of consecutive is in the matrix.

By doing this, we may expect that most rows will have just one block of consecutive is, while an occasional row might have four or five. Unfortunately, as already hinted above, these generalizations yield NP-hard problems. NP-hardness comes up again if we relax the assumption that probes must be unique, even if no errors are present.

Algorithmic Implications

Before describing algorithmic techniques for obtaining physical maps, we would like to explore a bit the constraints forced upon us by the difficulty of the problem. These considerations give a new light on what it means to *"solve"* a problem using an algorithm, and should therefore be of interest outside physical mapping as well. The first consideration is that what we are really trying to solve is a real-life problem, not an abstract mathematical problem. And the real-life problem is this: We want to know what the true ordering of the clones is in the target DNA. It is a combinatorial problem:

There are finitely many possible orderings, and one of them is the one we want. We will try to discover this true ordering by means of abstract models that give rise to optimization problems. These problems may be themselves very hard, notwithstanding the fact that they are abstractions of a yet harder problem. For example, the optimization versions of all problems are NP-hard. In general we will not be able to find optimal solutions to these problems quickly; and even if we did solve one of them to optimality within a reasonable time, we will have no guarantee that the solution found corresponds to the true solution. Thus it looks like we are in a terrible predicament. However, the situation is not as bleak as it seems.

First, since we have no guarantee that any optimal solution is the true solution, good approximate solutions may serve our purposes just as well. Second, we must remember that algorithms in molecular biology should be used in an iterative process. The scientist obtains experimental data and feeds it to an algorithm. Based on the solution or solutions given by the algorithm, the scientist can make more experiments and obtain more data and thus use the algorithm again. At each step we may expect that *"better"* (or *"closer to the truth"*) solutions will be found.

These observations suggest desirable features of a mapping algorithm, which we now list:

- It should work better with more data, assuming that the error rate stays the same.
- It should present a solution embedded in a rich framework of details, in particular showing how the solution was obtained, distinguishing *"good"* parts of the solution (groups of clones for which there was strong evidence for the ordering reported) from *"not so good"* parts. This greatly facilitates further experiments.
- If several candidate solutions meet the optimization criteria, all of them should be reported. If too many solutions are reported, the optimization criteria may be too weak (or the input data may contain too many errors). Conversely, if no solutions are reported, the optimization criteria may be too strong.

We would also like to have some guarantee that the true solution is among the set of approximate/optimal solutions found by the algorithm. In general this will not be possible. Instead, we try to obtain a few different optimization functions and respective algorithms whose optima share properties with the true solution. The hope is that the true solution will lie in the intersection of the solution sets found by each algorithm with high probability.

Note that instead of using several algorithms as suggested, we may try to design an algorithm that can optimize multi-objective functions. This is an important research area.

AN ALGORITHM FOR THE CIP PROBLEM

In this section we present a simple algorithm that determines whether an $n \times m$ binary matrix M has the CIP for rows. The goal of the algorithm is to find a permutation of the columns such that in each row all is are consecutive. For simplicity we assume all rows are different (that is, no two clones have the same fingerprit) and that no row is all zeros (that is, every clone is hybridized by at least one probe).

Understanding the algorithm requires that we look first at the way rows can relate to each other. For this we need a definition.

Definition 5.1

For each row i of M, let S_i be the set of columns k where $M_{i, k} = 1$.

With this definition we can now relate a row to another row. Given two rows i and j there situations can arise :

1. $S_i \cap S_j = \theta$
2. $S_i \subseteq S_j$ or $S_j \subseteq S_i$.
3. $S_i \cap S_j \neq \theta$, and none of them is a subset of the other.

We can clearly deal separately with rows that fall into the first of these situations, since the parmutation of columns corresponding to the elements of S_i will not interfere with those from S_j.

In such a case we can place i and j into different *components* and deal with each component separately. Now, how about rows that have nonempty intersection? (We are abusing language a bit here and speaking of rows instead of their respective sets S). Should they all be in the same component? Let us look at this case more carefully. Let us initially lump together in a component all rows that have nonempty intersection. Suppose there is a row k in this component such that S_k is either a subset of or is disjoint from S_j for all $i \neq k$ in this component.

It is clear that we can leave row k out (and any others with the same property) while dealing with this component. Row k will thus be in its own component, possibly sharing it with other rows. What exactly these other rows are will be made clear. Based on the intuition just given, we give a formal way to determine when two rows belong to the same component. We build a graph G_C using matrix M. In G_C each vertex will be a row from M. There is an undirected edge from vertices i to j if $S_i \cap S_j \neq \theta$ and none of them is a subset of the other. The components we want will be the connected components of G_C. We can now sketch the algorithm: Separate rows into components according to the rule above, permute the columns of each component, and then somehow join components together. This leaves us with two subproblems: how to find the right permutation for one component, and how to join components together. Let us first work on the column permutation inside a component.

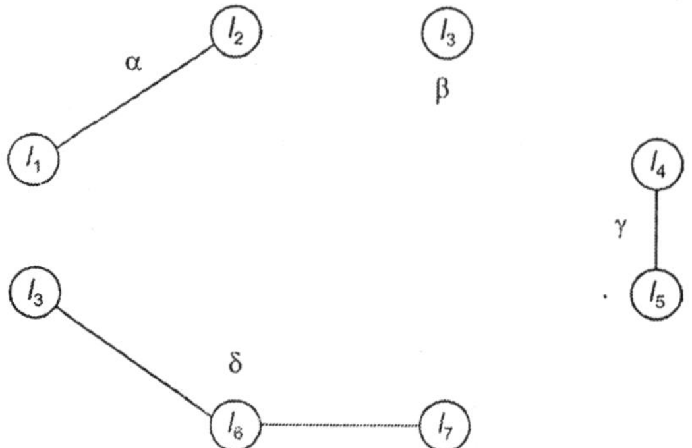

Fig. 5.7. Graph *Gc* corresponding to the matrix of Table 5.1. Connected components are inhdicated by Greek letters.

Table 5.1. A binary matrix.

	c_1	c_2	c_3	c_4	c_5	c_6	c_7	c_8	c_9
l_1	1	1	0	1	1	0	1	0	1
l_2	0	1	1	1	1	1	1	1	1
l_3	0	1	0	1	1	0	1	0	1
l_4	0	0	1	0	0	0	0	1	0
l_5	0	0	1	0	0	1	0	0	0
l_6	0	0	0	1	0	0	1	0	0
l_7	0	1	0	0	0	0	1	0	0
l_8	0	0	0	1	1	0	0	0	1

Taking Care of a Component

Let us look first only at row l_1. If it were the only row in this component, all we would have to do is permute the columns such that all 1*s* become consecutive. Note that if there are k 1*s* is this row, we will have $k!/2$ possible permutations. We are not counting reversals as distinct solutions, because they do not matter for keeping 1*s* consecutive. In the case of row l_1 the 1*s* are in columns 2, 7 and 8. Then one set of possible permutations is 2, 7, 8 or 7, 2, 8 or 2, 8, 7. We can encode all these solutions by associating a set of possible columns for each element equal to 1 in a given row (we will call these sets the *column sets*). For just one row this set is set S defined above, and we describe the resulting configuration as follows :

Table 5.2. A Section of Binary Matrix.

	c_1	c_2	c_3	c_4	c_5	c_6	c_7	c_8
l_1	0	1	0	0	0	0	1	1
l_2	0	1	0	0	1	0	1	0
l_3	1	0	0	1	0	0	1	1

			(2, 7, 8}	(2, 7, 8}	(2, 7, 8}		
l_1 →	...	0	1	1	1	0 ...

Now let us look at the following row, l_2. Because this row is in the same component as l_1 it must be the case that l_2 has 1*s* in some columns where l_1 also has 1*s*, and some other 1*s* where l_1 where l_1 does not. Now we have to place l_2 with respect to l_1 in a consistent manner. We have two choices for this placement: Either put the 1*s* of l_2 to the left of those of l_1 or to the right. This means placing column 5 either to the left or to the right of colums 2 and 7. Suppose we place 5 to the left. The possible solutions, encoded by the sets above each column, now become.

			(5}	(2, 7}	(2, 7}	(8}		
l_1 →	...	0	0	1	1	1	0	...
l_2 →	...	0	1	1	1	0	0	...

If these were the only two rows in this component we would have arrived at two possible permutations for its columns: 5, 2, 7, 8 and 5, 7, 2, 8. If we had chosen to place 5 on the right, we would obtain the reverse of these two permutations. This means that the direction we choose to place the second row does not matter, and this is a key observation. Now let us consider the third row, l_3. Because it is in the same component as the other two, we know that in graph G_C we should find edge (l_3, l_1), or edge (l_3, l_2), or both. In our example both edges are present. Let us now place l_3 with respect to l_2. It seems that again we have the choice of placing l_3 to the left or to the right of l_2. But now we have to take into account the relation between l_3 and l_1 as well. The way to do this is by considering the number of elements in the interesections between S_1, S_2, and S_3. Let $x \cdot y = |S_x \cap S_y|$, for any two sets x and y (the *internal product* of rows x and y). If $l_1 \cdot l_3 < \min(l_1 \cdot l_2, l_2 \cdot l_3)$, row l_3 must go in the *same* direction that l_2 was placed with respect to l_1. If $l_1 \cdot l_3 > \min(l_1 \cdot l_2, l_2 \cdot l_3)$, then we must place l_3 in the *opposite* direction used to place l_2 with respect to l_1 (we cannot have equality; why?). In either case, if the component is CIP, it must be the case that l_1 and l_3 are properly placed; conversely, if l_1 and l_3 are not properly placed, then the component is not CIP.

The reason for this lies on the fact that the only choice we made up until now was in the placing of l_2 with respect to l_1, and we know that both possibilities (left or right) result in the same solutions up to reversal.

In our example above, $S_3 = [1, 4, 7, 8]$. Then $1 \cdot 3 = 2$, $1 \cdot 2 = 2$, $3 \cdot 2 = 1$, which means we have to put l_3 to the right of l_2, as in the following figure:

				{5}	{2}	{7}	{8}	{1, 4}	{1, 4}		
l_1	→	...	0	0	1	1	1	0	0	0	...
l_2	→	...	0	1	1	1	0	0	0	0	...
l_3	→	...	0	0	0	1	1	1	1	0	...

We now know that the right permutation for columns 2, 5, 7 and 8 is 5, 2, 7, 8. Other columns may be determined as we consider more rows. But how do we handle additional rows? As we saw, in placing the third row we had no choice of placement. The same is true for all other rows in the same component.

All we have to do for a new row k is find two previously placed rows i and j such that there eixst edges (k, i) and (i, j) in G_C, and proceed as we did with the three-row case above. Note how the algorithm preserves possible permutations of a component in the sets above each column. considering that after placing the first two rows there are no other choices of placement for remaining rows, we see that this algorithm gives us *all* possible column permutations of a component having the CIP, up to reversal. In the example above, if the component had only three rows, the two possible permulations would be 5, 2, 7, 8, 1 4 and 5, 2, 7, 8, 4, 1. As we have already mentioned, keeping track of all possible solutions is a desirable feature in the case of DNA mapping. An implementation of the preceding algorithm is simple: Construct G_C and traverse it using depth-first search.

To check whether the algorithm has found a CIP permutation, we can verify the consistency of the column sets as we try each new placement, as indicated in the last line of procedure *place*. If we attempt to place a row and we find out that one of its columns should go, say, to the right, and the existing placement tells us that it should go to the left, then the component, and hence the matrix, does not have the CIP.

Algorithm *Place*

input: u, v, w, vertices of $G_C = (V, E)$ such that $(u, v) \in E$ and $(v, w) \in E$.
// *v and w can have value nil*
output: A placement for row u, if possible
if $v = nil$ and $w = nil$ then
  Place all 1*s* of row u consecutively
else if $w = nil$ **then**
  Let- or right-place the 1*s* of u with respect to the 1*s* of v
  Record direction used
else
  if $u \cdot w < \min(u \cdot v, v \cdot w)$ then
    Place u with respect to v using same direction used in v, w placement
    Record direction used
else
    Place u with respect to v using opposite direction used in v, w placement
    Record direction used
Check consistency of column sets.

Fig. 5.8. The procedure] to place a row given two others.

Let us analyze the running time of the algorithm. Building graph G_C takes $O(nm)$ time. We then process the n rows as sketched above, spending $O(m)$ per row to check consistency of column sets. Total time is thus $O(nm)$.

Joining Components Together

We now tackle the second part of the algorithm, component-joining. For this we need another graph, G_M, which will tell us how components fit together. Each component of the original matrix M will be a vertex in G_M. A directed edge will exist between vertex α and vetex β if the sets S_i for all $i \in \beta$ are contained in at least one set S_j of component α (we are making no distinction between a row j and its corresponding set S_j). The graph G_M corresponding to the components of the matrix.

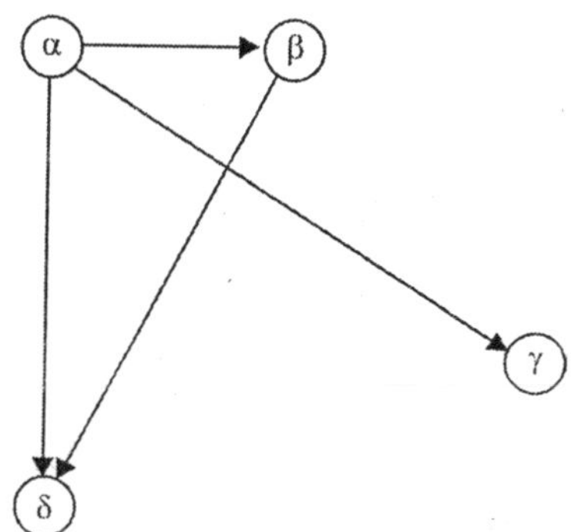

Fig. 5.9. Graph G_M corresponding to the components of the matrix from Table 5.1.

From the discussion in the beginning of this section we know that the containment relations between a set of β and sets of α are the same for all sets of β. Let us make this more clear. Suppose set S_j belonging to component β is contained in set S_j belonging to another component α. Because the rows i and j are in different components, there is no row k in component α

such that S_i is not contained in S_k and $S_i \cap S_k \neq \theta$. In other words, $S_i$is contained in some sets of α and is disjoint from the others. We claim that the exact same containments and disjunctions hold for all other sets from β. Suppose not and let us analyze one such exception. Suppose i, $l \in \beta$, $S_i \cap S_l \neq \theta, j \in \alpha$, and $S_i \subseteq S_j$, but $S_l \not\subset S_j$. In this case the only way l could be out of α is if $S_l \cap S_j = \theta$. But this cannot be, since S_i and S_l have at least one element in common and S_i is contained in S_j. We leave other cases to the reader. The joining of components depends on the way sets in one component contain or are contained in sets from other components. Intuitively, we should process first components that have sets that are not contained anywhere else. Because containment is given by edge direction, we have to process G_M in topological ordering. (We can only do this if G_M is acyclic. Is it? Yes, and this is not difficult to prove).

After having correctly permuted the columns of all components individually, we select all vertices in G_M without incoming edges, and freeze their columns. We then take the next vertex in topological order. Suppose we are following edge (α, β). This means that we must find some sort of *"reference column"* in component α that will tell us how to place the rows of β. Choose the row l from β that has the leftmost l, and call the column where this 1 is c_β.

We know that S_t is contained in some S_i of α but not in others. Find all rows *from* α that contain S_t, and find the leftmost column where all such rows have 1*s* (and call this column c_α). This is the reference column we are looking for, since we can now make c_α and c_β one and the same. Let us illustrate this process with the matrix and the graph, and let us consider its components in topological ordering. One such ordering is α,; β, δ, γ. Here are the rows of component α with columns permuted so that its is are consecutive :

			{1}	{2, 4, 5, 7, 9}					{3, 6, 8}			
l_1	→	...	1	1	1	1	1	1	0	0	0	...
l_2	→	...	0	1	1	1	1	1	1	1	1	...

We will now join the next component, which is β, a singleton row :

			{2, 4, 5, 7, 9}					
l_3	→	...	1	1	1	1	1	...

As should be clear, all we have to do is join l_3 with the rows from α, without any rearrangements :

			{1}	{2, 4, 5, 7, 9}					{3, 6, 8}			
l_1	→	...	1	1	1	1	1	1	0	0	0	...
l_2	→	...	0	1	1	1	1	1	1	1	1	...
l_3	→	...	0	1	1	1	1	1	0	0	0	

The next component is δ. Here are its rows, with consecutive 1*s* :

		...	{9, 5}		{4}	{7}	{2}	...
l_6	→	...	0	0	1	1	0	...
l_7	→	...	0	0	0	1	1	...
l_8	→	...	1	1	1	0	0	...

We can see that in component δ it is l_3 that has the leftmost 1 (in either column 5 or 9). In component α both rows l_1 and l_2 contain l_3, and it is the second column (representing a bunch of possible columns) that has all 1*s*. Therefore that second column should coincide with columns 9 or 5 from component δ. This is the result :

			{1}	{9, 5}		{4}	{7}	{2}	{3, 6, 8}			...
l_1	→	...	1	1	1	1	1	1	0	0	0	...
l_2	→	...	0	1	1	1	1	1	1	1	1	...
l_3	→	...	0	1	1	1	1	1	0	0	0	...
l_6	→	...	0	0	0	1	1	0	0	0	0	...
l_7	→	...	0	0	0	0	1	1	0	0	0	...
l_8	→	...	0	1	1	1	0	0	0	0	0	...

We finally glue in component *Y*. Its rows are

			{6}	{3}	{8}	
l_4	→	...	0	1	1	...
l_5	→	...	1	1	0	...

Joining with the rows above we obtain the final matrix with all 1*s* consecutive :

		{1}	{9, 5}		{4}	{7}	{2}	{6}	{3}	{8}
l_1	→	1	1	1	1	1	1	0	0	0
l_2	→	0	1	1	1	1	1	1	1	1
l_3	→	0	1	1	1	1	1	0	0	0
l_6	→	0	0	0	1	1	0	0	0	0
l_7	→	0	0	0	0	1	1	0	0	0
l_8	→	0	1	1	1	0	0	0	0	0
l_4	→	0	0	0	0	0	0	0	1	1
l_5	→	0	0	0	0	0	0	1	1	0

Notice that one of the column sets is not a singleton, indicating that we have found two solutions. In this particular case the reason is that column 5 and 9 are identical. But in general there may be multiple solutions that do not involve just the permutation of identical columns. For example, 1, 2, 7, 4, 9, 5, 8, 3, 6 is another solution for this same matrix. Multiple solutions may exist because G_M may allow different topological orderings, and this is compounded by the facts that each component may on its own have several solutions and each of them can be used in two ways (the permutation and its reversal). We now analyze the running time of the algorithm. Topological sorting of G_M takes time $O(n + m)$. It is possible to preprocess the entries of M so that the queries needed when traversing G_M take constant time. For example, we can store for each row the coloumn where its leftmost 1 is.

Such preprocessing takes at most $O(nm)$. Therefore this is the time needed for joining components. Total time for processing each component was seen to be $O(nm)$ as well, so this is the running time of the whole algorithm. Another algorithm exists that can find a CIP in time $O(n + m + r)$, where r is the total number of 1*s* in the matrix. In addition, the algorithm encodes in a compact way all possible solutions. This algorithm, however, is considerably more complicated than the algorithm we presented in this section. References are given in the bibliographic notes.

AN APPROXIMATION FOR HYBRIDIZATION MAPPING WITH ERRORS

In the previous section we studied an algorithm that solves the consecutive is problem. We saw that such a problem is a good model of hybridization mapping when there are no errors and when probes are unique. If errors are present, some other approach is needed, and that is the subject of this section. Let us first examine the effect errors can have on a clones × probes binary matrix M. Suppose M is presented to us with the true column permutation. Given one row, if there are no errors, all its 1*s* will be consecutive.

If a row corresponds to a chimeric clone, where two fragments were joined, we will see *two* blocks of 1*s* separated by some number of 0*s* (assuming no other errors are present in this row). We will call a consecutive block of 0*s* bordered by 1*s* a *gap*. Notice that this is different from the use of the term *gap* in other chapters of this book. We can thus say that a gap was created in this row because of the chimeric fragment. If, on another row, there is a false negative, the corresponding 0 may separate two blocks of 1*s*, creating another gap, as shown below :

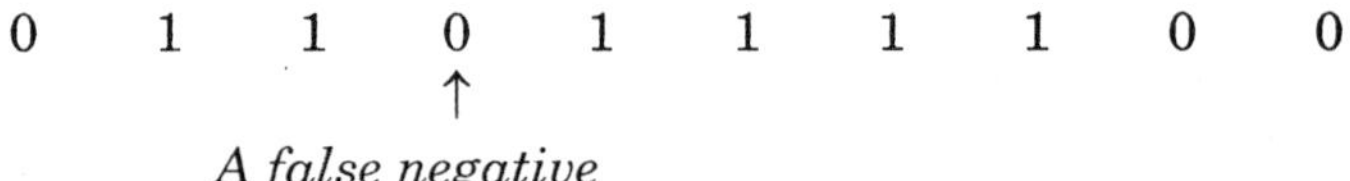

The gap will not be created if the probe was leftmost or rightmost for this clone. Finally, a false positive may split a block of 0*s* in two, thus possibly creating yet another gap.

In this way we see that there is a close correspondence between errors and gaps in the matrix. Given the basic assumption that we want to avoid explaining gaps by experimental error as much as possible, a reasonable approach is to try to find a permutation where the total number of gaps in the matrix is minimum. Such an approach has the desirable property that, if there is a CIP permutation, it will have the minimum number of gaps. In other words, gap minimization can be seen as a generalization of the consecutive 1*s* problem. We have already mentioned that some extensions to the CIP are NP-hard. Such is the case also with the gap minimization problem just sketched. However, for this particular NP-hard problem, we can use many special techniques to get approximate solutions that we can expect to be reasonably good, as we show next.

A Graph Model

It turns out that gap minimization is equivalent to solving a well-known graph problem. This is *traveling salesman problem* (TSP). This input to this version of the TSP is a complete undirected weighted graph G. The vertices of G correspond to columns of the clones × probes binary matrix M; that is, they correspond to probes. For reasons that will soon become clear, we also have to add an extra column to M filled with zeros, and G must have the corresponding vertex.

The weight on each edge of G is the number of rows where the two corresponding columns differ (this is also known as the *Hamming distance* between the rows). We will now argue that a minimum-weight cycle in G corresponds to a column permutation in M with the least number of gaps.

Table 5.3. A clones × probes matrix with added column p_6^*.

	p_1	p_2	p_3	p_4	p_5	p_6^*
c_1	1	1	1	0	0	0
c_2	0	1	1	1	0	0
c_3	1	0	0	1	1	0
c_4	1	1	1	1	0	0

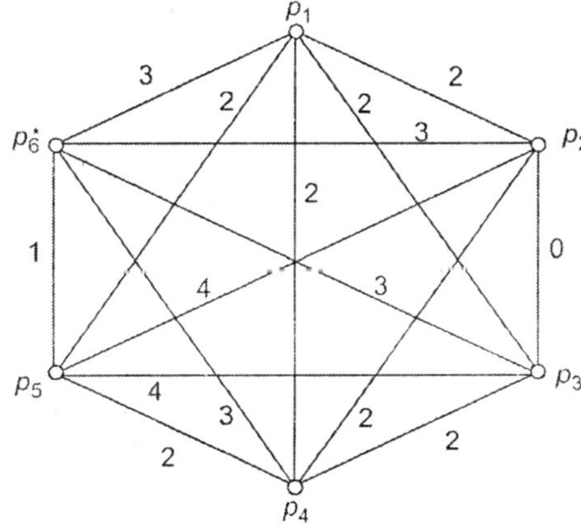

Fig. 5.10. TSP graph for matrix of Table 5.3.

To see this, note that given a permutation of the columns, a gap in a now means that at a certain point we have a transition from 1 to 0 and further on we have a transition from 0 to 1. Thus for each gap we have two transitions and each gap contributes exactly 2 to the weight of the cycle corresponding to the given column permutation. However edge weights may also be increased by *extremal transitions*, that is, transitions between elements in extremal (1 or m) columns, and these do not correspond to gaps.

To ensure that every row has a pair of extremal transitions we include an extra column of zeros in column $m + 1$. Without such a column, cycles in the graph correspond to permutations where we allow consecutive 1*s* to wrap around in each row, and we do not want this to happen. The relationship between cycles and permutations now becomes cycle weight = number of gap transitions + $2n$. This means that for a given n minimizing cycle weight is the same as minimizing the number of gaps. What we have just done is to show how we can reduce the gap minimization problem to the TSP. It is well known that the TSP is an NP-hard problem, so in principle we have not accomplished much.

However, a large array of techniques is available to solve or approximate traveling salesman problems, and these techniques can now be used in this context. The existence of such techniques is not in itself enough to give us confidence that solving traveling salesman problems will give us the true probe permutation. What we need are *guarantees* that such solutions are, in some sense, close to the true solutions. Before that, let us go back to the gap minimization problem. We have defined above a function that, given an input matrix, returns the total number of gaps in the matrix. We have further argued that obtaining a permutation that has the minimum value for this function (or approximately minimum) will help us find the true column permutation. Since we do not have a guarantee that the true permutation will be among the solutions we find, we should look for other functions that might also be helpful. The idea is that by carefully defining several such optimization functions and developing algorithms for them we will increase the likelihood of hitting upon the true solution.

In particular, it is reasonable to expect that the true solution will be in the intersection of all solution sets. This will only be the case, however, if each function represents one property that true solutions do have or are likely to have. Here is an example of another optimization function. One possible drawback of gap minimization is that in a permutation with a minimum value for this function one or a few rows may have many gaps, while others may have none. Having many gaps in one row is undersirable, since it would mean that one clone was subject to many more errors than other clones, which constradicts laboratory experience. Therefore we could try to minimize the number of gaps *per row*. We leave as Exercise 15 how to show that we can still use the preceding graph model. The resulting graph problem is known as *the bottleneck traveling saleman problem*, which is also NP-hard.

A Guarantee

In this section we give a formal proof that the TSP approach outlined in the last section will give us, with high probability, the true permutation. The proof we present depends on two basic assumptions: that the number of probes is sufficiently large, and that the mapping process obeys a certain mathematical model. This model seems to be a good representation of what actually happens in large mapping probjects. We describe the model next First we assume that the DNA molecule we are dealing with is so long that we may think of it as an interval on the real line, extending from 0 to N. The clones are subintervals of this long interval, and we assume that all of them have the same length; for convenience each clone is one unti long.

To simplify the exposition, we will speak of *clone permutations* rather than probe permutations. The unit length assumption makes them equivalent. This means that we will be looking for consecutive 1*s* in columns, not in rows, and that each vertex in the associated TSP will correspond to a clone. We use each *clone's* left endpoint as a clone locator. We, of course, do not know the precise position of each clone along the molecule; and because we are dealing with hybridization, all we will be able to determine is relative clone order. A critical feature of this model is clone distribution along the target DNA. We will assume that each clone's position is an independent random variable, that clone locators are distributed uniformly over $[0, N-1]$, and that the clones cover the interval $[0, N]$ (that is, for every subinterval I of $(0, N)$, there always exists at least one clone C such that $C \cap I \neq \theta$). Another important aspect of the model is probe distribution. We will *not* assume that each probe is unique; instead, we will assume that each probe occurs rarely along the target DNA. More formally, we will say that the occurrences of a given probe obey a Poisson process with rate λ. Moreover, the Poisson process of any one probe is independent of all the others. This part of the model lets us immediately obtain an expression for the probability of a specific probe hybridizing to a clone. The expression is

$$Pr\{\text{a given probe occurs } k \text{ times in a given clone}\} = e^{-\lambda}\frac{\lambda^k}{k!}. \tag{5.1}$$

This expression can be obtained from any textbook formula for Poisson processes, using the fact that our clones are unit length intervals.

This completes the model description. We shall now argue that, given a clones × probes binary matrix, the row permutation given by solving the associated TSP is a good approximation to the true permutation, in the following precise sense: The probability that both permutations are the same tends to 1 as the number of probes inreases. Note that the number of probes is *fixed* for a given instance of the problem. We are just claiming that in larger and larger instances of the problem (and we are measuring size here by number of probes) the TSP permutation will be the same as the true permutation with higher and higher probability. To prove this claim, we must argue in terms of graph weights, or more appropriately, clone distances.

As we mentioned, the weight of each edge of the graph associated with input matrix M is called the Hamming distance between its two endpoints (clones). We denote by h_{ij} the Hamming distance between clone i and clone j. We can also think of the *true* distance between clones. Denoting a clone's coordinates by l(left) and r(right) we can define this distance to be

$$t_{ij} = |l_j - t_i| + |r_j - r_i| = 2|l_j - l_i|, \qquad \text{...(5.2)}$$

given that clones are all of the same size.

Suppose now that we *knew* all true distances. Then it is clear that the largest such distance would give us the clones that are farthest apart, which is to say, the clones that occur at opposite ends of the interval [0, N]. The next largest such distance gives us another similar pair that occurs between the previous two, and so on. This means that given the true distance we are able to obtain the true clone permutation, which is not surprising.

However, we have distances h_{ij} and not distances t_{ij}. But because we are trying to obtain only the true *relative* order of clones, it would suffice if we could say that, given any four clones i, j, r and s, $h_{ij} < h_{rs}$ implies that $t_{ij} < t_{rs}$ and vice versa. The reason is that some notion of *order* between clone distances was all we needed to place clones relative to each order using the true distances. If we prove that the probability that $h_{ij} < h_{rs} \Leftrightarrow t_{ij} < t_{rs}$ tends to 1 as the number of probes increases, we will have the result we need. Let us look at a pair of clones i and j, and let us find the probability that a certain probe p contributes to their Hamming distance. This will happen if probe p hybridizes to i but not to j or vice versa.

Referring to Equation (5.1) we see that the probability that probe p does not occur on clone j is $e^{-\lambda}$, since $k = 0$. On the other hand, the probability of probe p occurring at least once in clone i is the same as the complement of the probability of probe p not occurring in that section of i that does not overlap j. If this overlap is z_{ij}, we obtain the following result :

$$Pr[p \text{ hybridizes to } i \text{ and not to } j \text{ or vice versa}] = p_{ij} = 2e^{-\lambda}(1 - e^{-\lambda(1-z_{ij})}). \qquad \text{...(5.3)}$$

This essentially means that there is a well-defined probability of this event happening. We can now take into account all probes, and each of them will have a certain probability of hybridizing to clones i and j. Assume we have m probes, and consider h_{ij}/m. This represents the mean contribution of each probe to the Hamming distance between clones i and j. (Note that a probe's contribution to the Hamming distance is either 0 or 1). If the number of probes is large, we can invoke the law of large numbers and say that h_{ij}/m approach p_{ij}. Or, for any fixed small positive real number $\in$, that

$$Pr\left\{\left|\frac{h_{ij}}{m} - p_{ij}\right| > \in\right\} \to 0$$

as $m \to \infty$. Note now that since clones are unit length, the true distance between two overlapping clones i and j with an intersection measuring z_{ij} is given by $t_{ij} = 2 - 2z_{ij}$. This allows us to substitute t_{ij} for z_{ij} in Equation (5.3) effectively showing us that $t_{ij} > t_{rs}$ if and only if $p_{ij} > p_{rs}$, for pairs of clones i, j and r, s. But if this is the case, then we can say that

$$Pr\{|t_{ij} - h_{ij}| > \in\} \rightarrow 0.$$

This implies that $h_{ij} < h_{rs} \Leftrightarrow t_{ij} < t_{rs}$, which is what we wanted to prove.

COMPUTATIONAL PRACTICE

In this section we present some actual results of computational tests with algorithms for the hybridization mapping problem. Before presenting these results, however, we must make considerations regarding the way the results of such tests can be interpreted. Some of these considerations are valid for many problems in computational biology. Initially we shall look at the input data using another graph.

By doing so it will become clearer what we can expect of any algorithm that tries to obtain the true probe permutation. This graph will also be used in the next section. We define the *hybridization graph H* as a bipartie graph (U, V, E) that is built using information from the hybridization matrix: Clones are the vertices of the U partition, and probes are the vertices of the V partition. There is an edge between two vertices if the corresponding probe hybridized to the corresponding clone.

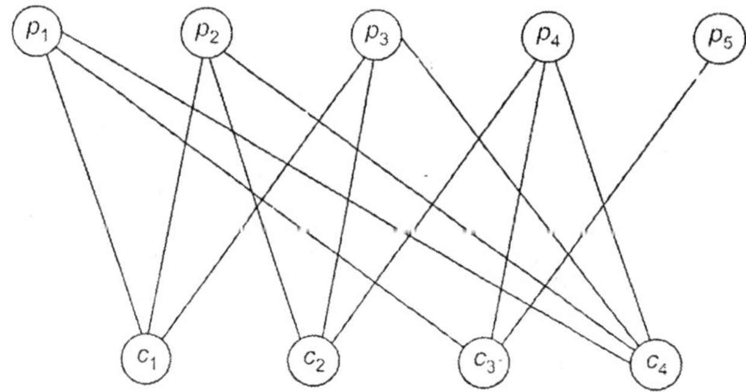

Fig. 5.11. Hybridization graph H corresponding to hybridization matrix from Table (5.3), without the added column.

The first thing to notice is that H may not be connected, even if all entries in the hybridization matrix are correct. If this is the case, then no matter how good our algorithm is, we will not be able to tell the relative order between probes that belong to different connected components; the information to do so is simply not present in the hybridization matrix. A connected component may be as simple as a singleton vertex, meaning that there is a probe that did not hybridize to any clone or a clone that was not hybridized to any probe. Another observation is that there may be *redundant probes*, or probes that hybridize to exactly the same set of clones.

This could happen if the probes, although different, hybridize to parts of the target DNA that are close together. It could also happen if certain clones for that particular DNA stretch are missing, leaving some probes without any positive hybridization results. Connected components of H show up when we solve the corresponding consecutive 1s problem. Redundant probes can also be easily seen in the hybridization matrix: They are columns that have exactly the same 1s and 0s. But if there are errors we may get wrong information regarding the number and structure of connected components of H and whether a probe pair is redundant. We say that errors may *mask* these properties. So notice the difficult situation that errors create:

The input matrix without errors may lack information necessary to find the true permutation (for example, the erroless H has several components), and we may have a lot of trouble just a recognize that lack of information (because our errorful H has only one component). Assuming matrix in Table (5.3) is errorless, probes p_2 and p_3 are redundant, but if there were a false positive between clone c_3 and prob p_3 we would not be able to recognize that. Given this situation, it is clear that evaluation of a mapping algorithm is a difficult task (in addition to mapping itself). We will now take a look at how we can evaluate such algorithms assuming that we somehow know the correct answer to any mapping problem.

This can be done, for example, if we use a computer program to generate artificial instances of mapping problems, simulating experimental errors. If such instances are faithful to real instances, it should be clear from the above discussion that the input matrix may lack information to enable an algorithm to determine the true probe order. Therefore we should try to evaluate how *"close"* the solution found by a particular mapping algorithm is to the true probe order. The question now becomes: How should we define closeness in this context? At the moment there is no consensus on how to do this. However, to give some idea of the performance of current mapping algorithms in practice, we will present as an example one definition that has appeared in the literature. it is a reasonable definition, but even if it becomes widely accepted, it may still undergo some refinements. We will measure a mapping algorithm by the fraction of *strong adjacencies* reported by it.

Strong adjacencies are defined in terms of the number b of blocks of consecutive 1*s* present in a hybridization matrix with a given probe permutation $\pi = p_1, p_2, .., p_m$. We analyze the effect of *translocations*, which are operations that reverse the order of a set of consecutive probes. We say that two adjacent probes p_i and p_{i+1} represent a strong adjacency if placing these probes apart by any translocation increases b in each that probes p_i and p_{i+1} should stay adjacent in all solutions.

Based on this concept, we can define the *strong adjacency coast* of a given permutation. This is given by the fomula

$$100\left(\frac{1}{m-1}\sum_{0}^{m-2}\delta_i\right), \qquad ...(5.4)$$

where $\delta_i = 1$ if p_i and p_{i+1} is a strong adjacency in the true permutation but these probes are not adjacent in the proposed permutation, and $\delta_i = 0$ otherwise. Note that the cost is given as a percentage. Good permutations should have low strong adjacency cost. With this definiton we are finally able to give the reader an idea of algorithm performance. Table (5.4) presents the strong adjacency cost of two algorithms. One of them is *"random"*: A random probe permutation is selected. The other is based on the TSP approach; this is how it works. Given the TSP graph the algorithm builds a cycle by choosing pair of vertices and making them adajcent on a path. The pair (u, v) chosen at each iteraction must fulfill the following conditions: If both u and v already belong to paths, and these paths are different, the paths can be joined only if u and v are endpoints in their respective paths (that is, they each have just one neighbor); and they must be the closest among all qualifying pairs. After all vertices are on the some path, that path is closed forming a cycle. This solution is then submitted to another alogrithm, which tries to improve the solution by applying another heuristic.

In Table (5.4) we can see that the TSP-based appraoch (called *"greedy"*) performs well compared to "random". The paper from which these results were obtained presents the performances of three other, much more sophisticated alogirithms, and the results are similar to those shown above for "greedy TSP". This can be seen as a point in favour of the TSP approach, but in a sense it is yet another measure of how difficult the mapping problem is.

The table also shows that in the presence of false negatives the solution of "greedy TSP" was fairly poor. It is fair to assume that results would be even worse if all kinds of errors were combined in the same instance. This motivates our next section, in which we present a heuristic that appears to be robust in the presence of false negatives.

Table 5.4. Strong adjacency corts for two algorithms on matrices with different kinds of errors. Error rates are indicated in the heading of each column (only one type of error per column). Coverage in all cases is 10, where coverage is the ratio betweent the total length of all clones and target DNA length.

	CIP *0*	*Chimerism* *0.5*	*False Positives* *0.04*	*False Negative* *0.32*
Greedy TSP	1.9	0.9	16.0	28.3
Random	86.4	89.7	94.4	94.9

HEURISTICS FOR HYBRIDIZATION MAPPING

As the previous sections have shown, mapping is a difficult problem, and no general and good algorithms for it have been found. As a consequence, what we see in practice is that researchers resort to various heuristics to help them arrive at a solution. In this section we present two such heuristics that have yielded good result in hybridization mapping projects.

Screening Chemeric Clones

As noted above, chimeric clones occur with high frequency in clone libraries, and their presence brings serious problems to any mapping algorithm. In this section we present a simple heuristic that tries to split chimeric clones into fragments. Such a heruistic is very useful as a *screening procedure*, which can be used as a preprocessing step before employing more sophisticated techniques. The idea is very simple: if a clone is a chimeric (and let us assume it is composed of only two fragments), the probes that hybridize to one of its fragments should not be related to the probes hybridizing to the other fragment. The key here is of course the concept of "relatedness".

This is made concrete by looking at the following graph. Take clone i and the set of probes that hybridize to it, P_i. We create a graph $H_i = (P_i, E)$ for every clone i. We create an edge between two probes from P_i if they hybridize to a clone other than i. If the resulting graph is connected, we say that i is not chimeric. If it has more than one component, we say that i is chimetric and we replace i by now "artichical" clones, each new clone given by a connected component of H_i. This method can be refined by requiring that an edge exists between probes p and q in H_i only if p and q hybridize to at least k other clones, where k is a parameter depending on the particular problem at hand.

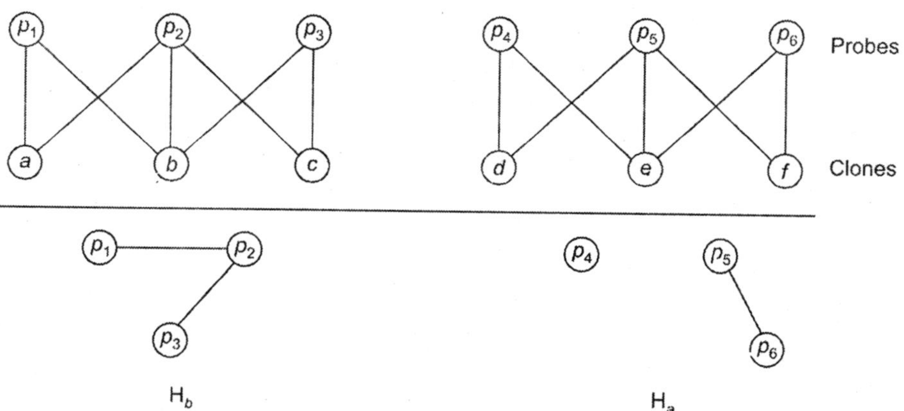

Fig. 5.12. Above the line we show the hybridization graph H for some clones × probes matrix. Below the line at left is graph H_b for clone b. This clone is probably not chimeric, because H_b is connected. At right is graph H_c for clone e. This clone could be chimeric, because H_c is not connected.

Obtaining a Good Probe Ordering

The heuristic presented here is more ambitious than the one we saw in the previous section. It aims at actually solving the problem, that is, obtaining a permutation of the probes. The idea is to estimate for every probe p the number of probes to its left and the number of probes to its right by looking at the hybridization graph H. We can then sort the probes using this estimate and thus obtain one "good" permutation.

Algorithm *Probe Permutation Heuristic*
Input: clones × probes hybridization matrix
output: a "good" permutation for the probes
Identify splitters as described in the text
for every splitter i do Determine the components A_i and B_i
for each probe p do Initialize l_p and r_p with zero
Select a pair (A_k, B_k) arbitrarily

    for each probe p do
        for each pair of components (A_i, B_i) do
            if $p \in A_i$ then
                if $|A_i \cap A_k| > |A_i \cap B_k|$ then
                    Increment l_p
        else
            Increment r_p
    else if $p \in B_i$ then
        if $|B_i \cap A_k| > |B_i \cap B_k|$
            Increment l_p
        else
            Increment r_p
Sort probes in decreasing order of $l_p - r_p$.

Fig. 5.13. Heuristic to obtain a "good" probe permutation.

Given a probe we will be able to count the number of probes to its left and to its right if we can somehow split the other probes into *two* separate components: one left component and one right component. Probes near the ends of the target DNA may not split the others in two; therefore, the probes for which we can obtain two components will be called *spliters*. The method for detecting splitters is described next. Given a probe p build a set of vertices S_p (whose elements are clones and probes) by including p in S_p and every other probe that shares a clone with p. Include in S_p all clones that are incident to any probe in S_p. Now remove all vertices in S_p from H. We will say that p is a splitter if the resulting graph has exactly two components. For each probe p, the algorithm keeps two counters, l_p and r_p, recording the number of left and right regions that contain p. These counters are initialized with zero and are incremented using the splitters. it is not obvious how to do this, because although we know the two components A_i and B_i of a splitter i, we cannot tell which one goes to the left or to the right. To solve this difficulty we rely on a fixed arbitrary splitter k that we use as a reference.

We assume that A_k lies to the left of B_k. Then, given another splitter i, a component X of i (either A_i or B_i) is the leftmost one when $|X \cap A_k| > |X \cap B_k|$. Once we have the final counts, we sort probes so that the ones with higher left bias $l_p - r_p$ come first. A simple improvement to this heruistic is in the choice of splitter k. The more "central" this splitter, the better should the results be. As with the heuristic of the previous section, experience has shown that this one performs well. In particular, it is relatively robust with respect to false negatives. In addition, one can envisage this heruistic also used as delivering its result to other algorithms or heuristics that could try to improve the solution.

RNA Secondary Structure Prediction

Ribonucleic Acid (RNA) is an important molecule that performs a wide range of functions in biological systems. In particular, it is RNA (not DNA) that contains the genetic information of viruses such as HIV and thereby regulates the functions of these viruses. RNA has recnetly become the center of much attention because of its catalytic properties (Cech and Bass 1988), leading to an increased interest in obtaining structural information. RNA molecules have two sets of structural information: first, the *primary structure* of RNA is a single strand made of the ribonucleotides A (adenine), C (cytosine), G (guanine), and U (uracil). Secondly, the ribounucleotide sequences fold over onto themselves to form double-stranded regions of base pairings, yielding higher order *tertiary structures*. It is well known that the structural features of RNAs are important in the molecular mechanisms involving their functions.

The presumption, of course, is that to a preserved function there corresponds a preserved molecular confirmation and, therefore, a preserved structure. The RNA *secondary structure* is a restricted subset of the tertiary structure that plays an important role between primary structure and tertiary structure, as the problem of inferring the tertiary structures of RNA molecules is often intractable. Based on a reliable secondary structure, the possible tertiary interactions that occur between secondary structural elements, as well as between these elements and single-stranded regions of RNAs, can be characterized. Curretly, the only completely accurate method of determining the folded structure of an RNA molecule is by X-ray crystallography; however, this is not only time consuming, but also expensive. The use of computational methods to predict RNA secondary structure began more than 30 years ago.

Many computational methods have been proposed in an attempt to predict RNA secondary structures. Although computational methods sometimes only provide an approximate RNA structural model, they facilitate the future study of RNA structures. To date, several apporaches have been established for predicting RNA secondary structure, most notably the phylogenetic comparative method, (Winker et al. 1990; Chan et al. 1991; Gutell et. al. 1992; Gutell et. al. 1994; Le et. al. 1995), the thermodynamic energy minimization method, Turner et. al. 1988; Zuker 1989; Rivas and Eddy 1999; Zuker 2000), and the stochastic context-free grammar method (Searls 1992, 1933; Sakakibara et. al. 1994; Knudsen and Hein 1999). There are other approaches for RNA secondary structure prediction, including the equilibrium partition function method (McCaskill 1990), gentic algorithms Shapiro and Wu 1996), algorithms that combine the phylogenetic and the thermodynamic methods (Wang and Zhang 1999), and method based on the hybridization of the thermodynamic and the phylogenetic methods as well as genetic algorithm (Chen et al. 2000). In this chapter we discuss algorithms/methods to predict RNA secondary structure.

Basic Definitions

Definition 1 (Primary Structure) Because an RNA sequence is composed of four possible bases, we can use a four-letter alphabet to represent an RNA sequence, $\Sigma = \{A, C, G, U\}$. This base sequence is usually referred to as *primary structure*. Formally, it is written as follows:

$$R = r_1, r_2, .., r_n, \qquad r_i \in \Sigma$$

Following convention, we denote the left end of the sequence as the 5′ end and the right side of the sequence as the 3′ end.

An RNA sequence folds by intramolecular base pairing and is stablized by the hydrogen bonds that result from that base pairing. Additionally, the stacking of base pairs in a helix stablizes the molecule and decreases the free energy of the folded structure, but the appearance of loops (see the definition below) destabilizes the molecule and increases the free energy of the RNA structure.

Definition 2 (Canonical Base Pairs) In an RNA secondary strucure, base paris are formed as one of the three kinds of pairs, C-G (G-C), A-U (U-A), and G-U (U-G). There are three hydrogen bonds between C-G (G-C) pairs, two between A-U (U-A), and one between G-U (U-G). Base pairs C-G (G-C) and A-U (U-A) are called Wastson-Crick base pairs. The base pair G-U (U-G) is referred to as a wobble base pair. These three types of pairings are referred to as *canonical base pairs*.

Definition 3 (Secondary Structure) We use $i \cdot j$ to represent the base pair formed by the ith base, r_i, and the jth base, r_j, where $1 \le i < j \le n$. Let S be a set of base pairs for sequence R, then set S is called *RNA secondary structure* if S satisfies the following conditions :

1. For any base pair $i \cdot j$ in S, the base pairing of r_i and r_j is a *canonical base paire* (see definition 2);
2. For any two base pairs $i_1 \cdot j_1$ and $i_2 \cdot j_2$, either $i_1 = i_2$ and $j_1 = j_2$ or $i_1 \ne i_2$, $i_1 \ne j_2$, $j_1 \ne i_2$ and $j_1 \ne j_2$;
3. If $h < i < j < k$, then S cannot contain both $r_h \cdot r_j$ and $r_i \cdot r_k$;
4. If S contains $r_i \cdot r_j$, then $|j - i| \ge 4$.

Following condition 2, a base r_i can be in one of the two states: either it is paired with another base, or it is unpaired. Condition 3 is usually referred to as the non-crossing condition. In reality, conditions 2 and 3 may not be true, which will result in "triples", "knots", and so on; however, all these are considered as features of the higher level structure (that is, the *tertiary structure*). Still, it is possible to violate condition 1, which will result in noncanonical pairs, such as G · A.

obviously, the RNA secondary structure is much more complicated than the RNA primary structure.

To facilitate the study of RNA secondary structure, the RNA secondary structures are analytically decomposed, using a process called K-loop decomposition, into five kinds of substructures, namely *stacked pairs, hairpin loops, bulge loops, interior loops*, and *multiple loops* (sometimes we group the latter four simply as *loops*).

Definition 4 (K-loop Decomposition) If $i \cdot j$ is a base pair and $i < k < j$, we say that k is accessible from $i \cdot j$ if there is no $i' \cdot j'$ such that $i < i' < k < j' < j$. Similarly, we say that the base pair $k \cdot l$ is accessible from $i \cdot j$ if both k and l are accessible from $i \cdot j$. The set of $(k - 1)$ base pairs and k' single-stranded bases accessible from $i \cdot j$ is called the k-loop closed by $i \cdot j$. The null k-loop, s_0, consists of those single bases and base pairs accessible from no base pair. They are also called the free bases and base pairs. Any secondary structure S decomposes the set $\{1, 2, .., n\}$ uniquely into k-loops $s_0, s_1, .., s_m$, where $m > 0$ if and only if $S \neq \varnothing$. This decomposition was first introduced by Sankoff *et al.* (1983). the present definition follows Zuker and Sankoff (1984) and Zuker (1986), where the closing base pair is not contained in the k-loop. A k-loop is sometimes also referred to as k-cycle.

Biochemists had developed their nomenclature for k-loops long before any formal definition was given. The various cases and subcases are as follows :

- $k = 1$: A 1-loop is called a hairpin loop.
- $k = 2$: Let $i' \cdot j'$ be the base pair accessible from $i \cdot j$. Then the 2-loop is called a
  1. stacked pair if $i' - i = 1$ and $j - j' = 1$, a
  2. bulge loop if $i' - i > 1$ or $j - j' > 1$, but not both, and an
  3. interior loop if both $i' - j > 1$ and $j - j' > 1$.
- $k > 2$: The k-loops are called multi-branched or multiple loops.

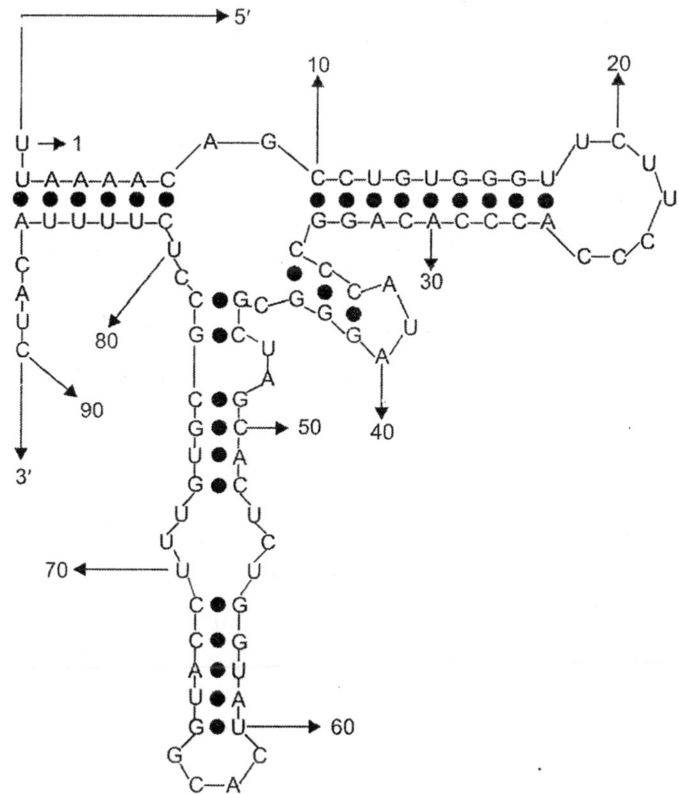

Fig. 6.1. An RNA secondary structure.

Figure 6.1 shows an RNA secondary structure. The nucleotides are laid out in such a way that paired bases are proximal with a indicating the base pairing.

- In figure 6.1, base pair 18 · 26 incloses a *hairpin loop*; all the bases that are surrounded by base pair 18 · 26 are not paired.

- Base pairs 2 · 86, 3 · 85, 4 · 84, 5 · 83, and 6 · 82 as a whole is called a *helix* or *stem*. Generally, only more than two consecutive pairs will be referred to as a helix or stem.
- The area closed by base pair 52 · 73 and 56 · 69 is called an *interior loop*. Note that all the bases are single bases.
- Base pairs 46 · 77 and 49 · 76 enclose a *bulge loop*; all the bases between 46 and 49 are not base paired.
- In this figure, a *multiple loop* is formed by the base pair 7 · 81, single bases 8–9, base pairs 10 · 34 and 35 · 43, single base 44, base pair 45 · 78, and single bases 79–80.
- The *(external) single-stranded region* consists of the single bases that are not surrounded by any base pairs in the structure, in this case single base 1 and single bases 87–90.

COMBINATORIAL ALGORITHM

In 1975, Pipas and McMahon published the *combinatorial method*, which is believed to be the first widely used method of predicting RNA secondary structures by using the free energy rules. The main idea of this algorithm is to form the structues by combining all potential helices in all possible ways. This algorithm works well when dealing with short RNA sequences, but is infeasible for long sequences. The algorithm consists of three steps.

In the first step, the *RNA sequence* is read and a bonding matrix B and a compatible matrix C are constructed.

A complete bonding matrix B for the sequence is set up by using the following rules:

- If base i is able to form a classical Watson-Crick hydrogen-bonding pair, C-G, G-C, A · U, or U · A, with base j, then the matrix element $B_{i,j} = 1$.
- If base i forms a G · U or U · G pair with base j, then $B_{i,j} = 2$.
- If base i and base j do not form the kind of base pairs discussed above, $B_{i,j} = 0$.

From this matrix, all stable helical regions can be determined. A stable helical region is defined as three or more consecutive base pairs ordered such that the strands are antiparallel. On the bonding matrix this corresponds to find a set, $\{B_{i,j}, B_{i+1,j-1}, \ldots, B_{i+m,j-m}\}$, where all elements are nonzero.

After the program has compiled the list of all possible helical regions derivable from the given primary sequence, it will set up a compatibility matrix (C-matrix) to indicate whether two regions can occur together in a given secondary structure. The elements of this matrix are defined as :

- Let i, j be two helical regions, if i and j are compatible, which means that these two helical regions can exist together in a given structure, then the matrix element will be $C_{i,j} = 1$, otherwise, $C_{i,j} = 0$.

The following two criteria are given in the paper. The first criterion excludes overlapping helices. The second criterion disallows pseudo-knots.

- Let i and j be two helical regions and let R_i and R_j be two sets whose elements are the bases forming region i and j respectively. Then $C_{i,j}$ will be assigned 1 if and only if $R_i \cap R_j = \varnothing$; otherwise $C_{i,j}$ will be given 0. This is used to avoid conflicting stems.
- Let i and j be two helical regions and let H_i be a set whose elements are the bases enclosed by region i. Let Q_i be a set whose elements are all bases not included in R_i and H_i. Then $C_{i,j}$ will be assigned 1 if and only if $R_j \cap H_i = \varnothing$ or $R_j \cap Q_i = \varnothing$; otherwise $C_{i,j} = 0$.

The second step is *permutation*. The main purpose of this step is to create all poosible structures that are obtainable from the given polynucleotide sequence. This is accomplished by generating all possible permutations of the nonzero (compatible) elements in the C-matrix. Here, a structure is defined as a set of three or more compatible helixal regions.

In the third step, all the *generated structures are evaluated*, base by base, and assigned a total free energy. The structures are then ordered by their free energies, and the best (*e.g.*, the most negative one) will be selected out and considered as the optimal structure from the given RNA primary structure. Favourable free energy contributions are assumed to be made by stacking interactions of stacked base pairs. The specific values assigned are empirical. Unfavourable free energy contributions are assigned to loops, as originally proposed by Tinoco *et al.* (1971).

This method can easily handle relatively short sequences, such as transfer RNAs. However, it is not very efficient for folding long RNA sequences: the time needed for long RNA sequences is at least proportional to 2^n, where n is the number of nucleotides in the sequence. Its major advantage is that it can predict a great many different RNA secondary structures.

ENERGY MINIMIZATION ALGORITHMS

The main idea of free energy minimization algorithms is that the whole structure can be considered as a collection fof substructures; thus, if one can obtain all the optimal substructures, the whole optimal structure can also be determined. Generally, an energy minimization emthod has two stages. One is the *filling algorithm* that computes the energy values of optimal structure of all fragments of a sequence. The other is the *trace back* algorithm that computes an optimal structure by searching systematically through the matrix of stored energy values.

Energy minimization algorithms all use dynamic programming. Dynamic programming methods for RNA structure began with Nussinov *et al.* (1978) and Waterman (1978). Since then many improvements have been made. Currently the most popular software for RNA folding. MFOLD, is based on Zuker's algorithm.

Base Pair Dependent Algorithm

Nussinov *et al.* (1978) were the first to apply the dynamic programming method to the folding problem by maximizing base pairings. Nussinov's algorithm is recursive. it computes the otpimal structure for small subsequences, and works its way up towards larger and larger subequences. The recurrence relation of Nussinov's algoirthm is as follows:

$$M_{i,j} = \max\left\{M_{i+1,j}, M_{i,j-1}, M_{i+1,j-1} + 1, \max_{i<k<j}\{M_{i,k} + M_{k+1,j}\right\}$$

where the $M_{i,j}$ represents the maximum number of base paris a secondary structure from subsequence $r_i, ..., r_j$ can possess. This equation states that the maximum number of base pairs between i and j is computed from the maximum number between $i + 1$ and j, the maximum number between i and $j - 1$, one (because i pairs with j) plus the maximum number between $i + 1$ and $j - 1$, and the maximum number between i to k and between $k + 1$ to j for some k.

The above equation is a very simple one, in which only the number of base pairs is considered, and the biological/chemical information is ignored. One improvement is to assign energy value to each base pair in a secondary structure. Suppose that there is a function e such that $e(r_i, r_j)$ is the energy value of a base pair $i \cdot j$. The above equation can be modified as :

$$W_{i,j} = \min\left\{W_{i+1,j} W_{i,j-1}, W_{i+1,j-1} + e(r_i, r_j), \max_{i<k<j}, \{W_{i,k} + W_{k+1,j}\right\}$$

Unfortunately, this simple method cannot fully reflect the detabilizing effects of various loops, or the nearest neighor interactions in helices and loops.

Loop Dependent Energy Minimization Algorithms

An RNA molecule tends to exist in a conformation characterized by the minimum free energy E. The exact calculation of E from basic physical and chemical principles for an arbitrary secondary structure S of an RNA is not feasible. However, free energy E of an RNA secondary structure can be approximated as the sum of individual contributions from stacked pairs and loops.

The weakenss of the base pair dependent energy minimization method is that it cannot easily be modified to incorporate stacking, stabilizing, and loop destabilizing energy rules.

The most sophisticated energy minimization algorithm for RNA secondary structure prediction is the Zuker algorithm. In Zuker's algorithm, the entire RNA secondary structure S is considered as a set of loops (K-loop decomposition), $s_0 \ldots s_m$, where $m > 0$. Note that stacked pairs are also referred as loops. Energies are assigned to the k-looops, and the energy of a structure S can be calculated as :

$$E(S) = \sum_{i=0}^{m} e(s_i)$$

Note that e is now a function of k-loops instead of a function of base pairs. In general, $e(s)$ is only negative when s is a stacked pair, because only stacked pairs contribute directly to the stability of the molecule. Hairpin loops, bulge loops, interior loops, and multiple loops all make positive energy contributions and therefore reduce the stability of the molecule. The goal of the algorithm is to find a secondary structure that minimuzes the free energy.

Energy parameters for various loops have been fitted to the results of experimental thermodynamic studies of small model RNAs (freier et. al. 1986; Turner et. al. 1987, 1988). They include parameters for stacking, hairpin loop lengths, bulge loop lengths, interior loop lengths, multiple loop lengths, single dangling bases, and terminal mismatches on stems.

For the exterior-loop, s_0, the energy is set to zero. We use eh(i, j) to represent the free energy of a *hairpin* loop closed by pair $i \cdot j$. We use ebi(i, j, i', j') to represent the free energy of the *bulge* or *interior loop* closed by the two base pairs $i \cdot j$ and $i' \cdot j'$. We use es(i, j) to represent the free energy of stacked pairs, $i \cdot j$ and $(i + 1) \cdot (j - 1)$.

A Simple Loop Dependent Energy Minimization Algorithm : We first discuss a simplified version of Zuker's algorithm where the energy contribution of multiple loops is set to zero and dangling single bases have no effect.

For $i < j$, let $W(i, j)$ be the minimum folding energy of all non-empty foldings on the subsequence $r_i, \ldots, r_j$ and $V(i, j)$ be the minimum energy of all foldings on the subsequence $r_i, \ldots, r_j$ that contains the base pair $i \cdot j$. Initially, $W(i, j) = V(i, j) = +\infty$, for those $|j - i| < 4$. The recurrence equation for W is as follows :

$$W(i,j) = \min\begin{cases} W(i+1,j), & (1)\\ W(i,j-1), & (2)\\ V(i,j), & (3)\\ \min\limits_{i \le k < j}\{W(i,k) + W(k+1,j)\} & (4) \end{cases}$$

This equation is obtained by considering the following cases. The first case is that, in the optimal structure, base i does not pair with any other base; therefore we have $W(i, j) = W(i + 1, j)$. The second case is that, in the optimal structure, base j does not pair with any other base, and therefore we have $W(i, j) = W(i, j - 1)$. The thrid case is that, in the optimal structure, both base i and base j are paired and they pair to each other, and therefore we have $W(i, j) = V(i, j)$. the fourth case is that, in the optimal structure, both base i and base j are paired but they do not pair to each other. Because in secondary structures there are no crossings, the optimal structure will come from two optimal substructures from subequences $i, ..., k$ and $k + 1, ..., j$ for some k. Therefore, we have

$$W(i,j) = \min_{i \le k < j} \{W(i,k) + W(k+1,j)\}$$

We now consider how to compute $V(i, j)$. Consider the optimal structure; there are several possibilities. Base pair $i \cdot j$ can be part of a hairpin, or can be part of a bulge or an interior loop, or can be stacked on base pair $(i + 1) \cdot (j - 1)$, or can be part of a multiple loop. This leads to the formula

$$V(i,j) = \min \begin{cases} \mathrm{eh}(i,j), & (1) \\ \mathrm{es}(i,j), & (2) \\ \mathrm{VBI}(i,j), & (3) \\ \mathrm{VM}(i,j) & (4) \end{cases}$$

where VBI(i, j) is for bulge or interior loop and VM(i, j) is for multiple loop.

The formulas for VBI(i, j) and VM(i, j) are the following :

$$\mathrm{VBI}(i,j) = \min_{\substack{i<i'<j'<j \\ i'-i+j-j'>2}} \{\mathrm{ebi}(i,j,i',j') + V(i',j')\}$$

$$\mathrm{VM}(i,j) = \min_{i+1<k<j-1} \{W(i+1,k) + W(k+1,j-1)\}$$

A Revised Loop Dependent Energy Minimization Algorithm : Zuker's algorithm considers some additional features in the folding process.

First, for the hairpin loop whose length is greater than 3, the so-called *terminal mismatched pairs* are considered; that is, if base pair $i \cdot j$ closes a hairpin loop, then the bases $i + 1$ and $j - 1$ have some effects on the free energy of the hairpin loop. In addition, specific tetraloops, that is, hairpin loops with four bases, are assigned enhanced stability. These tetraloops are known to be more stable or to be important in stabilizing tertiary structures. Furthermore, there are special hairpin loops—for instance, the hairpin with three single *Gs*, with their won energy rules. These considerations have been incorporated into the values of eh(i, j).

It has also been shown by experiments that the single bases also can stabilzie the RNA secondary structure. Suppose base i and base j form a base pair, and base $i + 1$ or base $j + 1$ is a single base, then base $i + 1$ or base $j + 1$ is called a *3′-dangling base;* if base $i - 1$ or base $j - 1$ is a single base, then base $i - 1$ or base $j - 1$ is called a *5′-dangling base.* Because single bases adjacent to helices are assumed to stack if the stacking is favourable, the algorithm considers that all dangling energies are nonpositive. The assumption that the contribution of a multiple loop is zero is too far from reality. Bilogists have shown that the multiple loops indeed play an important role in constructing RNA secondary structure.

A general energy function for multiple loops may require exponential computing time. Zuker' algorithm uses a linear energy function to calculate the energy of multiple loops,

$$e(k\text{-loop}) = a + bk' + ck, \qquad \text{where } k > 2$$

where a, b, and c are non-negative constants. k' is the number of single bases in the multiple loop.

To implement this equation, $W'(i, j)$ is used; the meaning of $W'(i, j)$ is alomost the same as $W(i, j)$, except that it treats the exterior loops of optimal structures as multiple loops with appropriate penalities. W' also considers the special rules for some hairpin loops and dangling stabilizing energies.

$$W'(i,j) = \min\begin{cases} W'(i+1, j) + b, & (1) \\ \text{ed}(i+1, j, i) + V(i+1, j) + b + c, & (2) \\ W'(i, j-1) + b, & (3) \\ \text{ed}(i, j-1, j) + V(i, j-1) + b + c, & (4) \\ V(i, j) + c, & (5) \\ \text{ed}(i+1, j-1, i) + \text{ed}(i+1, j-1, j) + V(i+1, j-1) + 2b + c, & (6) \\ \min\limits_{i \le k < j}\{W'(i, k) + W'(k+1, j)\} & (7) \end{cases}$$

where ed is the energy of the single dangling base: $\text{ed}(i, j, k)$ means that base k is dangling on base pair $i \cdot j$. In this new equation, three more cases are added due to the consideration of dangling bases, cases (2), (4), and (6). In actuality, these cases just refine the equation for $W(i, j)$. In addition, b and c are added as appropriate to match the linear penalty for multiple loop.

To understand this equation, consider the optimal structure of $r_i, \ldots, r_j$.

- In the optimal structure, if base i does not pair to any other base and does not stack on a base pair, then we have $W'(i, j) = W'(i + 1, j) + b$ where b is the penalty contribution of base i.
- In the optimal structure, if base i does not pair to any other base but stacks on base pair $(i + 1) \cdot j$, then we have $W'(i, j) = \text{ed}(i + 1, j, i) + V(i + 1, j) + b + c$, where b is the penalty contribution of base i and c is the penalty contribution of base pair $(i + 1) \cdot j$.
- In the optimal structure, if base j does not pair to any other base and does not stack on a base pair, then we have $W'(i, j) = W'(i, j - 1) + b$, where b is the penalty contribution of base j.
- In the optimal structure, if base j does not pair to any other base but stacks on base pair $i \cdot (j - 1)$, then we have $W'(i, j) = \text{ed}(i, j - 1, j) + V(i, j - 1) + b + c$, where b is the penalty contribution of base j and c is the penalty contribution of base pair $i \cdot (j - 1)$.
- In the optimal structure, if base i and base j form a base pair $i \cdot j$, then $W'(i, j) = V(i, j) + c$, where c is the penalty contribution of base pair $i \cdot j$.
- In the optimal structure, if base i and base j do not pair with any base and they both stack on base pair $(i + 1) \cdot (j - 1)$, then $W'(i, j) = \text{ed}(i + 1, j - 1, i) + \text{ed}(i + 1, j - 1, j) + V(i + 1, j - 1) + 2b + c$, where $2b$ are the penalty contribution of base i and base j and c is the penalty contribution of base pair $(i + 1) \cdot (j - 1)$.
- If none of the above is true, then the optimal structure will have at least two branches and we choose the structure with minimum energy.

Therefore we have

$$W'(i, f) = \min_{i \le k < j}\{W'(i, k) + W'(k + 1, j)\}$$

The equation for $V(i, j)$ is exactly the same as before.

$$V(i, j) = \min\{\text{eh}(i, j), \text{es}(i, j) + V(i + 1, j - 1), \text{VBI}(i, j), \text{VM}(i, j)\}$$

Recall that $\text{eh}(i, j)$ is for hairpin loops, $\text{es}(i, j)$ is for stacked pairs, $\text{VBI}(i, j)$ is for bulge or interior loops, and $\text{VM}(i, j)$ is for multiple loops. The computation for $\text{ch}(i, j)$, $\text{es}(i, j)$, and VBI$(i,$

j) has no change. However, the computation for VM(i, j) is now more complicated due to the stacking energy of single bases.

$$\text{VM}(i,j) = a + c + \min\{e_1(i,j) + e_2(i,j) + e_3(i,j) + e(i,j)\}$$

where
$$e_1(i,j) = \min_{i<k<j-1}\{W'(i+1,k) + W'(k+1,j-1)\}$$

$$e_2(i,j) = b + \text{ed}(i,j,i+1) + \min_{i+1<k<j-1}\{W'(i+2,k) + W'(k+1,j-1)\}$$

$$e_3(i,j) = b + \text{ed}(i,j-1) + \min_{i<k<j-2}\{W'(i+1,k) + W'(k+1,j-2)\}$$

$$e_4(i,j) = 2b + \text{ed}(i,j,i+1) + \text{ed}(i,j,j-1) + \min_{i+1<k<j-2}\{W'(i+2,k) + W'(k+1,j-2)\}$$

Because $i \cdot j$ closed a multiple loop, in the equation for VM(i, j) we have a penalty a for this multiple loop and a penalty c for base pair $i \cdot j$.

The functions $e_1(i, j)$, $e_2(i, j)$, $e_3(i, j)$, and $e_4(i, j)$ are obtained by considering various cases of dangling bases. When base $i + 1$ and base $j - 1$ are not stacked on $i \cdot j$, we have $e_1(i, j)$. When base $i + 1$ but not $j - 1$ is stacked on $i \cdot j$, we have $e_2(i, j)$. When base $j - 1$ but not base $i + 1$ is stacked on $i \cdot j$, we have $e_3(i, j)$. When both base $i + 1$ and base $j - 1$ are stacked on $i \cdot j$, we have $e_4(i, j)$.

The order to compute $W'(i, j)$ and $V(i, j)$ is for j to increase from 1 to n, and for i to decrease from j to 1. Once $W'(i, j)$ and $V(i, j)$ are computed, a linear array $W_5(i) = W(1, i)$, where $1 \le i \le n$, is computed. Finally, $W(1, n) = W_5(i)$ is necessary because $W'(1, n)$ treats exterior loops as multiple loops with penalty, whereas our energy rule for exterior loops is to assign them zero energy. The equation for $W_5(i)$ is obtained by considering that dangling bases will contribute to the stability of the structure and that the energy for an exterior loop is zero.

$$W_5(0) = W_5(n+1) = 0$$

$$W_5(i) = \min\{W_5(i-1), W_5^1(i), W_5^2(i), W_5^3(i), W_5^4(i)\}$$

where
$$W_5^1(i) = \min_{0\le k<i}\{W_5(k) + V(k+1,i)\}$$

$$W_5^2(i) = \min_{0\le k<i}\{W_5(k) + \text{ed}(k+2,i,k+1) + V(k+2,i)\}$$

$$W_5^3(i) = \min_{0\le k<i}\{W_5(k) + \text{ed}(k+1,i-1,i) + V(k+1,i-1)\}$$

$$W_5^4(i) = \min_{0\le k<i}\{W_5(k) + \text{ed}(k+2,i-1,k+1) + \text{ed}(k+2,i-1,i) + V(k+2,i-1)\}$$

Time Complexity : The time complexity of a direct implementation of this algorithm is $O(n^4)$, as we need $O(n^4)$ to compute VBI(i, j), the bulge or interior loop energy. Waterman and Smith (1986) and Lyng ϕ *et al.* (1999) have used alternative equations to compute the bulge or interior loop energy by considering the lengths of the loops, which reduced the computation time to $O(n^3)$. Therefore, the time complexity of Zuker's algorithm is $O(n^3)$. With this time complexity, this algorithm can easily handle relatively long sequences.

Suboptimal Structures : Another useful feature of Zuker's algorithm is that along with the optimal secondary structure, it can also produce subptimal secondary structures. The biologically correct structure is often not the calculated optimal structure, but rather a structure within a few percent of the calculated minimum energy. Therefore, suboptimal secondary structures will present alternative models for further investigation.

Suppose that E_{min} is the minimum folding energy. Let ΔE be a small energy increment; then, with some constraints, this algorithm can produce suboptimal secondary structures with energy value within ΔE of the optimal. Wuchty *et al.* (1999) proposed a different method to predict subptimal foldings.

PHYLOGENETIC COMPARATIVE METHODS

Given a single RNA sequence, Zuker's folding algorithm (Zuker 1989) can produce optimal and suboptimal structures according to minimum energy criteria. However, in some cases, these structures do not closely resemble the real structure. Biologists use phylogenetic comparative methods on a set of related RNA sequences to determine common structures that are closer to the real structure. One of the guiding principles in molecular biology is that structure is much more closely conserved than sequence. This principle also holds for RNA structures.

The implication for secondary structure is that secondary structure is conserved even though sequence drift occurs. Thus when one ase of a pair changes, we usually find that its partner also changes so as to conserve that base pair. This phenomenon is called a *compensatory base change* or *covariation*. This leads to the following observation: if we have a set of related (meaning evolutionarily close) RNA sequences, then they are expected to have similar secondary structures. If a helical region appears in all the sequences, then it is more reliable than those that only appear in a small number of sequences.

If a common secondary structure can be determined from the set, then it is probably more reliable than the one produced by using a thermodynamic algorithm on a single RNA sequence. This is the essence of the phylogenetic comparative method.

The framework of the phylogenetic comparative method is the following.

1. Select a set of related RNA sequences.
2. Perform multiple sequence alignment for the set of sequences.
3. Compute the covariation of any two columns of the alignment.
4. Determine the conserved helices and structures.

We need to select the sequences of homologous RNAs from different organisms among which the sequences vary. The choice of appropriate organisms for RNA structure comparisons requires a quantitative view of evolutionary relatedness.

The molecules compared must be different enough to provide numerous instances of sequence variation with which to test pairing possibilities, yet they must not differ so much that resideues cannot be aligned with confidence in the alignment step. In general, sequence similarities of 60–80 percent are favourable for inspecting the equivalent base pairings occurring in these sequences. In order to accurately evaluate the covariations, a good multiple alignment of the sequences is required.

Depending on the actual sequences, reliable alignments can be derived either from alignment software of from alignment software followed by manual adjustment. The result of this step is a multiple alignment of m RNA sequences :

$$
\begin{aligned}
R_1 &= r_1[1], r_1[2], r_1[3], .., r_1[n] \\
R_2 &= r_2[1], r_2[2], r_2[3], ..., r_2[n] \\
R_3 &= r_3[1], r_3[2], r_3[3], ... r_3[n] \\
\dots & \qquad \dots \qquad \dots \\
R_m &= r_m[1], r_m[2], r_m[3], ..., r_m[n]
\end{aligned}
$$

For any i, $1 \le i \le n$, the set of bases $r_k[i]$ with $1 \le k \le m$, is reffered to as a *column* of the alignment. Two columns of the alignment are said to be covaying when the variation in one column is echoed complementarily in the other column. The covariation of the two columns provides evidence for a conserved base pair.

The degree of covariation can be measured by several quantitative measures. One such measure is the *mutual information content* of two columns. For a single column i, let $f_i(r)$ be the frequency of the nucleotide r (where $r \in \{A, C, G, U\}$) in the ith column. Consider another column j, and let $f_{i,j}(r_1, r_2)$ be the joint frequency of the two nucleotides, r_1 from the ith column and r_2 from the jth column. If two columns vary independently, then $f_{i,j}(r_1, r_2)$ is roughly $f_i(r_1) \cdot f_j(r_j)$. Therefore we would expect $\log_2(f_{i,j}(r_1, r_2)/f_i(r_1) f_j(r_2))$ to be roughly zero. We are interested in measuring the divergence from independence. This leads to the definition of mutual information content $M(i, j)$, between two different columns, i and j:

$$M(i,j) = \sum_{r_1, r_2 \in \{A,C,G,U\}} f_{i,j}(r_1, r_2) \log_2 \frac{f_{i,j}(r_1, r_2)}{f_i(r_1) f_j(r_2)}$$

$M(i, j)$ is maximized when both columns are highly variable and also completly correlated. Two advantages of this measure are that any types of correlations can be found and that correlations that are quantitatively low, but still significant, can also be found. Once the covariation measure is computed, possible conserved helices can be calculated by searching the matrix $M(i, j)$.

Finally, a conserved common secondary structure can be constructed from conserved helices. One can apply Nussinov/Zuker algorithms for this step where the score being optimized is a function of covariation measure instead of the number of base pairs or of thermodynamic stacking energies. Nevertheless, this step normally requires manual intervention.

Phylogenetic comparative methods are currently considered the most effective methods, the advantage being that they produce reliable secondary structures.

The disadvantage is that they depend on the availability of high-quality multiple alignments and that they normally involve manual intervention. Phylogenetic comparative methods are currently considered the most effective methods, the advantage being that they produce reliable secondary structures.

The disadvantage is that they depend on the availability of high-quality multiple alignments and that they normally involve manual intervention. Phylogenetic comparative methods have also been used to determined RNA tertiary interactions.

STOCHASTIC CONTEXT-FREE GRAMMAR METHOD

A more recent approach uses a stochastic context-free grammar to model the common features of a set of RNA structures. We will now give a brief description of this method. One can view DNA and RNA sequences as sentences derived from a formal grammar. In the simplest kind of grammar, a regular grammar, sequences are derived from productions of the form $S \to aS$ and $S \to a$, where S is a non-terminal symbol and a is a terminal symbol. DNA sequences can be described by a regular grammar.

Base pairing in RNA secondary structure can be described by a context-free grammar (CFG). Context-free grammars are more powerful than regular grammars because they allow additional forms of productions, such as these of the form $S \to SS$ and $S \to aSa$. The process of generating a sequence by repeatedly applying productions is called derivation. For CFGs, a derivation can be arranged in a tree structure called a parse tree. Productions of the following forms can be used to model a RNA secondary structure: $S \to SS$, $S \to aSa$, $S \to aS$, and $S \to a$, where S is a non-terminal symbol and a is a terminal symbol.

The production $S \to aSa$ describes base pairings; $S \to aS$ and $S \to a$ describe unpaired bases; and $S \to SS$ describes branching of secondary structure. A parse tree of these productions represents an RNA secondary structure. to model a family of RNA sequences, addtional production forms will be needed. A stochastic context-tree grammar (SCFG) is a context-tree grammar such that each production is associated with a probability value. The probability of a derivation (parse tree) can be calculated as the product of the probabilities of the productions producing the derivation. The probability of a sequence S under SCFG G is the sum of probabilities over all possible derivations of G than produce the sequence.

$$\text{Prob}(s \mid G) = \sum_{\text{all derivations } d} \text{Prob}(S_0 \overset{d}{\Rightarrow} s \mid G)$$

$$= \sum_{\alpha_1,\ldots\alpha_n} \text{Prob}(S_0 \Rightarrow \alpha_1 \mid G) \times \text{Prob}(\alpha_1 \Rightarrow \alpha_2 \mid G) \times \ldots \times \text{Prob}(\alpha_n \Rightarrow s \mid G)$$

Because CFGs are ambiguous grammars, a sequence can be derived by more than one derivation or parse tree. For a single RNA, different parse trees usually represent different secondary structures. Because an SCFG assigns a probability to each parse tree, it has the advantage that the most likely parse tree for a given sequence can be determined. Intuitively, this most likely parse tree corresponds to the correct secondary structure if the SCFG G models the RNA sequences correctly.

In order to apply SCFG to RNA secondary structure prediction, we need to solve the following problems :

1. Given a set of example sequences/structures, determine the optimal SCFG. This includes the determination of the productions and the estimation of optimal probability parameters.
2. Given a SCFG G and an RNA sequence R, determine the most likely parse tree that corresponds to the correct RNA secondary structure.

The first problem can be solved by the inside-outside algorithm, which is an expectation maximization (EM) algorithm that calculates maximum likelihood estimates of an SCFG's parameters based on training data. The second problem can be solved by a dynamic programming procedure similar to the Viterbi algorithm for hidden Markov models (HMMs). Given a set of RNA sequences, Eddy and Durbin (1994) use multiple sequence alignment to construct a model (SCFG). They then iteratively reestimate the model by an EM algorithm.

With a new model generated, a new alignment can be constructed. This process is repeated until the model stabilizes. Sakakibara *et al.* (1994) proposed a method where an initial grammar instead of a multiple alignment is used. Their algorithm is based on a tree grammas and is more efficient than the inside-outside algorithm. One can vew SCFG method as an automated phylogenetic comparative method. Durbin *et al.* (1989) suggested that currently this aspect of the SCFG method is mostly theoretical rather than of practical interest.

CONCLUSIONS

RNA secondary structure prediction is an important problem in molecular biology. Thermodynamic energy minimization methods have been used to predict secondary structures from a single RNA sequence. Phylogenetic comparative methods have been used to determine secondary structures from a set of homologous RNAs whose sequences can be reliably aligned. One interesting problem is how to combine efficiently those two approaches, though there seems to be some inherent difficulties. Another challenging problem is to predict RNA tertiary structures. Although some initial attempts have been made, those problems have not been solved satisfactorily.

Protein Structure Prediction by Comparison: Homology-Based Modeling

Understanding the function and the physiological role of proteins is a basic requirement for the discovery of novel medicines (small molecules) and "biologicals" (protein-based products) with medical, industrial, or commodity applications. Although the draft sequence of the complete human genome is about to be ready, humankind is very far from understanding the function and the physiological role of the gene products it encodes. Indeed, being able to read the letters and the words is disconnected from understanding their meaning. Therefore, the attention of many biologists is now shifting to the functional analysis of the genome.

The term "functional analysis," however, merits a few short comments, as one can easily distinguish three levels of gene-related knowledge. First, once a gene sequence is known, one has to discover what transcripts it encodes. Indeed, genes can have several splice products yielding as many raw protein transcripts. Then, each transcript can be further altered by post-transcriptional modifications, which can be context specific (such as cell cycle). Second, proteins will often be attributed to a broad functional class (such as kinase, G protein-coupled receptor, etc.) by analogy with well-studied family members. However, in most cases, it is necessary to confirm these inferred functions through experimental approaches. By doing so, the knowledge base on protein will grow and functional predictions will become increasingly precise. Third, the physiological role of a protein has to be defined, meaning that the precise role of the protein in physiological processes has to be uncovered. This is a higher level of knowledge than the function.

Indeed, one can know the precise activity of an enzyme, including well worked-out reaction kinetics, but not understand how it contributes to the processes of life. This is very frequent, as one still discovers new roles for well-studied proteins. The "full picture" will thus take a very long time before it takes shape and represents the ultimate grand challenge in biology. Functional analysis, the first major step after genome sequencing and gene identification, must rely on a combination of technologies. Consequently, new experimental approaches, and their automation for large-scale applications, will need development. Concurrently, and in order to maximize the value of large data sets, one will witness the development of new data mining methods and mathematical models for biological processes simulation. A protein's function is tightly linked to its three-dimensional (3D) structure. As residues located far apart in the primary sequence can be very close in space, and only a few residues are generally responsible for a protein's function, insights into the 3D structure of a protein can represent a key component of the functional analysis process.

Consequently, an atomic-level 3D representation to assign roles to specific residues is a major asset, both for planning experiments and explaining observations. The "folding" process of a protein is very complex and as yet there is no objective and reliable way to determine it from the sole sequence. The scientific community is thus dependent on experimental protein structure elucidation.

The usual approaches, both X-ray diffraction and NMR (nuclear magnetic resonance), are both hampered by many technical hurdles and limitations. Consequently, several concerted "structural genomics" efforts are being launched in both private and the public sector to address these difficulties and increase the throughput of experimental structure elucidation. However, these will not be sufficient to elucidate the structure of all proteins of interest. Although today the protein sequence databases SWISS-PROT and trEMBL are populated with over 250,000 proteins, only approximately 10,000 of them have known 3D structures. Furthermore, SWISS-PROT and trEMBL hold fewer than 20,000 human proteins, meaning that at least another 80,000 will be added in the near future. The direct consequences of this is that only the "highly interesting" proteins should be elucidated experimentally in the foreseeable future. In this context, comparative modeling methods (homology based) have been developed and have matured to a point where many of the resulting models yield enough insights into a protein's 3D structure to be useful in functional analysis (Westhead and Thornton 1998).

WHAT IS COMPARATIVE PROTEIN MODELING?

Proteins from different sources and sometimes diverse biological functions can have similar sequences. It is generally accepted that high sequence similarity is reflected by structural similarity. Indeed, the relative mean square deviation (rmsd) of the alpha-carbon co-ordinates for protein cores sharing 50 per cent residue identity is expected to be around 1 Å. Thus the most reliable prediction methods, termed comparative protein modeling (also often called modeling by homology), consist of the extrapolation of the structure for a new (target) sequence from the known 3D structure of related family members (templates).

Identification of Modeling Templates

Comparative protein modeling requires at least one sequence of known 3D structure with significant similarity to the target sequence. In order to determine if a modeling request can be carried out, one generally compares the target sequence with a database of sequences derived from the Brookhaven Protein Data Bank (PDB). This can be performed using sequence similarity search tools such as FastA BLAST, and PSI-BLAST. Generally speaking, sequences with a FastA score 10.0 standard deviations above the mean of the random scores and a Poisson unlikelyhood probability $P(N)$ lower than 10^{-5} (BLAST) can be considered for the model building procedure.

The choice of template structures should be further restricted to those that share at least 25 per cent residue identity with 40 per cent of the target sequence as determined by sequence alignment algorithms such as SIM. The above procedure might allow the selection of several suitable templates for a given target sequence. The best template structure—the one with the highest sequence similarity to the target—should serve as the *reference.* Other selected templates can be used and must be optimally superposed onto it in 3D. As a result, each residue of the reference structure is then aligned with a residue from the other available templates. This yields a structurally corrected multiple sequence alignment.

Aligning the Target Sequence with the Template Sequence(s)

The target sequence can then be aligned with the template sequence or, if several templates are selected, with the structurally corrected multiple sequence alignment, using the best-scoring diagonals obtained by sequence alignment algorithms such as SIM. Residues, which should not be used for model building, for example those located in non-conserved loops, will be ignored during the modeling process. Thus, the common core of the target protein and the loops completely defined by at least one supplied template structure will be built.

Building the Model

Two very distinct classes of methods have been developed to build a model. One, implemented in *MODELLER*, is based on the satisfaction of spatial restraints derived from the alignment between the target sequence and its templates. The other, described below, is based on the building of an averaged framework from the coordinates of the templates. The coordinates of the model can then be built. Initially, an average framework for the target sequence will by calculated based on the multiple sequence alignment described above.

The spatial location for as many atoms of the target as possible will be derived from the positions of the corresponding atoms in the template structures. The coordinates will be calculated as a weight averaged position based on the corresponding atoms available from the templates. Each template will contribute to the average to an extent determined by its local degree of sequence similarity with the target. As the selected templates do not contain structural information about non-conserved loops and many of the side chains, these must be rebuilt in consecutive steps. Rebuilding non-conserved loops can be performed using a "*spare parts*" algorithm, as described by Jones and Thirup.

Although most of the known 3D structures available share no sequence or structure similarity with the target and templates, there might be similarities in the loop regions, which can be inserted in the protein model. Each loop is defined by its length and the geometry of its "*stems*," namely the coordinates of the alpha carbon (C^{α}) atom of the four residues preceding and following the loop. The fragments that correspond to the above loop definition are extracted from the PDB entries if the rmsd computed for their "*stems*" is lower than a specified cut-off value. Furthermore, only fragments that do not overlap with neighbouring parts of the structure are considered possible candidates.

The accepted spare parts are sorted according to their rmsd and their degree of sequence similarity with the target. The best fitting fragment is then added to the model. Because the spare parts algorithm does not always lead to convincing solutions, one can also use an approach based on a conformational space search driven by the satisfaction of stereochemical, distance, and steric constraints. Loops modeled with these methods are filtered according to criteria such as the surface exposure of hydrophobic moieties and relative conformational energies. In the final step of coordinate generation, it is necessary to determine the geometry of fully defined side chains, and build those that are incomplete or lacking. This can be achieved by using libraries of allowed side chain rotamers. Such tables are deduced from the highest resolution structures from the PDB, and each rotamer is ranked according to its frequency of occurrence.

More sophisticated tables provide the distribution of the side chain rotamers based on backbone conformations (Helix, Sheet, Turn, and Coil). Whereas the optimization of the fully defined side chain geometries is performed by replacing each one by its best fitting rotamer, the partially defined and lacking residues are built by searching for a combination of side chain conformations that minimizes the steric overlaps in the model structure.

Model Refinement

The final step of the coordinate generation process in the idealization of stereochemistry of the model; it consists mainly of the optimization of bond geometry and the removal of unfavorable non-bonded contacts. This step can be performed by energy minimization packages such as CHARMM or GROMOS.

Excessive energy minimization will cause the model to deviate markedly from the original model, which is not suitable and should be avoided. Indeed, experience has shown that the changes induced by force field computations do not improve the accuracy of the model with respect to a control experimental structure. Thus, one should keep the number of minimization cycles to a minimum, but sufficient to improve the stereochemistry of the model. A typical energy minimization procedure will use no more than three hundred cycles of energy minimization (a combination of 50–100 steps of steepest descent and 200–250 steps of conjugate gradient minimization) while imposing a harmonic force constant to the C^{α} carbon atoms.

AUTOMATED PROTEIN MODELING

By comparing all entries of the *SWISS-PROT/trEMBL* sequence database (release 36) with a non-redundant subset of all sequences of known 3D structure, we found that approximately 30 per cent of the database entries have at least one suitable modeling template for a portion of their sequence. This number increases every year by roughly 5 per cent, showing that the growth in available templates allows an ever increasing number of sequences to be modeled by comparative methods. Hence, we might speculate that comparative modeling procedures could be applied to more than 90 per cent of the protein sequences within the next 20 years. However, this projection does not address the issue that many protein structures do not cover the full length of the transcripts. Indeed, it is generally necessary to express only a domain or part of a protein in order to get crystals, so one often has only partial structural information for a protein sequence.

A typical example is the Fas ligand, which is composed of three domains: first, an intracellular domain for which there is an NMR-derived structure (PDB entry 1DDF); second, a membrane spanning domain for which there is no experimental information; and third, although there is no available experimental structure for the extracellular domain, a model can easily be built based on the tumor necrosis factor structure (PDB entry 1TNR). This means that in many cases, especially for multidomain proteins, several structure elucidation experiments have to be conducted before the entire sequence can be covered with structural information, and even then the overall structure will only result from an assembly of "pieces." Furthermore, the N-terminal and C-terminal parts of many protein structures are not resolved well enough and thus cannot be "*seen*" in the electron density map. Consequently, the average length of a protein model will be of 180 residues (ranging from 30 to 900). For the sequences that share a similarity of at least 30–35 per cent with a template, it is generally feasible to attempt comparative protein modeling using a completely automated approach with some limitations, which we will illustrate later in this chapter.

The emerging structural genomics efforts will make a major contribution to the broader application of comparative modeling as they will yield new modeling templates at a high rate. However, most large multidomain complex models cannot presently be built, mainly because no suitable modeling templates are available for such complexes and because the prediction of protein-protein contacts is inaccurate and still in its infancy. There is no doubt that the scrutiny of multiple sequence alignments is more informative than the analysis of an isolated sequence. This allows the identification of conserved and variable residues, and improves the rationalization of site-directed mutagenesis experiments.

Likewise, comparative structure analysis involving several members of a protein family, as opposed to single structure inspection, is expected to be invaluable in a number of situations. The understanding of the structural differences between residues occupying a similar portion of space in several family members or species variants will provide a strong basis for such applications as planning mutagenesis experiments and understanding the selectivity of compounds binding to active sites (*drug design*). Such a collection of structures can also be used during the rational design of combinatorial libraries of compound. It will thus be increasingly valuable to have models for as many members of a protein family as possible.

VERY LARGE-SCALE PROTEIN MODELING

Two years ago, we submitted all entries of the sequence databases *SWISS-PROT* and *trEMBL* to comparative protein modeling in a single experiment, called 3DCrunch. This followed several smaller experiments during which the full protein complement of bacterial and the yeast genome (Peitsch 1997; Peitsch *et al.* 1997) and others subjected to comparative modeling. In all cases we used the software framework underlying *SWISS-MODEL*. For this project, run in collaboration with Silicon Graphics Inc., we used a 64-processor Silicon Graphics *CRAY* Origin2000 server with 32Gb of memory. 3DCrunch needed 6828.4 cpu-hours to complete (4.4 days). This corresponds to 284.5 days of computing on a single processor machine. During the experiment, over 64,000 protein models were generated. This project is ongoing, as both the Protein Data Bank and *SWISS-PROT*/*trEMBL* are updated regularly and thus new sequences can be modeled on a monthly basis. To date, over 90,000 protein models have been generated and are available over the Internet at http:/www.expasy.ch/swissmod/. In order to provide information about the reliability of the modeling method implemented in 3DCrunch and *SWISS-MODEL*, we also generated twelve hundred models for proteins of known 3D-structure using modeling templates sharing between 25 and 95 percent sequence identity with the submitted sequences.

This is the first time that a modeling method was assessed at this scale, providing relevant information about its reliability. We then derived all degrees of identity between target and template sequences and the relative mean square deviation (base on C^{α} atoms) of the models from their corresponding experimental control structure. First, this allowed us to confirm a few observations:

1. We could confirm the observation that the common core of proteins sharing 50 per cent sequence identity deviates by approximately 1 Å relative mean square deviation. However, this deviation can be much increased if the protein structures are solved by NMR (also true for structures with identical sequences).
2. We could also confirm the long-known fact that the most reliable part of a protein model is the portion it shares with the modeling template, whereas the rebuilt non-conserved loop were a major contributor to model inaccuracy.

Furthermore, we made three observations that are particularly relevant in the context of automated comparative protein modeling in general and our approach in particular:

1. As shown in table 7.1, 63 per cent of the sequences sharing 40–49 per cent identity with their template yield a model deviating by less than 3 Å from their control structure. This number increases to 79 per cent for sequence identities ranging from 50 to 59 per cent. One can also see that below 30 per cent sequence identity, the accuracy of completely automated protein modeling is rapidly degrading.

2. Comparative protein modeling is guided by the alignment between target and template sequence. Error introduced by the alignment algorithm will have profound effects on the model. We have observed that these errors start to appear when the sequence identity between target and template sequence is lower than 30–35 per cent, which parallels and confirms the rule of thumb set forth by John Moult. It is noteworthy that less than 5 per cent of the models generated by *SWISS-MODEL* are inaccurate because of such errors.

   Analyzing several cases, we realized that visual inspection of the alignment and manual optimization of the alignment can in some cases improve the model markedly. This is not a general rule, as there is often no objective function able to distinguish between right and wrong in low similarity segments. Therefore, further benchmarks are needed to assess whether manual or fully automated alignment methods would yield the best results below the 30–35 per cent sequence identity threshold.

Table 7.1. Probabilities of *SWISS-MODEL* accuracy for target-template identity classes.

Percent sequence identity[a]	*Total number of models*[b]	*Per cent*[c] *models with rmsd < 1 Å*	*Per cent models with rmsd < 2 Å*	*Per cent models with rmsd < 3 Å*	*Per cent models with rmsd < 4 Å*	*Per cent models with rmsd < 5 Å*	*Per cent models with rmsd > 5 Å*
25–29	125	0	10	30	46	67	33
30–39	222	0	18	45	66	77	23
40–49	156	9	44	63	78	91	9
50–59	155	18	55	79	86	91	9
60–69	145	38	72	85	91	92	8
70–79	137	42	71	82	85	88	12
80–89	173	45	79	86	94	95	5
90–95	88	59	78	83	86	91	9

[a]*Range of sequence identity between target and template sequence.*

[b]*Total number of models in any given class of sequence identity. The table summarizes 1201 model-control structure pairs.*

[c]*Probability in per cent that the C_α atoms of a model, sharing X% sequence identity with its template, deviate by 1 Å or less from the corresponding experimental control structure. The following columns provide these probabilities for other relative mean square deviations.*

3. The choice of the "*right*" modeling template is crucial, as most high deviations between model and experimental control structures can be traced back to the selected modeling

templates. Indeed, it can be seen for instance, that 9 per cent of the models using templates with sequence identity of 90–95 per cent have an rmsd of 5 Å or more. After analysis of a number of those models, we realized that the template and experimental control structures were elucidated under sometimes very different conditions (including the use of NMR versus X-ray crystallography as an experimental method, presence or absence of ligands), which cause major structural differences between highly similar sequences. As set forth later in this chapter, the experimental conditions at the time of structure elucidation strongly impacts the resulting structure. It is thus crucial that the users be aware of the templates that are used by the modeling procedure or make their own informed choice.

Andrej Sali and co-workers at Rockefeller University are also addressing very large-scale protein modeling. Using the Saccharomyces cerevisiae genome as a starting point, they have developed a similar automated modeling pipeline using *PSI-BLAST* and *MODELLER* (Sali and Blundell). Their goal is to build models for as many protein sequences as possible. Although the modeling software cannot be used over the Web, one can easily search their model database (ModBase) and download the pre-computed models at http://guitar.rockefeller.edu/modbase. Today, their database contains close to seventeen thousand models from 10 complete genomes.

WEB-BASED AUTOMATED PROTEIN MODELING

In practice, using standard molecular graphics software, comparative protein modeling is a complex activity requiring an expert knowledge of the available methods and software functionality. Furthermore, most freely available software packages do not provide integrated visualization, analysis, and modeling capabilities optimized for protein modeling. Consequently, over the last several years we have concentrated on developing an automated comparative protein modeling server called *SWISS-MODEL* and its graphical front end Swiss-PdbViewer (Deep View) in order to provide the scientific community with a freely available sequence to structure workbench. Both server and client application are evolving systems, which are regularly updated to incorporate new algorithms and methods as we develop them. The major goal is to improve the accuracy of the models returned by the server, while providing an ever more flexible and user friendly graphical user interface.

NOTE ABOUT DETAILED MODELING

As stated above, it is unclear under which conditions detailed (manual) modeling yields better results than automated approaches. However, it is important to realize that homology modeling can be tricky, especially when the degree of similarity between target and template sequences drops below the 30–35 per cent range in some regions of the alignment. It is then generally advisable to further fine-tune the model manually. The resulting models can be of higher quality than their automated counterparts. The process we routinely apply to modeling projects is to build an initial automated model using the first approach mode of *SWISS-MODEL*. We then fine-tune this model using the interactive modeling features implemented in the Swiss-Pdb Viewer (Deep View). However, any combination of the above mentioned Web-based modeling servers and detailed modeling packages such as *MODELLER, COMPOSER, CONGEN*, or *WHATIF* (Vriend 1990) can be referred to follow a similar process.

WHAT DEFINES THE ACCURACY OF A MODEL

The quality of a model is determined by two distinct criteria, which will determine its applicability. First, the correctness of a model is dictated by the quality of the sequence alignment used to guide the modeling process. If the sequence alignment is wrong in some regions, then the spatial arrangement of the residues in this portion of the model will be incorrect. The first edition of the community-wide experiment known as critical assessment of protein structure prediction (CASP) already underscored that most severe modeling errors can be traced back to sequence alignment mistakes. This remains, despite many efforts to address this issue the main weakness of comparative protein modeling. Second, the accuracy of a model is essentially limited by the deviation of the used template structure(s) relative to the experimental control structure.

This limitation is inherent to the methods used, as models result from an extrapolation. As a consequence, the core C^{α} atoms of protein models that share 35–50 per cent sequence identity with their templates will generally deviate by 1.5 to 1.0 Å from their experimental counterparts, as do experimentally elucidated structures. However, one should not overlook the contributions of the templates to the model accuracy. The templates, which are obtained through experimental approaches, are subject to structural variations not only caused by experimental errors and differences in data collection conditions, such as the temperature, but also because of different crystal lattice contacts and the presence or absence of ligands.

Furthermore, X-ray crystallography and NMR generally yield 3D structures with an even broader rmsd spread. This is well illustrated by a typical example: the structure of interleukin-4 (IL-4), a cytokine consisting of a 130-residue four-helix bundle, was elucidated by X-ray crystallography as well as by NMR. The backbones of three IL-4 crystal structures (PDB entries 1RCB, 2INT, and 1HIK) show rms deviations of 0.4 to 0.9 Å, whereas those of three IL-4 NMR forms (PDB entries 1ITM, 1CYL, and 2CYK) deviated by 1.2 to 2.6 Å. These values illustrate the structural differences due to experimental procedures and the molecular environment at the time of data collection. It is thus crucial to know the experimental conditions under which the modeling templates were collected, as this has a direct impact on the accuracy of the derived models and thereby on their potential use. Almost every protein model contains non-conserved loops, which are expected to be the least reliable portions of a protein model. Indeed, non-conserved loops often deviate markedly from experimentally determined control structures.

In many cases, however, these loops also correspond to the most flexible parts of the structure as evidenced by their high crystallographic temperature factors (or multiple solutions in NMR experiments). On the other hand, the core residues—the least variable in any given protein family—are usually found in essentially the same orientation as in experimental control structures, although far larger deviations are observed for surface amino acids. This is expected as the core residues are generally well conserved and the rotamers of their side chains are constrained by neighbouring residues.

In contrast, the more variable surface amino acids will tend to show more deviations because there are few steric constraints imposed upon them. Some structural aspects of a protein model can be verified using methods based on the inverse folding approach. Two of them, namely the 3D profile based verification method and ProsaII, developed by Sippl, are widely used. The 3D profile of a protein structure is calculated by adding the probability of occurrence for each residue in its 3D context. Each of the 20 amino acids has a certain probability to be located in one of the 18 environmental classes (defined by criteria such as solvent-accessible surface, buried polar and exposed nonpolar area, and secondary structure) presently defined by Eisenberg and colleagues. In contrast, ProsaII relies on empirical pseudo-conformational energy potentials derived from the pairwise interactions observed in well-defined protein structures. These terms are summed over all residues in a model and result in a more (more negative) or less (more positive) favorable energy.

Both methods can detect a global sequence to structure incompatibility and errors corresponding to to-pological differences between template and target. They also allow the detection of more localized errors such as β-strands that are "out of register" or buried charged residues.

These methods are, however, unable to detect the more subtle structural inconsistencies often localized in non-conserved loops, and cannot provide an assessment of the correctness of their geometry. Finer details of a model, especially deviations from the ideal stereochemistry (bond length, dihedral angles, etc.), internal residue packing, and so on, can be easily analyzed with programs such as WhatCheck or *PROCHECK*. It is, however, crucial to realize that proper stereochemistry is not sufficient as a criterion for model correctness or value. It is possible, if not easy, to build models complying with all tests implemented in WhatCheck or *PROCHECK* but having little or no biological meaning. Nevertheless, these programs allow us to select the best model among a number of very closely related ones, especially when stereochemistry is the last distinguishing feature between a collection of models.

ABOUT THE USE OF PROTEIN MODELS

Protein models obtained with comparative modeling methods can be classified into three broad categories: (1) models that are based on incorrect alignments between target and template sequences. Such alignment errors, which generally reside in the inaccurate positioning of insertions and deletions, are caused by the weaknesses of the alignment algorithms and often cannot be resolved in the absence of a control experimental structure. It is, however, often possible to correct such errors by producing several models based on alignment variants and by selecting the most "sensible" solution.

Nevertheless, it turns out that such models are often useful as the errors are not located in the area of interest, such as within a well-conserved active site. (2) Models based on correct alignments are of course much better, but their accuracy can still be medium to low as the templates used during the modeling process have a medium to low sequence similarity with the target sequence. Such models, as the ones described above, are still very useful tools for the rational mutagenesis experiment design. They cannot be of great assistance during detailed ligand binding studies. (3) The last category of models comprises all those that were built based on templates that share a high degree of sequence identity (> 70%) with the target. Such models have proven useful during drug design projects and allowed the taking of key decisions in compound optimization and chemical synthesis.

For instance, models of several species variants of a given enzyme can guide the design of more specific non-natural inhibitors. However, nothing is absolute and there are numerous occasions in which models falling in any of the above categories could either not be used at all or in contrast prove to be more useful and correct than initially thought. In our experience, several applications of medium-accuracy models have proven successful. These can be classified into three categories.

Interpreting the Impact of Mutations on Protein Function: Potential Link to Diseases

One of the first uses one can make of a model structure is to interpret the impact a mutation can have on the overall function of a protein. Although the development of objective scoring functions has begun only recently, "visual inspection" associated with a good knowledge of the rules underlying protein structure has proven useful in defining the broad reasons for mutant malfunction. With the upcoming high-throughput production of single nucleotide polymorphisms (SNP), objective scoring functions will be crucial to make maximum use of the information. Indeed, a sizeable proportion of the SNPs will alter the translated protein sequences, and thus interpreting the potential functional effects of these mutants will be crucial to elucidate the molecular basis of human diseases.

Prioritization of Residues to Mutate to Determining Protein Function

As mentioned previously, the discovery of gene function in the genomic era will require a sustained experimental effort, including the creation of molecular mutants. The prioritization of residues to mutate will be greatly optimized by considering the 3D structure of the target protein.

Providing Hints for Protein Function

This is probably the broadest and least defined spectrum of potential applications for 3D models. The common feature of these applications is that models can be used to formulate a hypothesis around a protein, which can then be tested in experimental settings. It is well known that low, yet significant, degrees of sequence similarity are often not sufficient to attribute a function to a protein. In such cases, protein modeling can provide useful insights and help determine or confirm a potential functional assignment. Furthermore, one can use models to create hypothesis around potential enzymatic activities and possible ligand binding functions.

MODELING MEMBRANE PROTEINS

Membrane proteins still remain a class of proteins that represent an even greater challenge to modelers. G-protein coupled receptors in particular also represent a group of molecules of special interest to the pharmaceutical industry, as a very large proportion of today's medicines are modulators of their activities. Modeling such proteins has thus been attempted in many occasions by both de novo and comparative approaches. The two main steps along the path to a model have been automated: (1) algorithms have been developed to identify the transmembrane domains and (2) to generate 3D models using de novo approaches and comparative methods.

In all cases, however, the sequence analysis and coordinate generation steps were separated and could not be linked automatically due to the relatively low reliability of the first step. Consequently, this group of proteins is not yet amenable to high-throughput model building. This will of course dramatically change with the future availability of experimentally determined structure of one of their family members.

THE FUTURE

Over the next years we will focus on two main aspects of comparative protein modeling: (1) improving the sensitivity of the template identification and selection procedure in the sequence similarity "*twilight zone*," and (2) improving the model accuracy.

Increasing the Template Selection Sensitivity

The availability of iterative local similarity search algorithms allows for the identification of suitable template structures even at very low sequence similarity levels. In cases where the search results are not significant enough to choose a template, multiple models for a family of sequences based on several possible templates can be constructed. Evaluating the multiple model assembly will then enable the discrimination of compatible template folds. This would allow us to choose the correct template even beyond the limit of the search algorithms, provided a reliable function discriminating "*good*" from "*wrong*" models can be defined over a broad range of different types of errors.

Improving the Accuracy of Models

In order to improve both model correctness and accuracy, we will address two of the main causes of modeling errors. First, there is the sequence alignment, which guides the modeling procedure. Automated sequence alignment procedures often introduce errors by placing insertions and deletions ("*indels*") incorrectly. This phenomenon is mainly observed when the overall sequence identity between target and template sequences drops below 35 per cent.

The identification and subsequent multiple alignment of all members of a protein family results in more accurate "indel" placement. Second, side chain and loop rebuilding are also among the main causes of model inaccuracies. The increasing number of available high-resolution X-ray structures will improve the model accuracy by providing more high-quality templates. In addition, this will allow deriving improved secondary structure specific side chain rotamer libraries and broader loop databases. Both will contribute to the improvement of model accuracy.

Sequence Alignment

M. jannaschii is a methane-generating archaebacterium living at depths over 2.5 km on the ocean floor, near deep-sea thermal vents. From sequence comparisons of *M. jannaschii* with prokaryotes and eucharyotes, Bult *et al.* firmly established the validity of Woese's theory that *Archea* is a third life domain, along with the other two domains *Prokarya* and *Eukarya*. *M. jannaschii* has 1738 genes, and by performing sequence alignment with other genomes, it appears that archaebacteria share an evolutionary heritage with eukaryotes for transcription, transltion, and DNA replication, yet share certain other features with prokaryotes, since for instance archaebacteria have no nuclear membrane.

In this chapter, we will cover dynamic programming sequence alignment algorithms, and some of their applications. A good example of a recent commercial application of sequence alignment and homology testing was given by *Richard Roberts* in. Restriction enzymes appear to be certain bacteria's defense system against phases. Type II restriction enzymes (such as the well-known *ecoRI* endonuclease) recognize a small fragment of double-stranded DNA and cut both strands, either with a blunt cut, or leaving sticky ends. For instance, *ecoRI* recognizes the hexanucleotide palindrome *GAATTC*, and cuts, leaving sticky ends as indicated below :

```
                                  |AATTC . . .
                                   ‾‾‾‾|G . . .
. . . G|AATTC . . .
. . . CTTAA|G . . .   ──────►        and

                      . . . G|
                      . . . CTTAA|
```

In order to protect itself from the action of its own restriction enzymes, a bacterium methylates restriction enzyme recognition sites of its own genome. At the current time there are 3154 discovered restriction enzymes, some 5604 publications, and out of 16 (resp. 64) possible tetranucleotide (resp. hexanucleotide) palindromes (such as *AAGG* (resp. *GAATTC*)), all but 2 (resp. 2) have been shown to be recognition sites for certain restriction enzymes. Though there appears to be *no* recognizable sequence homology between different restriction enzyme genes, there *is* sequence homology between different recognition site methylation genes. Using this principle, the group of *R.* Roberts has successfully developed the *REBASE* software to locate likely candidates for potential restriction enzymes within genomic databases. We should like to define the distance between two sequences $a_1 \ldots a_n$ and $b_1 \ldots b_m$, allowing in exact matches, in producing an alignment.

Of course, the usual distance measures (Euclidean distance. Hamming distance, etc.) have little bearing on this problem, since the sequences have possibly different lengths (though there is work in mathematical evolution theory, where Hamming distance is used). For instance, given an initial sequence ACGTACGT of length 8, after 9540 generation, the seqeunce ACACGGTCCTAATAATGGCC was generated, assuming probability of deletion 0.0001, probability of insertion 0.001, probability of transitional substitution 0.00008, and probability of transversional substitution 0.00002. On a different program run with the same parameters, after 9540 generations, ACGTACGT evolved into the rather different sequence CAGGAAGATCTTAGTTC. The sequences ACACGGTCCTAATAATGGCC and CAGGAAGATCTTAGTTC are related in that both evolved from the same ancestor sequence ACGTACGT. What is their *true*, or historically accurate alignment? This question makes sense, because in the simulation program that generated the mutated sequences, the type and position of *edit* operation (delete. insertion, substitution) were taken into account. The true alignments are

– – ACG – T – A – – – CG – T – – – –

ACACGGTCCTAATAATGGCC

and

– – – AC – GTA – C – – G – T – –

CAG – GAAGATCTTAGTTC

so, by superposition, we obtain

– ACAC – GGTCCTAAT – – AATGGCC

CAG – GAA – G – AT – – CTTAGTTC – –

In contrast, Gotoh's algorithm with mismatch penalty of 3 and gap penalty function $g(k) = 2 + 2k$ for length k gap (discussed in this chapter), yields a completely different alignment:

ACACG – – GTCCTAATAATGGCC

– CAGGAAGATCT – – TAGTT – – C

There is recent work to determine the choice of mismatch and gap parameter, using in part the significance of an alignment score (significance is determined by how many random sequences having the same base composition have a better alignment score).

MOTIVATING EXAMPLE

In this section, we present a motivating example to give a more concrete understanding of some of the problems involved in sequence comparision. One of the purposes of sequence alignment is to detect *homologous proteins*. Two proteins (or protein domains) are *homologous* if they have evolved from a common predecessor. Usually, this implies that the two proteins have a similar structure and function. Homology is a very important concept in biology that is used as a fine-grained criterion to group together different proteins (with known structures). There are currently two principal databases for the hierarchical structural classification of proteins – namely CATH and *scop*. CATH classifies proteins according to Class (where the different protein classes are *mainly Alpha, Beta, Alpha and Beta,* and proteins having *few secondary structures*). Architecture (which is the description of the gross overall arrangement of the secondary structure elements).

Topology (which takes into account the overall shape as well as the connectivity of the secondary structure elements, using structural comparision algorithms), and Homology (where proteins are grouped together if they are believed to have a common ancestor). In deciding whether two proteins are homologous, sequence alignment is *one* of the techniques used *scop* (Structural Classification of Proteins) has a similar hierarchical structure, where proteins are grouped according to fold (major structural similarity), superfamily (probable common evolutionary origin) and family (clear evolutionary relationship). As an example of homologous proteins, we consider the family of reverse transcirptases.

Reverse transcriptases are used by retroviruses to replicate their own genetic information, which is stored in RNA. This RNA is translated back into DNA (for duplication) using reverse transcriptase. As an example, we take reverse transcriptase of Moloney murine leukemia virus, and the reverse transcriptase of HIV Type 1 virus. The homologous domains are classified as homologous by both the CATH database and the SCOP database. As already mentioned,

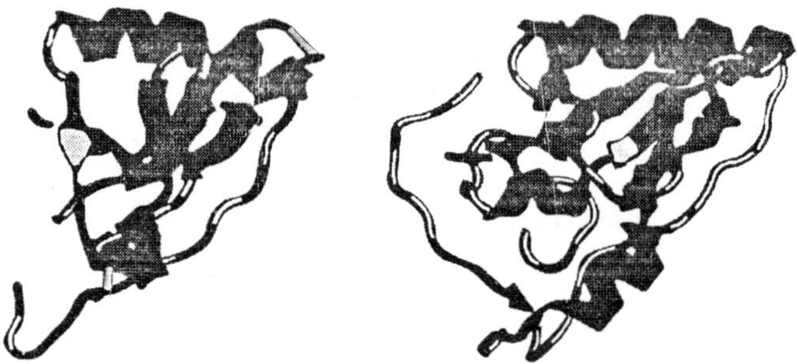

Fig. 8.1. Two homologous proteins of the reverse transcriptase family. The first protein is domain 1 of reverse transcriptase of the Moloney murine leukemia virus (PDB code IMML, and the second is domain 1 of the A chain of the HIV Type 1 reverse transcriptase (PDB code IRTH).

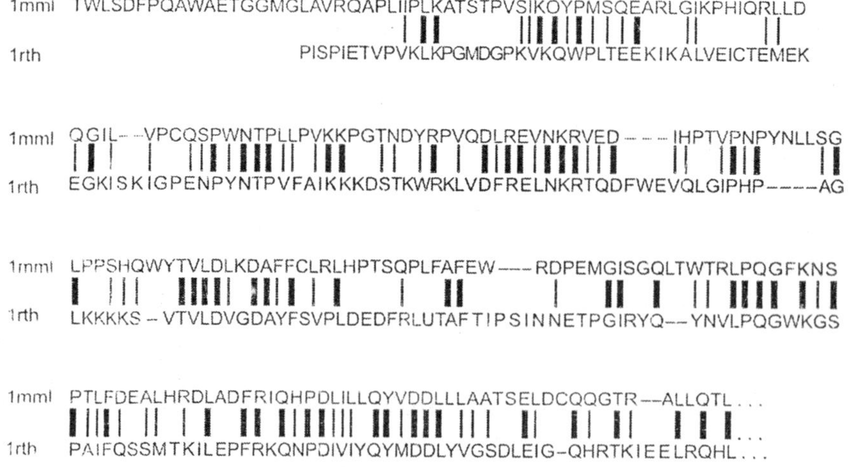

Fig. 8.2. Part of an alignment of Mmlv reverse transcriptase (PDB code IMM) and HIV-1 reverse transcriptase (PDB code IRTH). The alignments were produced using the FASTA package [PL88] with the Blosum 50 substution matrix [HH92]. The sequence of the first domains of both chains are printed in black, while all other amino acids are printed in gray. Thicker gray bars indicate an exact matching of amino acids, whereas the thinner gray bars indicate amino acids that are similar according to the Blossum 50 similarity matrix.

sequence alignment is one of the steps used to determine whether two proteins are homologous. Figure shows the sequence alignment for the amino acid sequences of both protein domains.

There are several problems involved in finding a good sequence alignment. First, the question is whether to align a complete protein sequence or only part of it (since it may be the case that only a part of the protein has been conserved). This is the difference between global and local sequence alignment. Second, major parts of the alignment are substitutions of amino acids by other amino acids, which have 'similar properties'. These substitutions are shown using thin gray bars (instead of thicker ones, which are used for exact matches). Now the question is how to weight the different substitutions (the weights are in fact measures of the similarity between amino acids). This will be discussed in the next section, which treats scoring matrices.

SCORING MATRICES

Alignments are used to reveal homologous proteins, or regions of proteins that are conserved in different proteins. Since evolution is a stochastic process, it is clear that there cannot be one single correct alignment. Instead, there are different possible alignments, and we have to choose the 'correct' ones (although we never know for sure which alignment is correct). To do this, one associates a *score* with every alignment, where the alignments with the highest score are the ones to be selected. Before we can choose an appropriate scoring function, we have first to fix the underlying evolutionary model. Usually, we have an evolutionary model where sequences can be mutated using insertions, deletions or substitutions.

Of course, this is a simplification, and there are approaches to incorporate other kinds of mutations as well. An example is tandem repeats, which will be discussed in this chapter. If we take insertion/deletions and substitution as possible evolutionary events, a scoring function can easily be based on scores for the different operations (since any alignment can be decomposed in a unique way in deletions/insertions and substitutions). Under the assumption that the differnet operations occured independently, the complete score can be taken to be the sum of scores associated with the different operations. We will show how to generate a scoring function for substitutions of amino acids.

The reason is that there is a well-established theory for generating scores for substitutions. In this case, the scores for the different substitutions are stored in a *substitution score matrix*, which contains for every pair (A, B) amino acids an entry s_{AB} (the score for aligning A with B). Clearly, the selection of an appropriate substitution score matrix is crucial for achieving good alignments. There are different manners in which a substitution score matrix can be derived. The first is an ad hoc approach – *i.e.*, a biologist can always set up a score matrix that produces good alignments. The second approach is to derive a score matrix from physical/chemical properties. The idea behind this approach is that an amino acid is more likely to be susbtituted by another if their properties are similar. The third and most often used approach is a statistical one. To be more precise, let s and s' be two amino acid sequence of length n, whose alignment score we would like to give, assuming that only substitutions (no insertions or deletions) are used in transforming s into s'. Using reliable amino acid substitution matrices allows one later to compute optimal *local* sequence alignments between two proteins of possibly different length; *i.e.* to find subsequences of the same length in the two proteins that are conserved.

Score Matrices Based on A Statistical Model

The score for aligning s with s' is generated by comparing two different hypothesis. The first hypothesis (the null hypothesis) is that s and s' *do not stem* from a common predecessor, *i.e.*, that we have an alignment by chance. This hypothesis assumes an underlying random model R that generates two sequences s and s' randomly. Assuming a probability q_A for producing and amino acid A in the model R (where we set q_A to be the relative frequency of the amino acid A in proteins), the probability $P(s, s' \mid R)$ for the null hypothesis is given by

$$P(s, s' \mid R) = \prod_{1 \le l \le n} q_{s_l} \prod_{1 \le l \le n} q_{s'_l} = \prod_{1 \le l \le n} q_{s_l} q_{s'_l}.$$

The second hypothesis (or the homologous hypothesis) is that s and s' *stem* from the same predecessor. The underlying model E is an evolutionary model. One assumes that there is an unknown ancestor sequence r of length n, from which s and s' are generated by random substitution of amino acids. Furthermore, we assume in the model E that p_{AB} is the probability that the amino acids A and B are aligned and have hence been derived from an ancestor amino acid C. How to determine these probabilities will be explained later. Given the probabilities p_{AB}, the probability $P(s, s' \mid E)$ for the homologous hypothesis is given by

$$P(s, s' \mid E) = \prod_{1 \le l \le n} p_{s_l s'_l}.$$

In principle, we want to determine what the chance (or what the odds are) is that the alignment of s and s' reflects the fact that we have aligned conserved region. Hence, a reasonable score for the alignment of s with s' is generated by comparing the probability of the homologous hypothesis with the null hypothesis. This is called the *odds ratio* :

$$\frac{P(s,s' \mid E)}{P(s,s' \mid R)} = \frac{\prod_{1 \le l \le n} p_{s_l s'_l}}{\prod_{1 \le l \le n} q_{s_l} q_{s'_l}} = \prod_{1 \le l \le n} \frac{p_{s_l s'_l}}{q_{s_l} q_{s'_l}}$$

Therefore, we can generate an appropriate score for the alignment of s and s' from scores $\frac{p_{s_l s'_l}}{q_s q_{s'_l}}$ for the amino acid pairings (s_l, s'_l) (with $1 \le l \le n$) in the alignment of s and s'. To achieve a scoring function that is additive instead of multiplicative, one defines the entries of the substitution socre matrix by *log odds ratios :*

$$s_{AB} = \log \frac{p_{AB}}{q_A q_B}. \qquad \text{...(8.1)}$$

In fact, it has been shown that any score matrix consists of log odds ratios under reasonable assumptions (See Alt 91).

Pam and Amino Acid Pair Probabilities

The next problem is to determine an appropriate evolutionary model E. To be more precise, the model E is a model $E((p_{AB})_{AB})$ that is parameterized in the pair probabilties p_{AB}, and we want to find these probabilities. In a statistical approach, these probabilities have to be estimated from data. This can be done if we have two sequences s and s' for which we ***know*** that they are homologous. Estimating p_{AB} from the known alignment of s with s' means that we want to find parameters p_{AB} that maximize.

$$P(E((p_{AB})_{AB}) \mid s, s').$$

To do this, we apply maximum-likelihood approach, *i.e.*, we compute parameters p_{AB} that maximize

$$P(s, s' \mid E(p_{AB})_{AB}).$$

A simple calculation using Lagrange multipliers shows that the maximum is achieved if we set

$$p_{AB} = \frac{n_{AB}(s, s')}{n},$$

where $n_{AB}(s, s')$ is the number of times the amino acids A and B are algined in one column in the alignment of s and s', and n is the length of s (and s'). Thus $\frac{n_{AB}(s,s')}{n}$ is the relative frequency of a pair (A, B) in the alignment of s and s'.

This approach was used for the first time by Dayhoff, Schwartz, and Orcutt (DSO 78) to generate the so-called PAM matrices, perhaps the most widely used substitution matrices. To construct the PAM matrix, Dayhoff *et al.*, were faced with the problem that the only way to find sequences s and s' believed to be homologous is to apply (local) sequence alignment. Hence, to find good substitution matrices for sequences alignment, one needs to apply sequence alingment, which sounds like a vicious circle. To overcome this problem, Dayhoff *et al.*, considered only very closely related sequences (*i.e.*, sequences that differ in at most 15% of their amino acids). The resulting alignment is thus very likely to reflect homology.

This approach yields a substitution score matrix (often simply called substitution matrix), which is valid only for sequences that are as closely related as the sequences used in generating the score matrix. On the other hand, one would like to align sequences that are more distantly related as well. To this end, Dayhoff *et al.* introduced an additional parameter t, which models the evolutionary time scale; *i.e.* they defined pair probabilities

$$p_{AB}(t) = P(A, B|t)$$

that depend on time t, rather than p_{AB} in our earlier discussion. How can one generate probabilities $P(A, B|t)$ for different values of t? Let

$$P(B|A, t)$$

denote the probability that amino acid A is substituted by B within evolutionary distance t. Assuming that time directionality of evolution can be ignored, we have

$$P(A, B|t) = P(B|A, t)\, P(A|t).$$

Assuming that the distribution of amino acids does not change during evolution, we have that $P(A|t) = P(A) = q_A$, hence $P(A, B|t) = P(B|A, t)\, q_A$, which implies that

$$P(B|A, t) = \frac{P(A,B \mid t)}{q_A}. \qquad \text{...(8.2)}$$

By equation $P(B|A, t)$ can be estimated from the relative frequency of the pair (A, B) in the known alignment of two sequences s and s' with distance t, and from the relative frequency q_A of the amino acid A.

The advantage of this approach is that from the matrix

$$M = P(B|A, t))_{AB}$$

of substitution probabilities for all pairs of amino acids for a specific evolutionary distance t, we can generate the matrix

$$M' = (P(B|A, kt))_{AB}$$

of substitution probabilities for any $k \in \mathbb{N}$, simply by setting

$$M' = M^k.$$

This allows the extrapolation of substitution probabilities over a longer time scale. From the substitution probabilities, a score matrix can be generated by equations (8.2) and (8.1).

How can the values $P(B|A, t)$ be adjusted to allow for arbitrary distances t? For this purpose, Dayhoff *et. al.* introduced the evolutionary time scale PAM, which is an acronym for 'point accepted mutation'. Two sequences s and s' have an evolutionary distance of 1 PAM is s was converted into s' by a *series* of accepted substitutions with an average of 1 accepted substitution per 100 amino acids. Since substitution can occur at the same amino acid site several times, a distance of n PAM between two sequences s and s' does not necessarily imply that exactly $n\%$ of the amino acid positions of s and s' differ. For this reason, it is possible to consider sequences whose PAM distance is greater than 100 (the PAM-250 substitution matrix is widely used). A substitution matrix

$$M = (P(B|A, t))_{AB}$$

is defined to be 1 PAM if the expected number of substitutions in a 'typical' protein using the substitution probabilities $P(A, B|t)$ is 1%. By the previous discussion, an n PAM matrix N can be generated from a 1 PAM matrix M by computing the nth power of M; *i.e.* $N = M^n$.

We can hardly expect to find homologous sequences s and s' having exactly 1 PAM distance. Dayhoff *et al.* therefore scaled the values $P(B|A, t)$ for $A \neq B$ by a value λ, such that the resulting substitution probabilities imply an expected substitution frequency of 1% (one amino acid among every 100 amino acids). More precisely, in the estimation of $P(B|A, t)$ from a known alignment of sequences s and s', Dayhoff *et al.* set

$$P(B|A, t) = \lambda \frac{n_{AB}(s, s')}{q_A}$$

for $B \neq A$, and defined

$$P(A|A, t) = 1 - \sum_{B \neq A} P(B \mid A, t),$$

where λ was chosen so that the resulting substitution matrix is 1 PAM.

The PAM matrices have been used with much success, their only problem being that since the entries in the 1 PAM matrix were estimated from very closely related sequences, it is therefore difficult to extrapolate to more distantly related proteins. For this reason, Hanikoff and Henikoff used highly conserved regions in multiple sequence alignments of several distantly related proteins, rather than pairwise alignment, in generating their BLOSUM matrices. Experimental results have shown that the BLOSUM matrices are better suited for aligning more distantly related proteins.

GLOBAL PAIRWISE SEQUENCE ALIGNMENT

Distance Methods

There are two different problems involved when comparing two similar sequences. The first one is to calculate the distance between the two different sequences (*i.e.*, are the sequences similar or not?). The second problem is to align the sequences in order to find conserved regions. Hence, we will give two different definitions for comparing sequences, namely *edit distance* and *alignment* (and *alignment distance*), and we will show under which conditions these two definitions yield the same result.

Let Σ be a finite alphabet. A word $a \in \Sigma^*$ is simply a sequence $a_1 \ldots a_n$, where $n \in \mathbb{N}$ and $\ldots$ for $1 \leq i \leq n$. The length of word a is denoted $|a|$. The empty word is denoted λ. With Σ^+ denote the set $\Sigma^* \backslash \lambda$. The gap symbol is—.

We first define the simplest notion of edit distance for two sequences a, b. The edit distance is interpreted as the number of evolutionary operations to transform a string a into b using mutation, insertion and deletion of one symbol as the only evolutionary operations. The edit distance, unlike the approach of Waterman *et al.* has no special treatment for gaps: *i.e.*, a size k gap is treated as k many single gaps (either insertions or deletion). Biologically, this is not realistic, so we retain the term edit distance for the approach of Needleman-Wunsch.

Definition (Edit Operation) : *Given an alphabet* Σ with $- \notin \Sigma$, *an* edit operation *is a pair*

$$(x, y) \in (\Sigma \cup \{-\} \times (\Sigma \cup \{-\}).$$

We *say that an edit operation* (x, y) *is a*

- substitution *if* $x, y, \in \Sigma$ *with* $x \neq y$,
- insertion *if* $x = -$ *and* $y \in \Sigma$,
- deletion *if* $x \in \Sigma$ *and* $y = -$.

Given $a, b \in (\Sigma \cup \{-\})^*$ *and an edit operation* (x, y) such that $x \neq y$, we write

$$a \rightarrow_{(x, y)} b$$

if b can be obtained from a by replacing one occurrence of x by y (if $x, y \in \Sigma$*), or by deleting one occurrence of x (if y = –), or by inserting one occurrence of y (if x = –). If* $S = s_1 \ldots s_n$ *is a sequence of edit operations, then we write*

$$a \Rightarrow_S b$$

if $a = a^{(0)} \rightarrow_{s_1} a^{(1)} \rightarrow_{s2} \ldots \rightarrow S_n\, a^{(r)} = b$ *for* words $a^{(0)} \ldots a^{(r)}$.

We will use *indels* as short for insertions or deletions. Note that the edit operations can be applied at every position of the string. *Context-sensitive* edit operations (*i.e.*, edit operations that for instance produce lethal mutations) are computationally more powerful than the edit operations defined above.

In evolution, some types of mutation happen more often than others. This is simulated in the definition of an edit distance by weighing the edit operations. Let $w(x, y)$ be a *cost function* assigning weights to the edit operations.

Definition (Edit Distance) : *Given a cost function* $w : (\Sigma \cup \{-\}) \times (\Sigma \cup \{-\}) \rightarrow \mathbb{R}$ *and two words* $a, b \in \Sigma^*$, *the cost of a sequence* $S = s_1 \ldots s_r$ *of edit operations is defined as* $\Sigma_{i=1}^{r} w(s_i)$. *The* edit distance of a, b *is defined* as

$$d_w(a, b) = \min\{m(S) \mid a \Rightarrow s\ b\}.$$

Usually, when w is implicit, we write $D(a, b)$ *rather than* $d_w(a, b)$.

One must be careful which kind of cost function can be used. For example, one must avoid the cost of the sequence (A, C) (C, T) being smaller than of the substitution (A, T) itself. This and other properties are combined in the notion of a metric.

Definition (Metric) : *A metric d must satisfy the following :*

1. $d(x, y) = 0$ *iff* $x = y$.
2. $d(x, y) = d(y, x)$ *(symmetry)*,
3. $d(x, z) \leq d(x, y) + d(y, z)$ *(triangle inequality)*

Proposition : *If w is a metric, then* d_w *is also a metric.*

For the calculation of edit distance, the straightforward definition of edit distance is not very useful. For this reason, we define the notions of alignment and alignment distance explicitly. We show that the definitions of alignment distance and edit distance agree in the case that the cost function is a metric.

Definition (Alignment and Alignment Distance) : *Let Σ be an alphabet with $- \notin \Sigma$. For every $u \in (\Sigma \cup \{-\})^*$ we define $u|_\Sigma$ to be the restriction of u to Σ (by deleting all occurrences of $-$ in u). An alignment is a pair (a°, b°) with $a^\circ, b^\circ \in (\Sigma \cup \{-\})^*$ such that*

$$|a^\circ| = |b^\circ|$$

and there is no position i such that

$$a_i^\circ = - = b_i^\circ$$

An alignment (a°, b°) is an alignment of (a, b) with $a, b \in \Sigma^*$ if

1. *$a^\circ /\Sigma = a$, and*
2. *$b^\circ/\Sigma = b$.*

Given a cost function w, we define the cost of an alignment by

$$w(a^\circ, b^\circ) = \sum_{i=1}^{|a^\circ|} w(a_i^\circ, b_i^\circ).$$

The alignment distance of a, b is

$$d_w^a(a,b) = \min\{w(a^\circ, b^\circ) \mid (a^\circ, b^\circ) \text{ alignment of } (a, b)\}.$$

Again we write $D(a, b)$ if w is clear from the context. The alignment (a°, b°) is optimal if $d_w^a(a,b) = w(a^\circ, b^\circ)$.

Proposition (Additivity of Alignments) : *Let (a°, b°) and (c°, d°) be two alignments. Then $(a^\circ c^\circ, b^\circ d^\circ)$ is also an alignment with*

$$w(a^\circ c^\circ, b^\circ d^\circ) = w(a^\circ, b^\circ) + w(c^\circ, d^\circ).$$

Proposition : *Let w be a metric cost function, and $a, b \in \Sigma^*$. Then for every alignment (a°, b°) of (a, b) there is a sequence S of edit operations such that $a \Rightarrow_s b$ and*

$$w(S) = w(a^\circ, b^\circ).$$

For every sequence S such that $a \Rightarrow_s b$, there is an alignment (a°, b°) of (a, b) such that

$$w(a^\circ, b^\circ) \leq w(S)$$

Proof : For every alignment (a°, b°), the sequence $S = s_1 \ldots s_{|a^\circ|}$ with

$$s_i = (a_i^\circ, b_i^\circ)$$

is a sequence of edit operations with $a \Rightarrow_s b$ and $w(a^\circ, b^\circ) = w(s)$.

The other direction is proven by induction on sequence length. Let S be a sequence of edit operations with $|S| = n + 1$. Then $S = S's$ and there is a $c \in \Sigma^*$ such that

$$a \Rightarrow_{s'} c \to_s b.$$

We prove the existence of an alignment (a°, b°) for (a, b) with $w(a^\circ, b^\circ) \leq w(S)$ for the case that s is a substitution. The other cases can be proven analogously.

By induction hypothesis, we know that there is an alignment (a°, c°) such that $w(a^\circ, c^\circ) \leq w(S')$. Note that we can use an arbitrary alignment with this condition. Let i be the position changed by $c \to_s b$(*i.e.*, i is the unique position with $c_i \neq b_i$) and let j be the corresponding position in c° (*i.e.*, the position of the ith Σ-letter in c°). We define (a°, b°) with

$$b^\circ = c_1^\circ \ldots c_{j-1}^\circ b_i c_{j+1}^\circ \ldots c_{|c^\circ|}^\circ,$$

which is an alignment of (a, b). Then

$$\begin{aligned} w(a^\circ, b^\circ) &= w(a^\circ, c^\circ) - w(a_j^\circ, c_j^\circ) + w(a_j^\circ, b_j^\circ) \\ &\overset{w \text{ metric}}{\leq} w(a^\circ, c^\circ) + w(c_j^\circ, b_j^\circ) \\ &= w(a^\circ, c^\circ) + w\,(\mathrm{s}) \\ &\overset{\text{Ind. Hyp.}}{\leq} w(S') + w(s) = w(S). \end{aligned}$$

Consider as an example the alphabet $\Sigma = \{A, C, G, T\}$ and the two words $a = ACCGGT\,A$ and $b = AGGCTG$. Then one possible alignment is

$$a^\circ = ACCGG - TA$$
$$b^\circ = A - - GGCTG.$$

A corresponding sequence S of edit operations with $a \Rightarrow_s b$ and $w(S) = w(a^\circ, b^\circ)$ is

$$S = (C, -)\,(C, -)\,(-, C)\,(A, G).$$

Remark : Note that Proposition is not true if w is not a metric.

Theorem (Needleman–Wunsch Edit Distance)

Let $w : (\Sigma \cup \{-\}) \times (\Sigma \cup \{-\}) \to \mathbb{R}$ *be a metric cost function for the given alphabet* Σ. *Let* $a, b \in \Sigma^*$ *with* $|a| = n$ *and* $|b| = m$. *Define the matrix* $(D_{i,j})$ *with* $0 \leq i \leq |a|$ *and* $0 \leq j \leq |b|$ *by*

$$D_{0,0} = 0,$$

$$D_{0,j} = \sum_{k=1}^{j} w(-, b_k),$$

$$D_{i,0} = \sum_{k=1}^{i} w(a_k, -),$$

$$\forall i, j > 0 : D_{i,j} = \min \begin{Bmatrix} D_{i,j-1} + w(-, b_j), \\ D_{i-1,j-1} + w(a_i, b_j), \\ D_{i-1,j} + w(a_i, -) \end{Bmatrix}. \quad \ldots(8.3)$$

Then $D_{i,j}$ *is the minimum global sequence alignment distance between sequences a and b, i.e.,*

$$D_{i,j} = D(a_1 \ldots a_i, b_1 \ldots b_i).$$

This theorem furnishes a dynamic programming algorithm (*i.e.*, fills in the distance matrix) with runtime $O(nm) = O(n^2)$ if $n \geq m$.

Proof : By induction on pairs (i, j), were $(k, l) < (i, j)$ if and only if $k < i$, or i, or $k = i$ and $l < j$ (lexicographic ordering). So assume as induction hypothesis that we have proven the theorem for all $(k, l) < (i, j)$.

We have to show that for all alignments (u°, v°) of $a_1 \ldots a_i, b_1 \ldots b_j$,

$$D_{i,j} \leq D(u^\circ, v^\circ),$$

and that there is an alignment (u°, v°) such that

$$D(u^\circ, v^\circ) \leq D_{i,j}.$$

For the first direction consider an arbitrary alignment (u°, v°) of $a_1 \ldots a_i$ and $b_1 \ldots b_j$. Let $r = |u^\circ| = |v^\circ|$. Then by Proposition 3.6 we have that

$$D(u_1^\circ \ldots u_r^\circ, v_1^\circ \ldots v_r^\circ) = D(u_1^\circ \ldots u_{r-1}^\circ, v_1^\circ \ldots v_{r-1}^\circ) + D(u_r^\circ, v_r^\circ). \qquad \ldots(8.4)$$

Since $u_r^\circ = v_r^\circ = -$ is excluded by the definition of an alignment, we have the three cases

$$\begin{array}{lll} u_r^\circ = a_i & u_r^\circ = a_i & u_r^\circ = - \\ v_r^\circ = - & v_r^\circ = b_j & v_r^\circ = b_j. \end{array}$$

Applying the induction hypothesis, we get for the different cases

$$\begin{array}{rlcl} & D_{i,j-1} & \leq & D(u_1^\circ \ldots u_{r-1}^\circ, v_1^\circ \ldots v_{r-1}^\circ), \\ \text{resp.} & D_{i-1,j-1} & \leq & D(u_1^\circ \ldots u_{r-1}^\circ, v_1^\circ \ldots v_{r-1}^\circ), \\ \text{resp.} & D_{i-1,j} & \leq & D(u_1^\circ \ldots u_{r-1}^\circ, v_1^\circ \ldots v_{r-1}^\circ), \end{array} \qquad \ldots(8.5)$$

With (8.4) and the recursion equation (8.3), this yields

$$D_{i,j} \leq D(a_1 \ldots a_i, b_1 \ldots b_j).$$

For the other direction, the case distinction shows how to produce an optimal alignment of $(a_1, \ldots a_i, b_1 \ldots b_j)$ given optimal alignments of $(a_1 \ldots a_i, b_1 \ldots b_{j-1})$, $(a_1 \ldots a_{i-1}, b_1 \ldots b_{j-1})$ or $(a_1 \ldots a_{i-1}, b_1 \ldots b_j)$.

With a small modification, we can also easily find all optimal alignments (*i.e.*, alignments whose distance is minimal). This is achieved via an additional matrix called the trace matrix. In principle, at every matrix cell we just store pointers that indicate which cells have been used in calculating the actual cell. The *trace matrix* $(tr_{i,j})$ for two words $a, b \in \Sigma^*$ is defined to be the matrix of elements $tr_{i,j} \subseteq \{\leftarrow, \uparrow, \nwarrow\}$ with

$$\begin{aligned} tr_0, 0 &= \theta, \\ tr_{0,j} &= \{\leftarrow\}, \\ tr_{i,0} &= \{\uparrow\}, \\ \forall i, j > 0 : \nwarrow \in tr_{i,j} &\Leftrightarrow D_{i-j} = D_{i-1,j-1} + w(a_i, b_j), \\ \uparrow \in tr_{i,j} &\Leftrightarrow D_{i,j} = D_{i-1,j} + w(a_i, -), \\ \leftarrow \in tr_{i,j} &\Leftrightarrow D_{i,j} = D_{i,j-1} + w(-, b_j), \end{aligned}$$

a *traceback* t is an element of $\{\leftarrow, \uparrow, \nwarrow\}^*$ that is a path in $(tr_{i,j})$ starting from the lower right corner of the trace matrix and following the pointer in $tr_{i,j}$ until we reach the upper left corner. The path is written left to right, *i.e.*, the right end corresponds to $tr_{|a|,|b|}$ and the left end to $tr_{0,0}$. The alignment (a°, b°) associated with t is generated as follows. Define $a^\circ \in (\Sigma \cup \{-\})^*$ from t by substituting a_i for the ith occurrence of $\nwarrow$ or $\uparrow$, and by replacing all occurrences of $\leftarrow$ with $-$. Similarly, we get b° by substituting b_j for the jth occurrence of $\leftarrow$ or $\nwarrow$, and by replacing all occurrences of $\uparrow$ with $-$.

Example : Consider the words $a = AT$ and $b = AAGT$ and a cost function where $\forall x \neq y$: $w(x, y) = 1$ and $w(x, x) = 0$. Then the distance matrix $(D_{i,j})$ for a, b is

		A	A	G	T
	0	1	2	3	4
A	1	0	1	2	3
T	2	1	1	2	2

and the corresponding trace matrix is

		A	A	G	T
	θ	{←}	{←}	{←}	{←}
A	{↑}	{↖}	{←, ↖}	{←}	{←}
T	{↑}	{↑}	{←}	{←, ↖}	{↖}

One possible traceback, indicated by black arrows, is

↖ ← ← ↖

The corresponding alignment is

$$a^\circ = A--T,$$
$$b^\circ = AAGT.$$

The other possible traceback, indicated by black and dark gray arrows, is

← ↖ ← ↖,

which produces the alignment

$$a^\circ = -A-T,$$
$$b^\circ = AAGT.$$

Figure shows the distance matrix together with the two tracebacks, which is the more conventional way of displaying tracebacks.

		A	A	G	T
	0	1	2	3	4
A	1	0	1	2	3
T	2	1	1	2	2

		A	A	G	T
	0	1	2	3	4
A	1	0	1	2	3
T	2	1	1	2	2

Fig. 8.3. Distance matrix and traceback.

By the definition of distance, the insertion (resp. deletion) of a gap of length n has exactly the same cost as the insertion of n gaps of length 1. This is often biologically unrealistic. So we refine the definition of alignment distance as follows. Given a word $u \in (\Sigma \cup \{-\})^*$, we say that u contains a *gap* of length k at position i if $u_i \dots u_{i+k-1} \in \{-\}^*$, and there is no other subword of u extending $u_i \dots u_{i+k-1}$ that is composed uniquely of $-s$. We define $\Delta_k(u)$ to be the number of different gaps of u that have length k.

Definition (Alignment with Gap Penalty) : A gap penalty *is a function* $g(k) : \mathbb{N} \to \mathbb{R}$ *that is subadditive, i.e.,*

$$\forall k, l : g(k+l) \leq g(k) + g(l).$$

A *gap penalty is called* affine *if there are* $\alpha, \beta \in \mathbb{R}$ *such that*

$$g(k) = a + k\beta.$$

Given a cost function w *on* $\Sigma \times \Sigma$, *the cost of an alignment* (a°, b°) *of length* $r = |a^\circ| = |b^\circ|$ *is defined as*

$$w(a^\circ, b^\circ) = \sum_{\substack{1 \leq i \leq r \\ a_i^\circ \neq -, b_i^\circ \neq -}} w(a_i^\circ, b_i^\circ) + \sum_{1 \leq k \leq r} \Delta_k(a^\circ) g(k) + \sum_{1 \leq k \leq r} \Delta_k(b^\circ) g(k).$$

$d_w^a(a, b)$ is defined as usual by

$$d_w^a(x, y) = \min\{ w(a^\circ, b^\circ) | a(a^\circ, b^\circ) \text{ alignment } (a, b)\}$$

Again we write $D(a, b)$ if w is clear from the context.

We have to redefine the notion of additivity for alignment with gap penalities. Then we present an $O(n^3)$ algorithm for arbitrary gap penalties, and finally an $O(n^2)$ optimization for affine gap penalties.

Proposition (Additivity with Gap Penalty) : *Given two alignments (a°, b°) with length* $r = |a^\circ| = |b^\circ|$ and (c°, d°). *If*

$$a_r^\circ \neq -V c_1^\circ \neq -$$

and $$b_r^\circ \neq V d_1^\circ \neq -,$$

then $(a^\circ c^\circ, b^\circ d^\circ)$ is also an alignment with

$$w(a^\circ c^\circ, b^\circ d^\circ) = w(a^\circ, b^\circ) + w(c^\circ, d^\circ).$$

Theorem (Waterman, Smith and Beyer [WSB76])

Let $g : \mathbb{N} \to \mathbb{R}$ be a gap penalty and w be cost function on $\Sigma \times \Sigma$. Let $a = a_1 \ldots a_n$ and $b = b_1 \ldots b_m$ be two word in Σ^*. We define $(D_{i,j})$ with $1 \leq i \leq n$ and $1 \leq j \leq m$ by

$$D_{0,0} = 0,$$
$$D_{0,j} = g(j),$$
$$D_{i,0} = g(i),$$

$$D_{i,j} = \min\begin{Bmatrix} \min\limits_{1 \leq k \leq j}\{D_{i,j-k} + g(k)\} \\ D_{i-1,j-1} + w(a_i, b_j), \\ \min\limits_{1 \leq k \leq i}\{D_{i-k,j} + g(k)\}\end{Bmatrix}.$$

The $D_{i,j} = D(a_1 \ldots a_i, b_1 \ldots b_j)$.

Proof: This is similar to the proof for Theorem 8.9, using the case distinction

$$u^\circ = \ldots\sigma\, a_{i-k+1} \ldots a_i \qquad u^\circ = \ldots a_i \qquad u^\circ = \ldots a_i - \ldots -$$
$$v^\circ = \ldots b_j - \ldots - \qquad v^\circ = b_j \qquad v^\circ = \ldots\sigma'\, b_{j-k+1} \ldots b_j$$

with $\sigma \in \{a_{i-k}, -\}$ and $\sigma' \in \{b_{j-k}, -\}$, together with the modified Additivity Proposition 8.12.

This theorem furnishes a dynamic programming algorithm with runtime $O(nm \cdot (n + m)) = O(n^3)$ if $n \geq m$. In the case of affine gap penalty, one can even apply an algorithm that has an $O(n^2)$ runtime. The main idea is to use three matrices instead of a single one. In the two additional matrices $(P_{i,j})$ and $(Q_{i,j})$, one stores the minimal distance for all alignments ending with a gap. That is, we define $P_{i,j}$ (resp. $Q_{i,j}$) to be the minmal distance for all alignments (u°, v°) for $(a_1 \ldots a_i, b_1 \ldots b_j)$ that are the form

$$u^\circ = \ldots a_i \qquad \qquad$$
$$v^\circ = \ldots - \qquad \left(\text{resp.} \quad \begin{matrix} u^\circ = \ldots - \\ v^\circ = \ldots b_j \end{matrix}\right).$$

Since we have affine gap penalties, it is easy to define the recursion equation for $P_{i,j}$ and $Q_{i,j}$. When considering $P_{i,j}$, we have two possible cases for the optimal alignment (u°, v°) (where optimal means optimal under the condition that a_i is aligned with a gap):

1. (u°, v°) extends an existing gap, *i.e.*, (u°, v°) is of the form

$$u^\circ = ..a_{i-1}\,a_i,$$
$$v^\circ = ..--.$$

Then we have to add only the gap extension cost β, *i.e.*, we get $P_{i,j} = P_{i-1,j} + \beta$.

2. (u°, v°) is creates a new gap, *i.e.*,

$$u^\circ = ..-a_i \qquad \text{or} \qquad u^\circ = -a_{i-1}\,a_i$$
$$v^\circ = ..b_j - \qquad\qquad v^\circ = ..b_j -.$$

Then we have to add the cost for a gap of length 1, *i.e.*, $P_{i,j} = P_{i-1,j} + g(1)$.

This is basic idea applied in the following theorem.

Theorem (Gotoh)

Let $g(k) = a + k\beta$ be an affine gap penalty, and let $w : \Sigma \times \Sigma \to \mathbb{R}$ be a cost function and a, $b \in \Sigma^*$. Define the matrices $(D_{i,j})$, $(P_{i,j})$ and $(Q_{i,j})$ recursively by

$$D_{0,0} = 0,$$
$$D_{0,j} = g(j),$$
$$D_{i,0} = g(i),$$

$$D_{i,j} = \min\begin{Bmatrix} D_{i-1,j-1} + w(a_i, b_j) \\ P_{i,j} \\ Q_{i,j} \end{Bmatrix},$$

with i, j ≥ 1, where for $1 \leq i \leq |a|$, $1 \leq j \leq |b|$,

$$P_{0,j} = \infty,$$

$$P_{i,j} = \min\begin{Bmatrix} D_{i-1,j} + g(1) \\ P_{i-1,j} + \beta \end{Bmatrix}$$

and

$$Q_{i,0} = \infty,$$

$$Q_{i,j} = \min\begin{Bmatrix} D_{i,j-1} + g(1) \\ Q_{i,j-1} + \beta \end{Bmatrix}.$$

Then

$$D_{i,j} = D(a_1 ..a_i, b_1 ..b_j). \qquad ...(8.6)$$

Proof: This is by induction over pairs (i, j). As an additional claim, we show that

$$P_{i,j} = \min\left\{ D(u^\circ, v^\circ) \,\middle|\, \begin{matrix}(u^\circ, v^\circ) \text{ alignment of } (a_1 \ldots a_i, b_1 \ldots b_j), \\ \text{with } u^\circ_{|u^\circ|} = a_i \text{ and } v^\circ_{|v^\circ|} = - \end{matrix} \right\}. \qquad ...(8.7)$$

$$Q_{i,j} = \min\left\{ D(u^\circ, v^\circ) \,\middle|\, \begin{matrix}(u^\circ, v^\circ) \text{ alignment of } (a_1 \ldots a_i, b_1 \ldots b_j), \\ \text{with } u^\circ_{|u^\circ|} = - \text{and } v^\circ_{|v^\circ|} = b_j \end{matrix} \right\}. \qquad ...(8.8)$$

Assume we have proven the claims for all $(k, l) < (i, j)$. We first prove that

$$P_{i,j} \leq \min\left\{ D(u^\circ, v^\circ) \,\middle|\, \begin{matrix}(u^\circ, v^\circ) \text{ alignment of } (a_1 \ldots a_i, b_1 \ldots b_j), \\ \text{with } u^\circ_{|u^\circ|} = a_i \text{ and } v^\circ_{|v^\circ|} = - \end{matrix} \right\}.$$

Let (u°, v°) with $r = |u^\circ| = |v^\circ|$ be an alignment of $(a_1 \,..a_i, b_1 \,..b_j)$ such that

$$u_r^\circ = a_i,$$
$$v_r^\circ = -. \qquad \text{...(8.9)}$$

If $i = 1$, then there is only one possible alignment given condition (8.9), namely

$$- \;\ldots - a_i,$$
$$b_1 \;\ldots b_j -.$$

Hence
$$\begin{aligned} D(u^\circ, v^\circ) &= g(j) + g(1) \\ &= D_{0,j} + g(1) \\ &= \min\begin{Bmatrix} D_{0,j} + g(1) \\ \infty \end{Bmatrix} = P_{1,j}. \end{aligned}$$

Otherwise, we know that (u'°, v'°) with $u'^\circ = u_1^\circ \ldots u_{r-1}^\circ$ and $v'^\circ = v_1^\circ \ldots v_{r-1}^\circ$ is an alignment of $(a_1 \ldots a_{i-1}, b_1 \ldots b_j)$. If $v_{r-1}^\circ \neq -$, then we get

$$\begin{aligned} D(u^\circ, v^\circ) &\overset{\text{Prop. 3.12}}{=} D(u'^\circ, v'^\circ) + D(a_i, -) \\ &= D(u'^\circ, v'^\circ) + g(1) \\ &\overset{\text{Ind. Hyp.}}{\geq} D_{i-1,j} + g(1). \end{aligned}$$

If $v_{r-1}^\circ = -$, then $D(u'^\circ, v'^\circ) \geq P_{i-1,j}$ by induction hypothesis. Let $k > 0$ be the length of the final gap of v'°. Then

$$\begin{aligned} D(u^\circ, v^\circ) &= D(u'^\circ, v'^\circ) - g(k) + g(k+1) \\ &= D(u'^\circ, v'^\circ) - \alpha - k\,\beta + \alpha + (k+1)\,\beta \\ &= D(u'^\circ, v'^\circ) + \beta, \\ &\overset{\text{Ind. Hyp.}}{\geq} P_{i-1,j} + \beta. \end{aligned}$$

It remains to be shown that there is an alignment (u°, v°) satisfyng (8.9) such that

$$D(u^\circ, v^\circ) \leq P_{i,j}.$$

If $P_{i,j} = P_{i-1} + \beta$, then take an alignment (u'°, v'°) of $a_1 \ldots a_{i-1}, b_1 \ldots b_j$ that is optimal under the condition that

$$u'^\circ = a_{i-1},$$
$$v'^\circ = -.$$

By induction hypothesis, this implies $D(u'^\circ, v'^\circ) = P_{i-1,j}$. Hence

$$\begin{aligned} D(u^\circ, v^\circ) &= D(u'^\circ a_i, v'^\circ -) \\ &= D(u'^\circ, v'^\circ) + \beta \\ &= P_{i-1,j} + \beta \\ &= P_{i,j}. \end{aligned}$$

If $P_{i,j} = D_{i-1,j} + g(1)$, then take an optimal alignment (u'°, v'°) of $a_1 \,..a_{i-1}, b_1 \,..b_j$. Then

$$D(u'^\circ, v'^\circ -) = D_{i-1,j}$$

by induction hypothesis, and hence

$$D(u'^{\circ} a_i, v'^{\circ} -) \leq D_{i-1,j} + g(1).$$

Again we can achieve an optimal alignment using trace matrices and tracebacks. Since we are using different matrices, we must use a different pointer set P, namely

$$\mathcal{P} = \{^D\nwarrow, {}^Q\bullet, {}^P\bullet, {}^D\leftarrow, {}^Q\leftarrow, {}^D\uparrow, {}^P\uparrow\}.$$

Thus, a traceback pointer points to a matrix and a cell in this matrix. Instead of taking one trace matrix, we define three matrices (tr^D), (tr^P), and (tr^Q), where

$$tr^D_{0,0} = \theta$$

$$tr^D_{0,j} = \{^D\leftarrow\},$$

$$tr^D_{i,0} = \{^D\uparrow\},$$

and

$$\forall i,j>0: {}^D\nwarrow \in tr^D_{i,j} \Leftrightarrow D_{i,j} = D_{i-1,j-1} + w(a_i, b_j),$$

$${}^Q\bullet \in tr^D_{i,j} \Leftrightarrow D_{i,j} = Q_{i,j},$$

$${}^P\bullet \in tr^D_{i,j} \Leftrightarrow D_{i,j} = P_{i,j};$$

$$\forall i,j>0: {}^P\uparrow \in tr^P_{i,j} \Leftrightarrow P_{i,j} = P_{i-1,j} + \beta,$$

$${}^D\uparrow \in tr^P_{i,j} \Leftrightarrow P_{i,j} = D_{i-1,j} + g(1);$$

$$\forall i,j>0: {}^Q\leftarrow \in tr^Q_{i,j} \Leftrightarrow Q_{i,j} = P_{i,j-1} + \beta,$$

$${}^D\leftarrow \in tr^Q_{i,j} \Leftrightarrow Q_{i,j} = D_{i,j-1} + g(1)$$

A traceback is defined using intuitive functions on pairs of integers for the pointers $\{^D\nwarrow, {}^Q\bullet, {}^P\bullet, {}^D\leftarrow, {}^Q\leftarrow, {}^D\uparrow, {}^P\uparrow\}$, where we use triples (M, i, j) to name matrix elements:

$$\forall i\geq 0, j\geq 0: \quad {}^M\bullet(i,j) = (M,i,j) \quad \text{for } M\in\{Q,P\},$$

$$\forall i\geq 0, j>0: \quad {}^M\leftarrow(i,j) = (M,i,j-1) \quad \text{for } M\in\{D,Q\},$$

$$\forall i>0, j\geq 0: \quad {}^M\uparrow(i,j) = (M,i-1,j) \quad \text{for } M\in\{D,P\},$$

$$\forall i>0, j>0: \quad {}^D\nwarrow(i,j) = (D,i-1,j-1).$$

An element $t \in \{^D\nwarrow, {}^Q\bullet, {}^P\bullet, {}^D\leftarrow, {}^Q\leftarrow, {}^D\uparrow, {}^P\uparrow\}^*$ is a *traceback for* (a, b) if there is sequence of matrix cells

$$(M_0, i_0, j_0) \ \ldots (M_{|t|}, i_{|t|}, j_{|r|})$$

such that

1. $(M_{|t|}, i_{|t|}, j_{|t|}) = (D, |a|, |b|$ and $(M_0, i_0, j_0) = (D, 0, 0)$:
2. for all $1 \leq k \leq |t|$:

$$t_k \in tr^{M_k}_{i_k, j_k} \text{ and } t_k(i_k, j_k) = (M_{k-1}, i_{k-1}, j_{k-1}).$$

The corresponding alignment is defined analoigously to the case of alignmenht without gap penalty.

Alignment with Tandem Duplication

In the previous alignment algorithms, we considered edit operations of the forms insertion, deletion and substitution. Benson considered an additional operation, tandem duplication, which generates tandem repeats. The corresponding evolutionary model is called the DSI model (duplications, substitutions and indels). The SI model (substitutions and indels) is the underlying evolutionary model of standard alignment algoirthms.

A tandem duplication is an operaiton that replaces a subsequence w in a sequence a by w^k. An example of the application of tanem duplication is

$$\text{ACGAG}\,\boxed{\text{CCGTAGAA}}\,\text{TACCG}$$
$$\downarrow$$
$$\text{ACGAG}\,\boxed{\text{CCGTAGAA}}\boxed{\text{CCGTAGAA}}\boxed{\text{CCGTAGAA}}\boxed{\text{CCGTAGAA}}\,\text{TACCG}.$$

Here, 3 copies of CCGTAGAA have been inserted. Benson considered the problem of tandem repeats under two assumptions :

1. Tandem duplication occurs before the other types of operations.
2. There is no removal of copies of a tandem repeat region.

Both assumptions together imply that one could can generate the alignment using the usual alignment algorithm if one knew the regions that have been duplicated.

Given the above assumptions, Benson showed that one can extend Gotoh's algorithm for affine gap penalty to incorporate also duplications. Let a and b be two sequences that should be aligned under the DSI model (with affine gap penalities). Let g be an affine gap penalty, $d : \mathbb{N} \rightarrow \mathbb{R}$ be an affine duplication cost (*i.e.*, $d(l) = \gamma + \delta l$, where γ is the duplication initiation cost and δ is the duplication extention cost) and $w : \Sigma \rightarrow \mathbb{R}$ be a cost function on Σ. Then we define matrices $(D_{i,j})$, $(P_{i,j})$ and $(Q_{i,j})$ as follows. $(P_{i,j})$ and $(Q_{i,j})$ are defined and $(D_{i,j})$ is defined by

$$\begin{aligned}
D_{0,0} &= 0,\\
D_{0,j} &= g(j),\\
D_{i,0} &= g(i),\\
D_{i,j} &= \min\begin{Bmatrix} D_{i-1,j-1} + w(a_i, b_j) \\ P_{i,j} \\ Q_{i,j} \\ Dup[i,j] \end{Bmatrix},
\end{aligned}$$

where $Dup[i, j]$ is defined as follows. For a word w, let $w_{r\,..s}$ be the subsequence $w_r \ldots w_s$ of w. Then

$$Dup[i,j] = \min_{r,s}\begin{cases} D_{i-r,j-s} + \min_{k\in N}\{d(k) + D(a_{i-r+1..i}, (b_{(j-s+1)..j})^k), \\ D_{i-r,j-s} + \min_{k\in N}\{d(k) + D((a_{i-r+1..i})^k, b_{j-s+1...j})\}. \end{cases}$$

Here, the notation $(b_{(j-k+1)..j})^l$ is a shortcut for the word consisting of l copies of $b_{(j-k+1)...j}$, *i.e.*, the word

$$\underbrace{b_{(j-k+1)...j\,...}\,b_{(j-k+1)...j}}_{l\text{ copies}},$$

and similarly for $(a_{i-h+1...i})^l$. Thus, the remaining problem is to calculate

$$\min_{k\in N}\{d(k) + D(a, b^k)\}.$$

We consider only the problem of aligning a word a with an unknown number k of copies of a word b, ignoring the duplication cost. The *alignment distance with tandem repeat* $D^{\text{tandem}}(a, b)$ is defined by

$$D^{\text{tandem}}(a, b) = \min\{D(a, b^k) \mid k \in \mathbb{N}\}$$

As we will see, the alogrithm for calculating $D^{\text{tandem}}(a, b)$ can easily be extended to incorporate the duplication cost, too. An *alignment with tandem repeats* is defined to be a pair $(a^\circ, b_1^\circ \ldots b_k^\circ)$, where $(a^\circ, b_1^\circ \ldots b_k^\circ)$ is an alignment as usual with

- $a^\circ|\Sigma = a$, and
- for all $1 \le l \le k : b_l^\circ|\Sigma = b$.

In the following, we consider only an alignment distance D with an affine gap penalty $g(l) = \alpha + l\beta$ and a cost function w. The principle idea is to consider all possible alignment matrices $(D^k_{i,j})$ for the usual alignment of a with b^k for $k \in \mathbb{N}$. Each matrix is then divided into k submatrices of size $|a| \times |b|$, and all these sub-matrices are overlayed onto a single matrix using minimization.

This general idea is fomrulated as follows. Let $(D^k_{i,j}), (P^k_{i,j})$, and $(Q^k_{i,j})$ be the matrices for alignment distance between a and b^k as defined by Gotoh's algorithm. The *overlay matrices* $(D_{i,j})$, $(P_{i,j})$, and $(Q_{i,j})$ of size $|a| + 1 \times |b| + 1$ are defined by

$$D_{0,0} = 0,\ D_{0,j} = g(j),\ D_{i,0} = g(i),$$

$$Q_{i,0} = \infty,\ P_{0,j} = \infty,$$

and

$$\begin{aligned} D_{i,j} &= \min\{D^k_{i,l|b|+j} \mid k, l \in \mathbb{N} \wedge l < k\}, \\ \forall i>0, j>0 : P_{i,j} &= \min\{P^k_{i,l|b|+j} \mid k, l \in \mathbb{N} \wedge l < k\}, \\ Q_{i,j} &= \min\{Q^k_{i,l|b|+j} \mid k, l \in \mathbb{N} \wedge l < k\}, \end{aligned} \quad \text{...(8.10)}$$

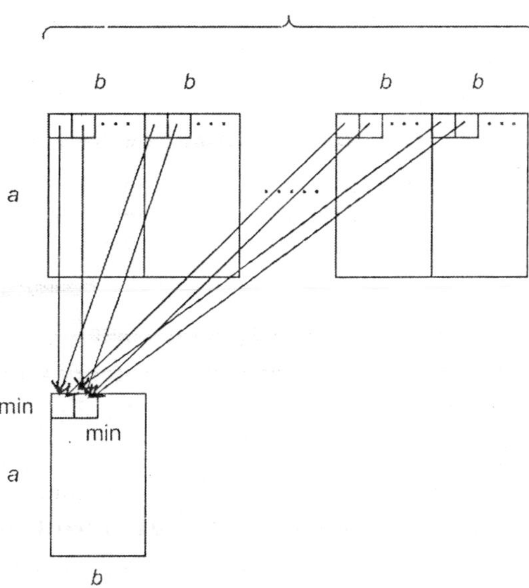

Fig. 8.4. Overlay matrix for tandem repeats.

Proposition : *For all* $1 \le i \le |a|$, $1 \le j \le |b|$,

$$\exists k \in \mathbb{N} : D_{i,j} = D(a_1 \ldots a_i, b^k b_1 \ldots b_i).$$

Proof : This follows directly from the definition of $D_{i,j}$.

Now we want to find a recursive definition for $D_{i,j}$ that can be used for a dynamic programming algorithm similar to Gotoh's algorithm. The main idea is again to simulate the Gotoh's algorithm on k copies of b using a single copy of b, except that we have a special treatment in the case that the dynamic programming algorithm hits the boundary. This is called the wraparound step (see Figure 8.5). The following two lemmas give a recurrence equation for the matrices $(D_{i,j})$, $(Q_{i,j})$ and $(P_{i,j})$ for matrix elements that do not depend on matrix elements beyond a boundary in Lemma 8.16 and for the matrix elements that depend on matrix elements beyond a word boundary in Lemma 8.17.

Lemma : *Let* $(D_{i,j})$, $(Q_{i,j})$, *and* $(P_{i,j})$ *be defined as in (8.10). Then*

$$\forall i > 0, j, > 1: D_{i,j} = \min\begin{Bmatrix} D_{i-1,j-1} + w(a_i, b_j) \\ P_{i,j} \\ Q_{i,j} \end{Bmatrix}, \qquad \ldots(8.11)$$

$$\forall i > 0, j > 0: P_{i,j} = \min\begin{Bmatrix} D_{i-1,j} + g(1) \\ P_{i-1,j} + \beta \end{Bmatrix}, \qquad \ldots(8.12)$$

$$\forall i > 0, j > 1: Q_{i,j} = \min\begin{Bmatrix} D_{i,j-1} + g(1) \\ Q_{i,j-1} + \beta \end{Bmatrix}. \qquad \ldots(8.13)$$

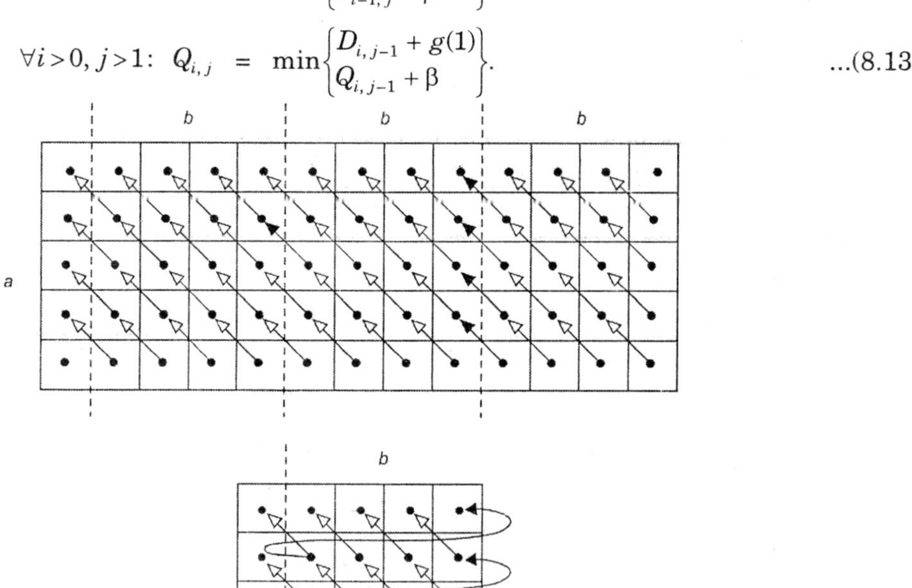

Fig. 8.5. Wraparound step. The first picture shows which matrix elements of $(D_{i,j})$ are evaluated in Gotoh's algoirthm to calculate a specific matrix element D_{kl} for the alignment a with three copies of b. Note that we have ignored the $(P_{i,j})$ and $(Q_{i,j})$ matrices. The second picture shows which matrix elements are evaluated in the wrapround dynamic programming. All arrows in the first picture that cut a boundary between two successive copies of b (indicated in gray) can be thought to be superimposed onto one wraparound step (indicated in gray) in the second picture.

***Proof*:** By Theorem, we get

$$D_{i,j} = \min_{k,l<k}\{D^k_{i,lb+j}\}$$

$$= \min_{k,l<k}\left\{\min\left\{\begin{matrix} D^k_{i-1,\,l|b|+j-1} + w(a_i, b_j) \\ P^k_{i,\,l|b|+j} \\ Q^k_{i,\,l|b|+j} \end{matrix}\right\}\right\}$$

$$= \min\left\{\begin{matrix} \min_{k,l<k}\{D^k_{i-1,l|b|+j-1}\} + w(a_i, b_j) \\ \min_{k,l<k}\{P^k_{i,l|b|+j}\} \\ \min_{k,l<k}\{Q^k_{i,l|b|+j}\} \end{matrix}\right\}$$

$$\underset{=}{(j-1>0)} \left\{\begin{matrix} D_{i-1,\,j-1} + w(a_i,\, b_j) \\ P_{i,\,j} \\ Q_{i,\,j} \end{matrix}\right\}.$$

The proofs for the other matrices are analogous.

Now we have to consider the special case $D_{i,\,1}$ and $Q_{i,\,1}$. For example, we know that in the calculation of $D_{i,\,1}$, the left or diagonal step either marks the beginning of the alingment (*i.e.*, takes into account the cell $D_{i,\,0}$ or $D_{i\,-\,1,\,0}$), or it is the start of a new tandem repeat. In the latter, the costs for the previous alignment are stored in $D_{i,\,|b|}$ or $D_{i\,-\,1,\,|b|}$.

Lemma *: Let* $(D_{i,j})$, $(Q_{i,j})$ *and* $(P_{i,j})$ *be defined as in. Then*

$$\forall i > 0 : D_{i,1} = \min\left\{\begin{matrix} D_{i-1,0} + w(a_i, b_j) & \\ D_{i-1,|b|} + W(a_i, b_j) & \text{(wraparound)} \\ P_{i,1} & \\ Q_{i,1} & \end{matrix}\right\}, \qquad \text{...(8.14)}$$

$$\forall i > 0 : Q_{i,1} = \min\left\{\begin{matrix} D_{i,0} + g(1) & \\ D_{i,|b|} + g(1) & \text{(wraparound)} \\ Q_{i,0} + \beta & \\ Q_{i,|b|} + \beta & \text{(wraparound)} \end{matrix}\right\}. \qquad \text{...(8.15)}$$

Proof by Theorem, we get again

$$D_{i,1} = \min_{k,l<k}\left\{d^k_{i,lb+1}\right\}$$

$$= \min_{k,l<k}\left\{\min\left\{\begin{matrix} D^k_{i-1,l|b|} + w(a_i, b_j) \\ P^k_{il|b|+1} \\ Q^k_{i,l|b|+1} \end{matrix}\right\}\right\}$$

$$= \min\left\{\begin{matrix} \min_{k,l<k}\left\{D^k_{i-1,l|b|}\right\} + w(a_i, b_j) \\ \min_{k,l<k}\left\{P^k_{i,l|b|+1}\right\} \\ \min_{k,l<k}\left\{Q^k_{i,l|b|+1}\right\} \end{matrix}\right\}$$

$$= \min\begin{Bmatrix} \min_{k,l<k}\{D^k_{i-1,l|b|}\} + w(a_i, b_j) \\ p_{i,1} \\ Q_{i,1} \end{Bmatrix}$$

The problem here is that there is no definition for $\min_{k,l<k}\{D^k_{i-1,l|b|}\}$ in. But we can write the following :

$$\min_{k,l<k}\{D^k_{i-1,l|b|}\} + w(a_i,b_j) \quad = \quad \min\begin{Bmatrix} \min_k\{D^k_{i-1,0}\} \\ \min_{k,1\le l<K}\{D^k_{i-1,l|b|}\} \end{Bmatrix} + w(a_i,b_j)$$

$$= \min_k\{D^k_{i-1,0}\}$$

$$= \quad \min\left\{\min_{k,0\le l<k-1}\{D^k_{i-1,l|b|+|b|}\}\right\} + w(a_i,b_j).$$

By Theorem we have $D^k_{i-1,0} = g(i-1) = D_{i-1,0}$, and therefore

$$\min_k\{D^k_{i-1,0}\} = D_{i-1,0}.$$

Further more we know that given $k \le k' \in \mathbb{N}$, then for all $l < k$ and $j \le |b|$ we have

$$D^k_{i-1,l|b|+j} = D^{k'}_{i-1,l|b|+j}.$$

This implies

$$\min_{k,0\le l<k-1}\{D^k_{i-1,l|b|+|b|}\} = \min_{k-1,0\le l<k-1}\{D^{k-1}_{i-1,l|b|+|b|}\} = D_{i-1,|b|}.$$

Hence,

$$\min_{k,l<k}\{D^k_{i-1,l|b|}\} + w(a_i,b_j) = \min\begin{Bmatrix} \min_k\{D^k_{i-1,0}\} \\ D_{i-1,|b|} \end{Bmatrix} + w(a_i,b_j),$$

which proves the claim. The proof for the $(Q_{i,j})$ matrix is analogous.

The last two lemmas give a recursion equation that is satisfied by $((D_{i,j}), (P_{i,j}), (Q_{i,j}))$ defined by. The remaining part is to show that every triple $((D_{i,j}), (P_{i,j}), (Q_{i,j}))$ satisfying the recursion equation also satisfies.

Lemma : *Let a, b be given such that* $|b| > 1$. *Furthermore, let the gap penalty be* $g(k) = \alpha + \beta k$ such that $\alpha \ge 0$ and $\beta > 0$. Then every triple $((D_{i,j}), (P_{i,j}), (Q_{i,j}))$ *of matrices satisfies.*

Proof : Let $(D_{i,j})$, $(P_{i,j})$, and $(Q_{i,j})$ satisfy (3.11) – (3.15). We have to show that for all $j > 0$,

$$M_{i,j} = \min\{M^k_{i,l|b|+j} \mid k,l \in \mathbb{N} \wedge l < k\}$$

for $M \in \{D, P, Q\}$. Thus, we have to show that for all $j > 0$,

$$\forall k,l \in \mathbb{N} : M_{i,j} \le M^k_{i,l|b|+j} \qquad ...(8.16)$$

and

$$\exists k,l \in \mathbb{N} : M_{i,j} \ge M^k_{i,l|b|+j}. \qquad ...(8.17)$$

- For the first direction, assume that does not hold. Let (k^*, l^*, i^*, j^*) with $1 \le j^* \le |b|$ be a minimal quadruple (under the lexicographic ordering) such that (8.16) is not satisfied, *i.e.*, such that

$$M_{i^* j^*} > M^{k^*}_{i^*, l^*|b|+j^*}.$$

for some $M \in \{D, P, Q\}$. Note that $l^* < k^*$.

We derive a contradiction for $M = Q$.The contradiction for the other matrices can be proven analogously. Since the matrices $(D^k_{i,j}), (P^k_{i,j})$, and $(Q^k_{i,j})$ satisfy Gotoh's recursion equations, we get

$$Q^{k^*}_{i^*, l^*|b|+j^*} = \min\begin{Bmatrix} D^k_{i^*, l^*|b|+j^*-1} + g(1) \\ Q^{k^*}_{i^*, l^*|b|+j^*-1} + \beta. \end{Bmatrix}.$$

We have two cases :

1. $j = 1$. If $l = 0$, then

$$D^{k^*}_{i^*, 0\cdot|b|+0} = D_{i^*, 0} \quad \text{and} \quad Q^{k^*}_{i^*, 0\cdot|b|+0} = Q_{i^*, 0}$$

by definition of the matrices. If $l > 1$, then we get again by minimality of (k^*, l^*, i^*, j^*) that

$$D^{k^*}_{i^*, l^*|b|+1-1} \overset{l \geq 1}{=} D^{k^*}_{i^*, (l^*-1)|b|+|b|} \geq D_{i^*}, |b|$$

and
$$Q^{k^*}_{i^*, l^*|b|+1-1} \overset{l \geq 1}{=} Q^{k^*}_{i^*}, (l^*-1)|b| + |b| \geq Q_{i^*, |b|}.$$

Hence,

$$\begin{aligned} Q^{k^*}_{i^*, l^*|b|+j^*} &\geq \min\begin{Bmatrix} D_{i^*,0} + g(1) \\ D_{i^*,|b|} + g(1) \\ Q_{i^*,0} + \beta \\ Q_{i^*,|b|} + \beta \end{Bmatrix} \\ &= Q_{i^*, 1}, \end{aligned} \qquad \text{(by (8.15))}$$

which is a contradiction.

2. $j > 1$. This is similar to the above case.

- For the other direction, assume that is not satisfied. Let (i^*, j^*) be the minimal quadruple such that

$$\forall k \in \mathbb{N}, l < k : M_{i^*, j^*} < M^k_{i^*, l|b|+j^*}. \qquad \text{...(8.18)}$$

If $M = P$, then by the recursion equation (8.12) we get

$$P_{i^*, j^*} = D_{i^*-1, j^*} + g(1) \qquad (\text{resp. } P_{i^*, j^*} = P_{i^*-1, j^*} + \beta).$$

By the minimality of (i^*, j^*), we get that there are k, l such that

$$D_{i^*-1, j^*} \geq D^k_{i^*-1, l|b|+j^*} \qquad (\text{resp. } P_{i^*-1, j^*} \geq P^k_{i^*-1, l|b|+j^*}).$$

Since $(D^k_{i,j})$ (resp. $(P^k_{i,j})$) satisfy Gotoh's recursion equation, this gives us $P_{i^* j^*} \geq P^k_{i^*, l|b|+j^*}$ in both cases, and therefore contradiction.

A similar argument can be applied if $M = D$ or $M = Q$ and $j^* \neq 1$.

The remaining case is $M = D$ or $M = Q$, and $j^* = 1$. The main idea is to apply a similar argument by generating a traceback in the row i^* starting at M_{i^*, j^*}. The cells are connected via the recursion equation. If the traceback went back to the previous row, then the same traceback could be done in some $((D^k_{i,j}), (P^k_{i,j}), (Q^{k)}_{i,j}))$, which would yield a direct contradiction to. If this is not the case, then we will show that the recursion equations are not satisfied.

For the formal definition, let $(M(1), j(1) = (M, j^*)$, and let the tuple

$$(M(r+1), j(r+1))$$

be generated from $(M(r), j(r))$ as follows :

$$(M(r+1), j(r+1)) = \begin{cases} (P, j(r)) & \text{if } M(r) = D \text{ and } D_{i^*, j(r)} = P_{i^*, j(r)} \\ \text{else} \quad (Q, j(r)) & \text{if } M(r) = D \text{ and } D_{i^*, j(r)} = Q_{i^*, j(r)} \\ \text{else} \quad (D, j(r)-1) & \text{if } M(r) = Q \\ & \text{and } Q_{i^*, j(r)} = D_{i^*, j(r)-1+g(1)} \\ \text{else} \quad (Q, j(r)-1) & \text{if } M(r) = Q \\ & \text{and } Q_{i^*, j(r)} = Q_{i^*, j(r)-1} + \beta \\ \text{else} \quad (D, |b|) & \text{if } j(r) = 1, M(r) = Q \\ & \text{and } Q_{i^*,1} = D_{i^*,|b|} + g(1) \\ \text{else} \quad (Q, |b|) & \text{if } j(r) = 1, M(r) = Q \\ & \text{and } Q_{i^*,1} = Q_{i^*,|b|} + \beta \\ \text{else} \quad \textit{undefined.} & \end{cases}$$

We have the following cases :

1. The traceback is finite. Let r be the maximal number such that $(M(r), i(r))$ is defined, and let $M = M(r)$. By definition of the tuples, there must be a matrix M' and a constant $c \in \mathbb{R}$ such that either $j(r) \neq 1$ and

$$M_{i^*, j(r)} = M'_{i^*, j(r)} + c \text{ or } M_{i^*, j(r)} = M'_{1^*-1, j(r)-1} + c,$$

or $j(r) = 1$ and

$$M_{i^*, 1} = M'_{i^*-1, |b|} + c.$$

In any case we know by the minimality of (i^*, j^*) that there are k, l such that

$$M_{i^*-1, j(r)} \geq M^k_{i^*-1, l|b|+j(r)},$$

$$\text{resp.} \quad M_{i^*-1, j(r)-1} \geq M^k_{i^*-1, l|b|+j(r)-1},$$

$$\text{resp.} \quad M_{i^*-1, |b|} \geq M^k_{i^*-1, (l-1)|b|+|b|},$$

By the recursion equation for $((D_{i,j}), (P_{i,j}), (Q_{i,j})$ and for $((D^k_{i,j}), (P^k_{i,j}), (Q^k_{i,j}))$, we immediately get

$$M_{i^*, j(r)} \geq M^{k.}_{i^*, l|b|+j(r)}.$$

Proceeding this way, one get that there are k, l such that $M_{i^*, j^*} \geq M^k_{i^*, l|b|+j^*}$, which is a contradiction.

2. The traceback is infinite. Then the traceback must be cyclic, *i.e.*, there are $r < r'$ such that

$$M(r) = M(r') \quad \text{and} \quad j(r) = j(r').$$

By definition of the traceback and the conditions $|b| > 1$, $\alpha \geq 0$ and $\beta > 0$, there must be a positive value $v \in \mathbb{R}$ with $v > 0$ such that

$$M(r)_{i^*, j(r)} = M(r')_{i^*, j(r')} + v,$$

which is a contradiction to $(M(r), j(r)) = (M(r'), j(r'))$.

The previous does not directly lead to a dynamic programming (DP) algorithm. The problem is with the $(Q_{i,j})$ matrix, where we need to know $Q_{i,|b|}$ before we can define $Q_{i,1}$. Hence, we have to modify the DP algorithm to work into two passes. An application example, comparing the matrix yielded by the standard Needleman-Wunsch algorithm [NW70] applied to two copies of b, and the matrix as generated by the wraparound Dynamic Programming.

In the following, we want to use $D_{i,j}^{(m)}$ to indicate the values of $D_{i,j}$ after the pass m, where $m = 1$ or $m = 2$. We can show that if the last row of the matrices in the first and second passes agree, then we have found the final matrices. If this is not true, then we need additional passes.

Lemma : *If for all* $1 \le i \le |a|$

$$\begin{aligned} \exists 0 < j <| b \mid D_{i,j}^{(1)} &= D_{i,j}^{(2)}, \\ \exists 0 < j <| b \mid P_{i,j}^{(1)} &= P_{i,j}^{(2)}, \\ \exists 0 < j <| b \mid Q_{i,j}^{(1)} &= Q_{i,j}^{(2)}, \end{aligned} \quad \text{...(8.19)}$$

then

$$D_{i,j}^{(m)} = D_{i,j}$$

Algorithm Wraparound Dynamic Programing

Initialization :

{

for $i = 0$ to $|a|$ {

$D_{i,0} = g(i)$

$Q_{i,0} = \infty$

}

for $j = 0$ to $|b|$ {

$D_{0,j} = g(j)$

$P_{0,j} = \infty$

}

}

pass 1 : {

for $i = 1$ to $|a|$

for $j = 1$ to $|b|$ {

$$P_{i,j} = \min \begin{Bmatrix} D_{i-1,j} + g(1) \\ P_{i-1,j} + \beta \end{Bmatrix}$$

$$Q_{i,j} = \min \begin{Bmatrix} D_{i,j-1} + g(1) \\ P_{i,j-1} + \beta \end{Bmatrix}$$

$$D_{i,j} = \min \begin{Bmatrix} D_{i-1,j-1} + w(a_i, b_i) \\ \text{if } j > 1 \text{ then } \infty \text{ else } D_{i-1,|b|} + w(a_i, b_i) \text{ end if} \\ P_{i,j} \\ Q_{i,j} \end{Bmatrix}$$

}

}

Pass 2 : {
 for i = 1 to $|a|$
  for j = 1 to $|b|$ {

$$P_{i,j} = \min\begin{Bmatrix} D_{i-1,j} + g(1) \\ P_{i-1,j} + \beta \end{Bmatrix}$$

$$Q_{i,j} = \min\begin{Bmatrix} D_{i,j-1} + g(1) \\ \text{if } j > 1 \text{ then } \infty \text{ else } D_{i,|b|} + g(1) \text{ end if} \\ P_{i,j-1} + \beta \\ \text{if } j > 1 \text{ then } \infty \text{ else } P_{i,|b|} + \beta \text{ end if} \end{Bmatrix}$$

$$D_{i,j} = \min\begin{Bmatrix} D_{i-1,j-1} + w(a_i, b_j) \\ \text{if } j > 1 \text{ then } \infty \text{ else } D_{i-1,|b|} + w(a_i, b_j) \text{ endif} \\ P_{i,j} \\ Q_{i,j} \end{Bmatrix}$$

  }
}

		A	C	C	G	A	C	C	G
	0.00	1.00	2.00	3.00	4.00	5.00	6.00	7.00	8.00
A	1.00	0.00	1.00	2.00	3.00	4.00	4.00	6.00	6.00
T	2.00	1.00	1.00	2.00	3.00	4.00	5.00	6.00	7.00
C	3.00	2.00	1.00	1.00	2.00	3.00	4.00	5.00	6.00
G	4.00	3.00	2.00	2.00	1.00	2.00	2.00	4.00	5.00
G	5.00	4.00	3.00	3.00	2.00	2.00	3.00	4.00	4.00
G	6.00	5.00	4.00	4.00	3.00	3.00	3.00	4.00	4.00
A	7.00	6.00	5.00	5.00	4.00	3.00	4.00	4.00	5.00
C	8.00	7.00	6.00	5.00	5.00	4.00	3.00	4.00	4.00
T	9.00	8.00	7.00	6.00	6.00	5.00	4.00	4.00	6.00
G	10.00	9.00	8.00	7.00	6.00	6.00	5.00	5.00	5.00
T	11.00	10.00	9.00	8.00	7.00	7.00	6.00	6.00	5.00

		A	C	C	G
	0.00	1.00	2.00	3.00	4.00
A	1.00	0.00	1.00	2.00	3.00
T	2.00	1.00	1.00	2.00	3.00
C	3.00	2.00	1.00	1.00	2.00
G	4.00	2.00	2.00	2.00	1.00
G	5.00	5.00	3.00	3.00	2.00
G	6.00	3.00	3.00	4.00	3.00
A	7.00	3.00	4.00	4.00	4.0
C	8.00	4.00	3.00	4.00	5.00
T	9.00	5.00	4.00	4.00	5.00
G	10.00	5.00	5.00	5.00	4.00
T	11.00	5.00	6.00	6.00	5.00

Alignment: a = ATCGGGACTGT
bb = A–C–CGACCG–

Fig. 8.6. Comparison of Needleman-Wunsch and Wraparound Needleman-Wunsch. The cost is 1 for substitutions, and 1 for indels. The task is to find the minimal distance between a = ATCGGGACTGT and an unknown number of copies of b = ACCG. The optimum is found by two copies.the figure shows the distance matrix for Needleman-Wunsch applied to a and bb, and the distance matrix for Wraparound Needleman-Wunsch applied to a and b. One can see that the traceback of the first matrix is just folded into the second matrix. Note that the number of wraparound steps in the traceback determines the number of copies of b used.

Proof: After pass 2, we know that $(D^{(2)}_{i,j})$ staisfies and holds, then especially

$$D^{(1)}_{i,|b|} = D^{(2)}_{i,|b|},$$
$$P^{(1)}_{i,|b|} = P^{(2)}_{i,|b|},$$
$$Q^{(1)}_{i,|b|} = Q^{(2)}_{i,|b|}.$$

By the definition of the algorithm, this implies

$$Q^{(2)}_{i,1} = \min\begin{Bmatrix} D^{(1)}_{i,0} + g(1) & \\ D^{(1)}_{i,|b|} + g(1) & \text{(wraparound)} \\ D^{(1)}_{i,0} + \beta & \\ D^{(1)}_{i,|b|} + \beta & \text{(wraparound)} \\ D^{(2)}_{i,0} + g(1) & \\ D^{(2)}_{i,|b|} + g(1) & \text{(wraparound)} \\ Q^{(2)}_{i,0} + \beta & \\ Q^{(2)}_{i,|b|} + \beta & \text{(wraparound)} \end{Bmatrix}$$

This gives and, with a similar argument, we also get. The same can be done for the $(D^{(2)}_{i,j})$ matrix. By Lemma, this proves the claim.

The cost for duplications can easily be.incorporated by adding the duplication extension cost every time a wraparound step is performed.

Similarity Methods : Let $a = a_1 \ldots a_n$ and $b = b_1 \ldots b_m$ be words over a finite alphabet Σ. Rather than minimizing the *distance* between a, b, we *maximize* the similarity between a, b. Instead of a cost function $w(x, y)$, we consider a similarity function $s(x, y)$, where generally $s(x, y)$ is positive if $x = y$, and negative if $x \neq y$ (the score for gap $s(x, -)$ and $s(-, y)$ may be different than the socre $s(x, y)$ of a mismatch). *Transition* is the substitution of a purine (adenine or guanine) for a purnine, or a pyrimidine (cytosine or tymine) for a pyrimidine. *Transversion* is the substitution of a purine for a pyrimidine or vice versa. Such information can be incorporated into the similarity function s. The best scoring global alignment between a, b (without special treatment of gap, *i.e.*, without context sensitivity) must have score.

$S(a, b)=$

$$\max\left\{\sum_{i=1}^{L} s(a_i^*, b_i^*) \,\middle|\, \text{over all } a^*, b^*, \text{provided that for no } i \text{ are both } a_i^*\ b_i^* \text{ eqal to '–'}\right\}.$$

Here, $L \leq n + m$, and a^*, b^* are sequences in $\Sigma \cup \{-\}$, where – represents a gap (– in the a sequence represents insertion into the b sequence, while – in the b sequence represents deletion from the a sequence; this is clear if one imagines a sequence of edit operations which produce b from a).

A dynamic algorithm to compute $S(a, b)$ goes as follows. Let

$$S_{i,j} = S(a_1, .., a_i, b_1, .., b_j).$$

Letting δ be the gap penalty, we consider the linear gap function $g(n) = \delta(n)$ Then

$$S_{0,0} = 0,$$
$$S_{0,j} = j\delta,$$
$$S_{i,0} = i\delta,$$
$$S_{i,j} = \max\{S_{i-1,j-1} + s(a_i, b_j), S_{i-1,j} + \delta, S_{i,j-1} + \delta\}.$$

The similarity score computation takes time $O(nm)$, and either defining *pointers* or using *tracebacks* produces the sequence alignment with this socre.

T. Smith and M. Waterman modified the previous algorithm to produce *local alignments.* Define the local similarity measure.

$$H(a, b) = \max\{S(a_k \ldots a_i, b_l \ldots bj) \mid 1 \leq k \leq i \leq n, 1 \leq l \leq m\}.$$

In computing the similarity score for local alignments, it is necessary that the expected local similarity score between random sequences is negative (this condition is guaranteed for instance by the PAM matrices.) Otheriwse, a local alignment could always be extended to a longer local alignment and one could expect a higher score. This condition is considered in detail Durbin et al. To determine $H(a, b)$, compute

$$H_{i, j} = \max \{0, S(a_k \ldots a_i, b_l \ldots b_j) \mid 1 \le k \le i, 1 \le l \le j\}$$

by dynamic programming, where

$$\begin{aligned} H_{0,0} &= 0, \\ H_{0,j} &= 0, \\ H_{i,0} &= 0, \\ H_{i,j} &= \max\{0, H_{i-1,j-1} + s(a_i, b_j), H_{i-1,j} + \delta, H_{i,j-1} + \delta\}. \end{aligned}$$

Then the score of the best socring subsequence of a aligned with subsequence of b is

$$H(a, b) = \max\{H_{i,j} \mid 1 \le i \le n, 1 \le j \le m\}.$$

Multiple Sequence Alignment

The problem of multiple sequence alignment is, given k sequences of length at most n, to determine an optimal alignment of all sequences. Suitably formulated and with k, n arbitrary, this problem is known to be NP-complete. There are many approaches towards multiple sequence alignment, of which we mention only a few.

For the fomal description of multiple sequence alignment, let Σ be the alphabet, and $\Sigma' = \Sigma \cup \{-\}$ be Σ extended by the gap character –. An alignment for k sequences $S_1 \ldots S_k$ is given by a character matrix

$$A = (A_{ij})_{1 \le i \le k, 1 \le j \le K}$$

over the alphabet Σ' with the property that S_i can be obtained from $A_{i1} \ldots A_{iK}$ by removing the gaps. Thus, the matrix A represents the alignment of the k sequences $S_1 \ldots, S_k$ in K columns.

The cost function for multiple sequence alignment is a straightforward extension of the cost function for pairwise sequence alignment. In both cases, the cousts are given to the individual columns of the alignment. Thus, a cost function for aligning sequences $S_1 \ldots S_k$ is a function $w : \Sigma^k \to \mathbb{R}$. The *cost* $D(A)$ of an alignment A with K columns is given by

$$D(A) = \sum_{1 \le j \le K} w(A_{1j}, \ldots, A_{kj}).$$

Different multiple sequence alignment problems can be formulated by sepcifying how the function $w(c_1, \ldots, c_k)$ is composed.

Dynamic Programming : In a straightforward manner, it is possible to extend the dynamic programming techniques from pairwise alignment to the alignment of k sequences. For instance, when trying to algin three sequences, $a_1 \ldots a_n$, $b_1 \ldots b_m$ $c_1 \ldots c_e$, drawn from the finite alphaet Σ, define

$$D_{i,j,k} = \min \begin{Bmatrix} D_{i-1,j-1,k-1} + w(a_i, b_j, c_k) \\ D_{i,j-1,k-1} + w(-, b_j, c_k) \\ D_{i-1,j,k-1} + w(a_i, -, c_k) \\ D_{i-1,j-1,k} + w(a_i, b_j, -) \\ D_{i-1,j,k} + (a_i, -, -) \\ D_{i,j-1,k} + w(-, b_j, -) \\ D_{i,j,k-1} + w(-, -, c_k) \end{Bmatrix}.$$

where $w(x, y, z)$ is the cost of comparing $x, y, z \in (\Sigma \cup \{-\})$. This situation corresponds to a linear gap function, and one could define $w(x, y, z)$ to be the sum of pairwise costs $w(x, y) + w(x, z) + w(y, z)$ or alternatively define a new measure, such as

$$w(x, y, z) = \begin{cases} 0 & \text{if } x = y = z, \\ 1 & \text{if 2 of the 3 symbols are the same,} \\ 2 & \text{if all 3 symbols are distinct.} \end{cases} \quad \text{...(8.20)}$$

The determination of $D_{n, m, e}$ then requires $O(nme) = O(n^3)$ time if $n \geq m, e$. Similary, multiple alignment for k sequences by dynamic programming would take n^k time, where n is the maximum sequence length, when using a linear gap function.

Gibbs Sampler : In an interesting application of the method of Gibbs sampler was developed for local multple sequence alignment. The generic idea concerns alignment without gaps, although a modification of the method allows for gaps.

Suppose that we have m sequnces, of possibly differing lengths. Let w be a fixed window size. After convergence, Algorithm determines a local sequence alignment of all windows. In Algorithm, various window sizes are tried, and the optimal scoring local alignment is output.

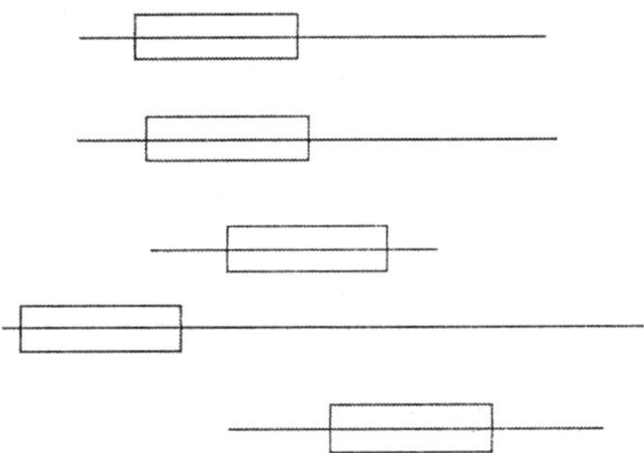

Fig. 8.7. Multiple sequence alignment using Gibbs sampler.

Algorithm, Lawrence, Altschul, Bogurki, Liu, Neuwald, Wootton

1. Identify random size w segments, or windows, from m amino acid sequences.
2. Choose one of the m proteins at random and temporarily call it P, where $r_1, .., r_n$ is its amino acid sequence.
3. Define a $20 \times w$ frequency matrix Q where

$$Q_{r,j} = \frac{N_{r,j}}{m-1}$$

and $N_{r,j}$ is the number of occurrences of resideue r at position j among the $m - 1$ remaining proteins. (To avoid a probabilitty of 0 in the case than $N_{r,j} = 0$, *pseudocounts* are added – see for details).

4. For $i = 1$ to $|P| - w + 1$ compute

$$p_i = \prod_{j=1}^{w} Q_{r_{i+j},\, j},$$

which is the probability that the subsequence of P of size w beginning at position i is generated by Q.

5. Choose the starting position of the window in P randomly according to probability p_i. In other words, define cumulative probabilities q_i by

$$q_0 = 0,$$
$$q_1 = p_1,$$
$$q_i = q_{i-1} + p_i = \sum_{j \le i} p_j$$

for $1 \le i \le n - w + 1$. Generate random $z \in (0, 1)$, determine i_0, the smallest $1 \le i \le n - w$ for which $z \le q_i$, and set the new starting position for the window in P at position i_0.

6. If convergence has occurred, then stop, else return to Step 2.

Maximum Weight Trace

Kececioglu introduced a special formulization of multiple sequence alignment, the *complete maximum-weight trace* (*CMWT*) formalization. The CMWT formalization handles a subclass of multiple sequence alignment problems, which can roughly be described as multiple sequence alignment problems that merges pairwise sequence alignments. The most prominent instance of this subclass is the sum-of-pirs multiple sequence alignment problem, which was introduced by Carrillo and Lipman. We will first define the sum-of-pairs multiple sequence alignment problem. Later, we define the CMWT.

In the *sum-of-pairs* multiple sequence alignment problem with linear gap penalties, an alignment for n sequences $S_1, .., S_n$ is given as usual by a character matrix

$$A = (A_{i,j})_{1 \le i \le n,\, 1 \le j \le K}$$

over the alphabet Σ' with the property that S_i can be obtained from $A_{i1} .. A_{iK}$ by removing the gaps. The cost function is composed of pairwise cost function $w_p : \Sigma' \times \Sigma' \rightarrow \mathbb{R}$. The *cost* $D(A)$ of an alignment A is given by

$$D(A) = \sum_{i<i'} \sum_{1 \le j \le K} w_p(A_{ij}, A_{i'j}).$$

Of course, this is equivalent to $D(A) = \Sigma_{1 \le j \le K} \Sigma_{i < i'} w_p(A_{i,j}, A_{i'j})$, and is hence a special case of the general multiple sequence alignment problem, where the cost for a column is given by $w(a_1 .., a_n) = \Sigma_{i<i'} w_p(A_{ij}, A_{i'j})$.

The sum-of-pairs problem (with linear gap penalties) is one of the motivations for the CMWT problem. In this problem, the letters of the strings $S_i = s_{i1}, ...s_{in}$, are considered to be the set of vertices $V = V_1 \uplus ... \uplus V_n{}^5$ of a complete n-partite graph $G = (V, E)$ (*i.e.*, G satisfies that for every $v \in V_i$ and $v' \in V_j$, we have $e = (v, v') \in E$ if and only if $i \neq j$). G is called the *complete alignment graph* for the sequences $S_1, ..., S_n$. It represents all possible multiple sequence alignments for the sequences $S_1, ..., S_n$. An *alignment graph* G' is a subgraph of the complete alignment graph. Alignment graphs can be used to restrict the search for a multiple sequence alignment to a subset of all possible alignments. For example, let S_1 be AACG and S_2 be AGG. Then the complete alignment graph for AACG and AGG is the 2-partite graph.

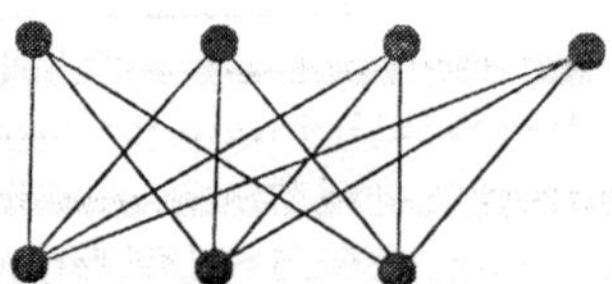

With every edge $e \in E$, there is a positive weight $w(e)$ associated. an alignment A for the sequences $S_1, .., S_n$ *realizes* an edge $e = (s_{ij}, s_{i'} j') \in E$ of an alignment graph $G(V, E)$ for the seqeucnes $S_1, .., S_n$ if the jth character of S_i and the j'th character of S_i' are aligned in A. For example, consider the alignment

A A C G
A – G G .

Then this alignment realizes three edges, indicated by straight lines :

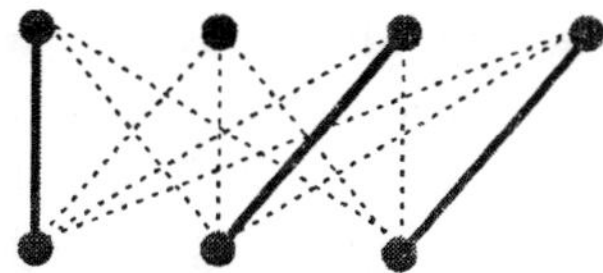

Given an alignment A, the set of all eges realized by A is called the *trace* of A. A set $T \subset E$ of edges is called a *trace* if it is the trace of some alignment A. Given the weight function w, the *weight* of a trace T is $\Sigma_{e \in T} w(e)$.

Problem (Complete) Maximum-Weight Trace : Let $S_1, ..S_n$ be sequnces, let $G = (V, E)$ be the complete alignment graph for $S_1, .., S_n$, and let w be a weight function.The complete maximum-weight trace problem is to find a trace $T \subseteq E$ that has maximal weight (undor w). The *maximum-weight trace problem* is defined analogously for an alignment graph $G = (V, E)$ for $S_1, ..., S_n$.

A remaining problem is that not any subset of edges is a trace (*i.e.*, not every subset of E corresponds to a real alinment). Consider again the two sequences AACG and AGG, and consider the following subset of edges indicated by straight lines:

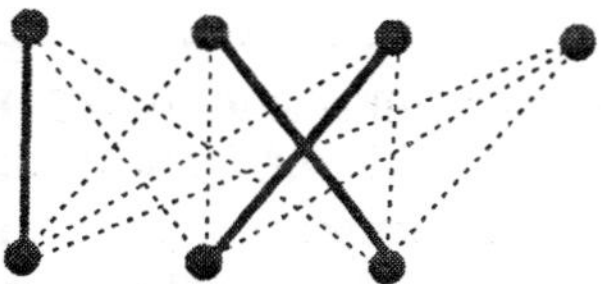

The problem are the two crossing edges indicated in gray above.

In order to characterize set of edges realized by alignments, we must define extended alignment graphs. We need some definitions.A binary relation $\leq$ is a *partial order* if it is (1) *reflexive* (*i.e.*, $a \leq a$), (2) *antisymmetric* (*i.e.*, $a \leq b$ and $b \leq a$ implies $a = b$), and (3) *transitive* (*i.e.*, $a \leq b$ and $b \leq c$ implies $a \leq c$). A binary relation $<$ is a *strict partial order* if it is (1) *irreflexive* (*i.e.*, $a \not< a$), and (2) *transitive* (*i.e.*, $a < b$ and $b < c$ implies $a < c$).

Given sequences $S_1, .., S_1$ with $S_i = s_{i1} ... s_{in_i}$, one defines the *extended alignment graph* $G = (V, E, \prec)$ for $S_i, ..., S_n$ to be a triple such that (V, E) is an alignment graph for $S_1, .., S_n$, and $\prec$ is defined by

$$\prec \; = \; \{(s_{ij}, s_{ij+1}) \mid 1 \leq i \leq n \wedge 1 \leq j < n_i\}.$$

With $\prec^*$, we denote the transitive colour of $\prec$, *i.e.*,

$$\prec^* \; = \; \{(s_{ij}, s_{ij'}) \mid 1 \leq i \leq n \wedge 1 \leq j < j' \leq n_i\}.$$

Note that $\prec^*$ is a strict partial order of V.

Using the extended alignment graph, one can characterize traces. A *connected component* of a graph $G = (V, E)$ is a $\subseteq$-maximal set $V' \subseteq V$ such that for all vetices $v, v' \in V'$ there is a path of edges in E connecting v and v'. For any two subsets $X, Y \subseteq V$, we define

$$X \triangleleft Y \text{ if and only if } \exists v \in X \, \exists v' \in Y : v \prec v'.$$

We define $\triangleleft^*$ to be the transitive closure of $\triangleleft$. Hence, $X \triangleleft^* Y$ if and only if $\exists v \in v \in X \, \exists v'$

$\in Y : v \prec^* v'$. For the sequences AACG and AGG the extended complete alignment graph is where we have indicated the edges for $\prec$ by arrows.

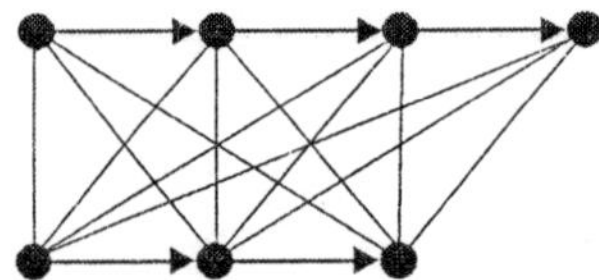

Where we have indicated the edges for $\prec$ by arrows.

Theorem (Kececioglu) : *Let* $S_1, ..., S_n$ *be sequences, and let* $G = (V, E, \prec)$ *be the extended alignment graph for* $S_1, ..., S_n$*. Then a subset* $T \subseteq E$ *is a trace if and only if* $\lhd^*$ *is a strict partial order on the connected components of* $G' = (V, T)$.

Proof : For the first direction, let T be the trace of an alignment A with K columns. By the definition of an alignment, we know that $(s_{ij}, s_{kl}) \in T$ if an only the jth position of S_i and the lth position of S_k are aligned in one column. Hence, any connected component of $G' = (V, T)$ consists of the vertices that are aligned in one column, *i.e.*, $V = X_1 \uplus .. \uplus X_K$, where X_r is the connected component of $G' = (V, T)$ corresponding to the rth column of the alignment A. We have to show that $X_r \lhd^* X_s$ implies $r < s$, since then $\lhd^*$ must be a strict partial order on $X_1, ..., X_K$. If $X_r \lhd^* X_s$, then there is a s_{ij} in X_r such that there is a $j' > j$ with $s_{ij'}$ in X_s, which implies by the definition of an alignment that column s must appear after column r.

For the second direction, let T be a subset of E such that $\lhd^*$ is a strict partial order on the connected components of $G' = (V, T)$. Let $\lhd^*_{tot}$ be any strict total order of the connected components of $G' = (V, T)$ extending $\lhd^*$ (which must exist since $\lhd^*$ is a strict partial order). Let $X_1, ..., X_K$ be an enumeration of the connected components of $G' = (V, T)$ such that $r < s$ implies $X_r \lhd^*_{tot} X_s$. Then we define the alignment A of $S_1, ..., S_n$ in K rows by

$$A_{ij} = \begin{cases} s_{il} & \text{if } s_{il} \in X_j, \\ - & \text{otherwise.} \end{cases}$$

We have to show that A is an alignment. First, we have to show that A is properly defined; *i.e.*, we have to show that $s_{ij} \in X_j$ implies that there is no other $s_{il'}$ such that $s_{il'} \in X_j$. Suppose that there were $s_{il} \in X_j$ implies that there is no other $s_{il'}$ such that $s_{il'} \in X_j$. Suppose that there were $s_{il} \in X_j$ and $s_{il'} \in X_j$ with $l \neq l'$. Without loss of generality, we can assume that $l < l'$, which implies that $s_{il} \prec^* s_{il'}$ and therefore $X_j \lhd^* X_j$, which is a contradiction to the assumption that $\lhd^*$ is a strict partial order. Second, we have to show that we achieve the string S_i if we ignoring the gap symbol – in the string $A_{i1} ... A_{iK}$. Let w be the string that results from ignoring the gap symbol – in $A_{i1} ... A_{iK}$. Since every s_{ij} must occur in one of the X_r, we know the word w is composed of some permutation of the symbos in S_i. Hence, we have to show that the permutation is the identity, *i.e.*, we have to show that the symbols occur in correct order.

So assume that there is a s_{il} such that there is some s_{ik} with $l < k$ that appears earlier in the string $A_{i1} ... A_{iK}$ than s_{il}. By the definition of A, we know that there are X_r and X_s such that $s_{ik} \in X_r$, $s_{il} \in X_s$, and $r < s$. This implies $X_r \lhd^* X_s$. On the other hand, $l < k$ implies $s_{il} \prec^* s_{ik}$ and therefore $X_s \lhd^* X_r$. But $X_s \lhd^* X_r$ and $X_r \lhd^* X_s$ is contradictory to the assumption that $\lhd^*$ is a strict partial order.

In the complete maximum weight trace problem was generalized and coded in the framework of integer linear programming.

Hidden Markov Models : In a later chapter, we will describe hidden Markov models (HMM), a widely used machine learning technique, based on solid mathematical foundations of maximum likelihood and expectation maximization. In HMMs have been described for multiple sequence alignment; moreover HMM multiple sequence alignment software has been independently developed by various groups, including *P* Baldi, Y. Chauvin and V. Mittal-Henkle S. Eddy and R. Hughey, K. Karplus *et al*.

The underlying idea is that by viewing the sequences to be aligned as a *training set* of observations, which deviate from an ancestor sequence by a stochastic process, a stochastic model is computed, which is most likely to have generated the sequences in the training set. This approach leads to an algorithm for multiple sequence alignment that scales linearly in the number of sequences, rather than exponentially; however, determining the alignment with *global maximum* likelihood, rather than *local maximum*, can require exponential time.

Steiner Sequences : In a problem somewhat related to multiple sequence alignment is considered. In the laboratory technique of *single-molecule DNA sequencing*, single-stranded DNA is cut, a single base at a time. This base then flows down a microscopic tube at high speed past an optical sensor, which detects the type of base. This technique is error-prone, due to sputtering, especially at the beginning and end of the DNA molecule being cut. By repeating the process many times, we produce a collection of N erroneous copies of an original DNA sequence $s_1, ..., s_n$, which is to be determined. A heuristic to determine $s_1, ..., s_n$ is given in Algorithm.

The algorithm's runtime is of order

$$\frac{N}{3}n^3 + \frac{N}{3^2}n^3 + \frac{N}{3^3}n^3 + \ldots + \frac{N}{3^{\log_3 N}}n^3 \leq \frac{3}{2}Nn^3 = \Omega(Nn^3).$$

Algorithm 3.3 can be appropriately modified to produce a general multiple sequence alignment, by constructing a tree with branching factor 3 and depth $\log_3 N$, where the N sequences to be aligned are at the leaves, and intermediate nodes are Steiner sequences of the children. From the tree, it is clear how to produce the multiple sequence alignment. The resulting multiple sequence alignment could, however, have the following undesired property. Suppose that $s_1, ..., s_N$ are the N sequences to be aligned, and that s_i and s_j are identical (or almost identical), where $1 \leq i \ll J \leq N$. Then it could happen, because of grouping together of every 3 adjacent sequences in producing the intermediate Steiner sequences, that s_i and s_j are misaligned in the resulting multiple sequence alignment. A heuristic remedy might be to first apply pairwise sequence alignment in order to sort the sequences, thus producing $s_{\sigma(1)}, ..., s_{\sigma(N)}$ for some permutation $\sigma \in S_N$, with the property that adjacent sequences in the sorted order are similar to each other.

How well does the heuristic perform in determining the original sequence $s_1, ..., s_n$ from

Algorithm Kececioglu, Li, Tromp

1. Give N erroneous copies of an original DNA sequence to be determined, let $M = N$.
2. Partition the M sequences into $M/3$ groups of three sequences.
3. Apply dynamic programming with cost function $w(x, y, z)$ given in to determine an optimal alignment of the three sequences U, V, W in each group. From this alignment, define the *consensus sequence* S obtained by taking the majority symbol in each column and then removing all occurrences of '–' (*i.e.*, it can happen that '–' is the majority symbol

of a column, in which case this is removed from S). This consensus sequence S is a *steiner sequence* for U, V, W, meaning that $S \in \{A, C, G, T\}^*$ satisfies.

$$D(S, U) + D(S, V) + D(S, W) = \min\{D(T, U) + D(T, V) + D(T, W) : T \in \{A, C, G, T\}^*\}.$$

4. The previous step yields $M/3$ Steiner sequences. If $M = 3$, then stop and output the resulting sequence. Else let $M = M/3$, and return to Step 2.

erroneous copies produced by single-molecule DNA sequencing, and how many erroneous copies are required to determine $s_1, ..., s_n$ with high probability? Kececioglu, Li and Tromp answer these questions, subject to the hypotheses that

1. the original DNA sequence $s_1, ..., s_n$ is a random sequence (technically *Kolmogorov random*, or *incompressible*, essentially related to maximal entropy). and
2. insertion, deletion and substitution are equally likely.

The former assumption is biologically unrealistic, especially for DNA sequences of interest (genes, promoter sequences, etc.), while the latter assumption seems reasonable, given the physical device used for single-molecule DNA sequencing. Under these assumptions, it is shows in that Step 3 of Algorithm reduces an original error rate of $\in$ to almost $\in^2$.

It suffices to reduce the error rate to less than $1/n$, since the expected number of errors in a sequence of length n will be less than $n \cdot 1/n = 1$, and the original sequence will have been determined. To this end let c be such that $\in = 1/c$, and note that the least k such that $E^{2^k} = \frac{1}{c^{2^k}} < 1/n$ is $[\log_2 \log_2 n - \log_2 \log_2 c] = O(\log_2 \log_2 n)$. Assuming that the resulting $M/3$ Steiner sequences in each pass of the algorithm are random (this has no analytic justification), Kececioglu, Li, and Tromp then argue that it suffices to take $N = 3^k = N = 3^k = 3^{\log_2 \log_2 n} = (\log_2 n)^{\log_2 3} \approx \log_2^{1.5849} n$, which is a small number of erroneous copies.

Genomic Rearrangements

Using dynamic programming methods for sequence alignment, we have shown how genetic distance between two species can be measured using sequence alignment (global or local, using distance or similarity methods) of two homologous DNA or amino acid sequences. *Chromosomal rearrangement* events are much rarer than pointwise mutations, and hence can possibly be used to determine similarity between vastly different organisms.

1. *Intra*-chromosomal events: *inversion* of a contiguous segment of genes (from $5' \rightarrow 3'$ to $3' \rightarrow 5'$, or vice versa); *duplication*, possibly caused by transposons, leading to pseudogene families such as ALU; *transposition*, where a portion of the chromosome is broken and placed elsewhere in the same chromosome.
2. *Inter*-chromosomal events: *reciprocal translocation*, where the end segments (not including the centromere) of two chromosomes break off and exchange positions (this can happen during crossover); *chromosomal duplication*, where the number of chromosomes in the genome is doubled this happened in the evolution of a wild grass into wheat); *fission*, where one chromosome is broken into two; *fusion*, where two chromosomes fuse into one.

An even higher level of abstraction was considered by Ferretti, Nadeau, and Sankoff in picturing a chromosome as an *unordered set* of genes. Define a *synteny* as a partition of the genome into distinct sets of genes, or chromosomes. Two genes are said to be syntenic if they both lie on the same chromosome. From physical genome maps, current knowledge about the placement of genes on chromosomes is far greater than more detailed knowledge about the order of genes, their orientation, etc.; hence the algorithmic study of synteny seems particularly appropriate.

If we consider that synteny sets (chromosomes, as unordered sets of genes) can be transformed only by the operations of fission, fusion, and reciprocal translocation, then the following questions immediately come to mind.

- Given synteny sets of current organisms, can one construct likely synteny sets of ancestral organisms?
- How many chromosomes did ancestral species have?
- Which genes were on which chromosomes of ancestral species?
- Do phylogenetic trees constructed from comparing synteny sets of different species resemble those constructed from sequence alignment data?

In order to answer such questions, we must compute the *syntenic distance* between two species, *i.e.*, the minimal number of fission, fusion and reciprocal translocation events necessary to transform one synteny set into another.

For example, suppose the genome $\mathcal{G}$ consists of genes a, b, c, d, e, f, g, h and is partitioned into two chromosomes A, B, where $A = \{a, b, c, d\}$ and $B = \{e, f, g, h\}$. Suppose the genome $\mathcal{H}$ consists of the same genes a, b, c, d, e, f, g, h and is partitioned into three chromosomes C, D, E where $C = \{a, b\}$, $D = \{c, e, f\}$, and $E = \{d, g, h\}$. Then a fission of A into $\{a, b\}$ and $\{c, d\}$, followed by a translocation of $\{c, d\}$ and $\{e, f, g, h\}$ into $\{c, e, f\}$, and $\{d, g, h\}$ transforms genome $\mathcal{G}$ into genome $\mathcal{H}$.

As another example, suppose the genome $\mathcal{G}$ consists of genes a, b, c, p, q, r, x, y and is partitioned into three chromosomes C_1, C_2, C_3, where $C_1 = \{x, y\}$, $C_2 = \{p, q, r\}$, and $C_3 = \{a, b, c\}$. Suppose the genome $\mathcal{H}$ consists of (different) genes a, b, p, q, r, x, y, z and is partitioned into two chromosomes D_1, D_2, where $D_1 = \{p, q, x\}$ and $D_2 = \{a, b, r, y, z\}$. The common genes between genomes $\mathcal{G}$ and $\mathcal{H}$ are a, b, p, q, r, x, y, and the syntenic distance between $\mathcal{G}$ and $\mathcal{H}$ is 2, since a translocation of $\{p, q, x\}$, $\{a, b, r, y\}$ produces $\{x, y\}$, $\{a, b, p, q, r\}$, and a fission of $\{a, b, p, q, r\}$ produces $\{p, q, r\}$, $\{a, b\}$ thus transforming the genome $\mathcal{H}$ into $\mathcal{G}$. Note that we only consider those genes common to both genomes.

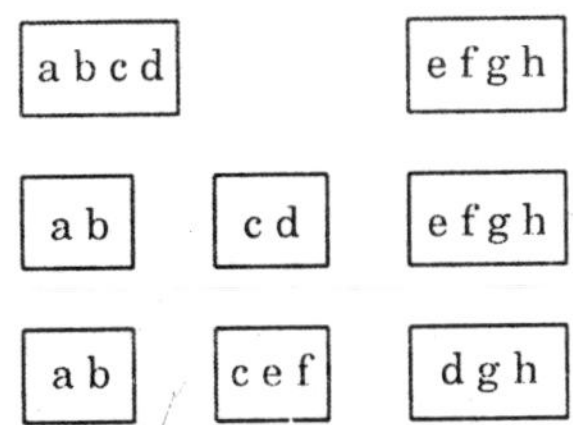

Fig. 8.8. Fission and translocation of synteny sets.

It is known that the problem of determining syntenic distance is NP-complete. Certain other measures of chromosomal rearrangement distance have been shown to the NP-complete, where, however, a variant of the problem can sometimes be shown to be polynomial-time-computable. In some cases, using heuristics, branch-and-bound, or other means, efficient algorithms for approximate distance have been developed.

Locating Cryptogenes and Guide RNA

A recent application of sequence alignment methods was given by A. von Haeseler *et al.* DNA is transcribed into pre-edited mRNA, whereupon the intron transcriptions are removed. In some instances, before producing the final mRNA, the pre-edited mRNA is *edited* – pointwise, certain nucleotides are added or removed. The original genes, called *cryptogenes*, may thus be significantly different than the cDNA obtained by retranscribing the edited mRNA back into DNA. One editing operation can shift the entire reading frame, so it is truly amazing that nature has evolved such an intricate mechanism to produce proteins.

There are many documented cases of editing found in plants, fungi, and animals, but for mitochondrial mRNA in the case of kinetoplastid protozoa (such as trypanosomes), *guide* RNA(gRNA) has been found that forms, over a small region, a (reverse complementary) template for the desired nucleotide sequence. The gRNA forms a short *anchor region* with the pre-edited mRNA downstream of the edited region (*i.e.*, on the 3′ end of the pre-edited mRNA) and forms a perfecte hybrid (allowing G, U base pairs) with the mRNA. the anchor region is terminated by a mismatch, which starts the *guiding region* of the guide RNA. This region guides the editing of the cryptogene.

There is low sequence similarity of the guiding region with the corresponding region of the pre-edited mRNA of the cryptogene, but a perfect match (allowing G, U base pairs) with the edited mRNA. More complicated still, there are cases of *chaining* of guide RNAs in the sense that a first gRNA anchors itself on the pre-edited RNA, permits the editing over a small region, and the newly edited portion is reverse complementary to a second gRNA responsible for further changes. This permits a linear application of desired editing. In algorithms derived from Smith-Watermam local sequence alignment are given for finding potential gRNA genes from a given cryptogene, and for finding potential cryptogenes from similar proteins. Since Smith-Waterman alone does not place real gRNA genes among the highest scoring sequences found, certain rules concerning the formation of an *anchor* region between gRNA and pre-edited mRNA are given.

Later a Markov chain model is developed, for which the expected length of gRNA is computed. This computation agrees fairly well with the data. The following algorithm accounts

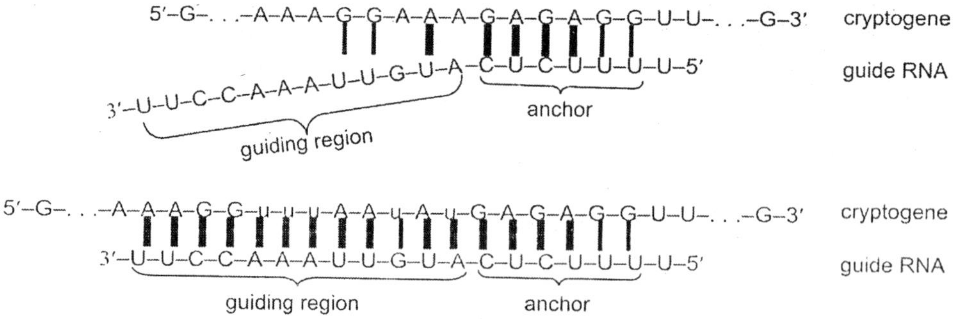

Fig. 8.9. Cryptogene and guide RNA before and after editing. The inserted uracils are indicated by lower-case us. Usual Watson–Crick base pairs are indicated by thick lines, G–U bonds by thin lines.

for pointwise uracil additions, as documented in the trypanosome data. For the application to gRNA, we match the cryptogene with the gRNAk gene. For simplicity, we consider the matching of the RNA transcripts of both the gRNA gene and the cryptogene (instead of the DNA genes themselves; of course, one has to use the DNA version for a database search). Recall that in

RNA,. there are no steric interferences between guanine and uracil, and so we consider the weaker G–U bond in addition to the Watson–Crick pairs A–U with 2 hydrogen bonds and C–G with 3 hydorgen bonds.

Consider the pre-edited cryptogene transcript as the a sequence, and the gRNA as the b sequence. Thus, the pairs $(-, A)$ and $(-, G)$ corresponds to insertions in the cryptogene a. Since U base-paris with both A and G, this correspond to the insertion of U as described above. This suggests the similarity function.

$$\sigma(x, y) = \begin{cases} c_1 & \text{if } (x, y) \in \{(A, U), (U, A)\}, \\ c_2 & \text{if } (x, y) \in \{(C, G), (G, C)\}, \\ c_3 & \text{if} (x, y) \in \{(G, U), (U, G)\}, \\ c_4 & \text{if} (x, y) \in \{(-, A), (-, G)\}, \\ -\infty & \text{if} (x, y) \in \{(-, U), (-, C), (A, -), (C, -), (G, -)\}, \\ c_5 & \text{if } (x, y) \in \{(U, -)\}, \\ -\infty & \text{otherwise.} \end{cases}$$

Note that c_4 corresponds to the editing instance of adding uracil to the pre-edited mRNA, while c_5 corresponds to the editing instance of deleting uracil from the pre-edited mRNA. If one models no uracils being deleted, then one sets $c_5 = -\infty$. This model *disallows* bulges and interior loops, since we do not account for any pairing of bases that do not form a Watson–Crick or G–U bond. Thus, the last line in definition of σ is $-\infty$.

For gRNA search in the organism *L. tarentolae* (4 known cryptogenes at the time of the similarity function was defined by

$$\sigma(x, y) = \begin{cases} 500 & \text{if } (x, y) \in \{(A, U), (U, A)\}, \\ 1000 & \text{if } (x, y) \in \{(C, G), (G, C)\}, \\ 1 & \text{if } (x, y) \in \{(G, U), (U, G)\}, \\ 0 & \text{if } (x, y) \in \{(-, A), (-, G)\}, \\ -\infty & \text{if } (x, y) \in \{(-, U), (-, C), (A, -), (C, -), (G, -)\}, \\ -\infty & \text{if } (x, y) \in \{(U, -)\}, \\ -\infty & \text{otherwise.} \end{cases}$$

The algorithm of [vHBS+92] is reported to be robust, in that setting $c_1 = 1$, $c_2 = 3$, $c_3 = 1$ does not change the results of search.

Anchor and Periodicity Rules. **:** Since known gRNAs did not have the highest scores for the algorithm, the following rules were introduced in. There rules concern formation of an *anchor region* (where gRNA binds to pre-edited mRNA) and the preriodicity of uracil inserts. Only insertion of uracil was considered.

Rules

1. There are at least 5 base pairs at the 3′ end of cryptogene region, including G, U pairs. This is the *ancor* rgion.
2. There can be no more than 3 G, U base pairs in the anchor.
3. There must be at least 4 contiguous watson–Crick base pairs in the anchor.
4. No more than 8 contiguous Us may be inserted.

5. There are no more than 3 base pairs between successive U inserts.
6. There must be at least 1 base pair at the 5′ end of the cryptogene (end of editing).

The rule applications were incorporated into the dynamic programming algorithm, providing better (but still not optimal) results.

Search for Cryptogenes : Until now, knowledge of the cryptogene has been assumed, and we have searched for gRNA genes. One can search for cryptogenes by finding similarity between unknown cryptogenes and homologous proteins $y_1 \ldots y_m$ (y_i s are amino acids) of closely related organisms. Here, similarity between two amino acids is given by the Dayhoff PAM 250 matrix, δ_N is the cost for a nucleotide gap, while δ_A is the cost for an amino acid gap. Again, we use the RNA transcript of the cryptogene for simplicity. Since we compare a nucleotide sequence $a_1 \ldots a_n$ with an amino acid sequence $y_1 \ldots y_m$, we have always to compare a nucleotide codon with an amino acid in the matching case. Note that this implies that we explicity model the reading frame (and the shifting of the reading frame caused by insertion of U_s). Hence, define

$$H_{i,j} = \max \left\{ \begin{array}{lll} 0 & & \\ H_{i-1,j} & - & \delta_N \\ H_{i,j-1} & - & \delta_A \\ H_{i-3,j-1} & - & s(a_{i-2}a_{i-1}a_i, y_i) \\ H_{i-2,j-1} & - & s(ua_{i-1}a_i, y_j) \\ H_{i-2,j-1} & & s(a_{i-1}ua_i, y_i) \\ H_{i-2,j-1} & - & s(a_{i-1}a_i u, y_j) \\ H_{i-1,j-1} & - & s(ua_i u, y_j) \\ H_{i-1,j-1} & - & s(a_i uu, y_j) \\ H_{i,j-1} & - & s(uuu, y_j) \end{array} \right\}$$

Again, we use a lower-case u to indicated insertion of U_s in the transcript of the cryptogene.

Expected Length of gRNA in Trypanosomes

The problem with the previous approaches is that they do not rank known guide RNAs among the highest scoring ones. For this reason, developed a Markov chain to model guide RNAs that respect the rules (with the exception of the rule concerning 8 contiguous uracils) in order to estimate the expected maximum length of gRNA from given parameters. It would be interesting to write a program to output the expected length without going through the following statistical analysis. The general concept is as followed. Since a sequence is a guide RNA only if there is a corresponding cryptogene, one generates the guide RNA together with the cryptogene.

The simplest way to describe the construction of guide RNAs is to start with the anchor and proceed into the guiding region. Since we want to use the standard left-to-right direction for strings, this implies that we write down alignments of cryptogene and guide RNAs in the reverse direction where we have also specified the editing events).

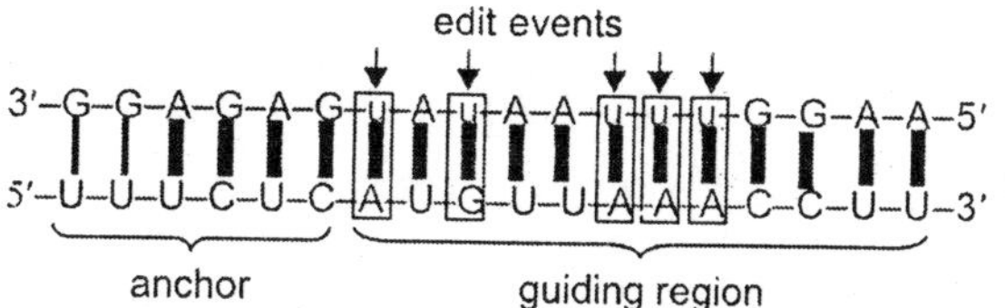

Fig. 8.10. Reverse alignment (together with edit events).

Recall that $a = a_1 \dots a_n$ is the transcript of the cryptogene, and $b = b_1 \dots b_m$ the guide RNA. Sequences are modeled as being generated from independent, identically distributed nucleotides. Set

$$p_a = P_r[a_i = a]$$

and

$$q_a = P_r[b_i = a]$$

for $a \in \{A, C, G, U\}$ (since we use the RNA transcripts again). Let

$$p_1 = p_A q_U + p_U q_A + p_C q_G + p_G q_C$$

be the probability of a Watson–Crick base pair, and

$$p_2 = p_G q_U + p_U q_G$$

be the probability yielding a G, U base pair. Then the probability of a non-pairing is

$$Pr[\text{mispair}] = 1 - p_1 - p_2.$$

Let L_a be the length of the anchor region, and L_e the length of the edited region. The length of the guide RNA is clearly $L_a + L_e$. Elementary probability yields an estimate for maximum length of L_a, while the method of Perron-Frobenius yields an approximation for the expected maximum length of L_e.

The anchor must begin after a mismatch, then consist of a region with base pairs, allowing at most three G, U base pairs, and requiring at least 4 contiguous Watson–Crick base pairs. Here is an example of a mismatch followed by an anchor :

$$XWWWNWWWNWWWNWWWW,$$

where X denotes mismatch (before start of anchor), W denotes Watson–Crick base pair, N denotes G, U base pair. Note there are 3 G, U base pairs in the anchor (the maximum allowed), and that the anchor length is 16. If the length of the guide RNA is shorter than 16, then not all possible strings containing less than 3 Ns have at least 4 contiguous Ws (Watson–Crick base pairs). For example, the string of $XWWNWWNWWNWW$ of length 11 does not satisfy this condition, whereas the string $XWWNNWWWWNWW$ of the same length has 4 contiguous Ws. If, on the other hand, the anchor has length at least 16, then we know that the condition 'less than 3 Ns' implies that there are at 4 contiguous Ws. This implies that for anchors of length at least 16,

$$Pr[L_a \geq l] \geq (1 - p_1 - p_2) \sum_{k=0}^{3} \binom{e}{k} p_1^{l-k} p_2^{k}.$$

From the binomial theorem, we have the approximation

$$(p_1 + p_2)^l = \sum_{k=0}^{l} \binom{e}{k} p_1^{l-k} p_2^{k} \geq \sum_{k=0}^{3} \binom{l}{k} p_1^{l-k} p_2^{k},$$

so that

$$Pr[L_a \geq l] \approx (1 - p_1 - p_2)(p_1 + p_2)^l;$$

hence

$$Pr[L_a = l] \approx (1 - p_1 - p_2)^2 (p_1 + p_2)^l$$

where the extra $(1 - p_1 - p_2)$ term comes from having a mismatch at the end. In this case, the mismatch signals the start of editing.

We continue with the description of the guiding region of the gRNA, which starts with the mismatch that ends the anchor region. note that, by definition, a RNA sequence is a gRNA if there is a corresponding cryptogene whose edited mRNA aligns to the gRNA. Hence, we have to generate (at random) alignments of gRNA to cryptogenes.

We have to consider the edit event in more detail first. An edit event has two basic properties: (1) editing can take place only if the gRNA has G or A (since these can base-pair with U), and (2) there is a mismatch if editing is not performed. Hence, the possible edit events are

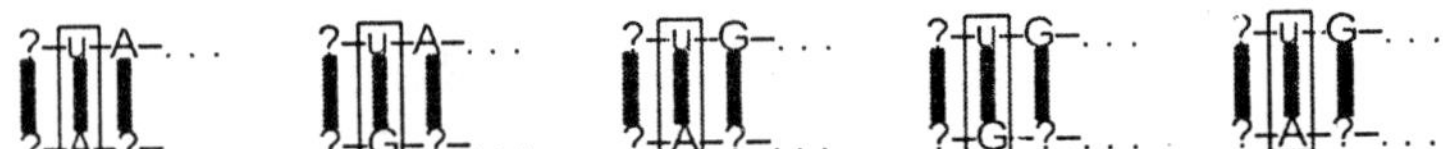

these events corresponds to the following mismatches with the pre-edited mRNA of the cryptogene :

respectively. Thus, when generating at random an edit event, we have to take into account that there is no base-pairing with the next nucleotide in the cryptogene. For this reason, we add the next nucleotide that has to appear in the (pre-edited) cryptogene to the edit event. The states

characterize edit events, where the following nucleotides in the cryptogene are A, A, G, G, C, respectively. The states

represent base-pairing in the alignment. They are used to count the number of matches after an edit event. The Moore automaton for generating alignments of gRNA and cryptogene is given in Figure 8.11. Recall that a Markov model is a stochastic Moore automation. Note that the automation restricts the number of successive matches to 3, which accounts for Rule 5.

For example, one possible sequence of states generated by the automation is

This sequence of states represents a set of possible guiding region alignments. For example, this set includes the alignment but excludes the alignment.

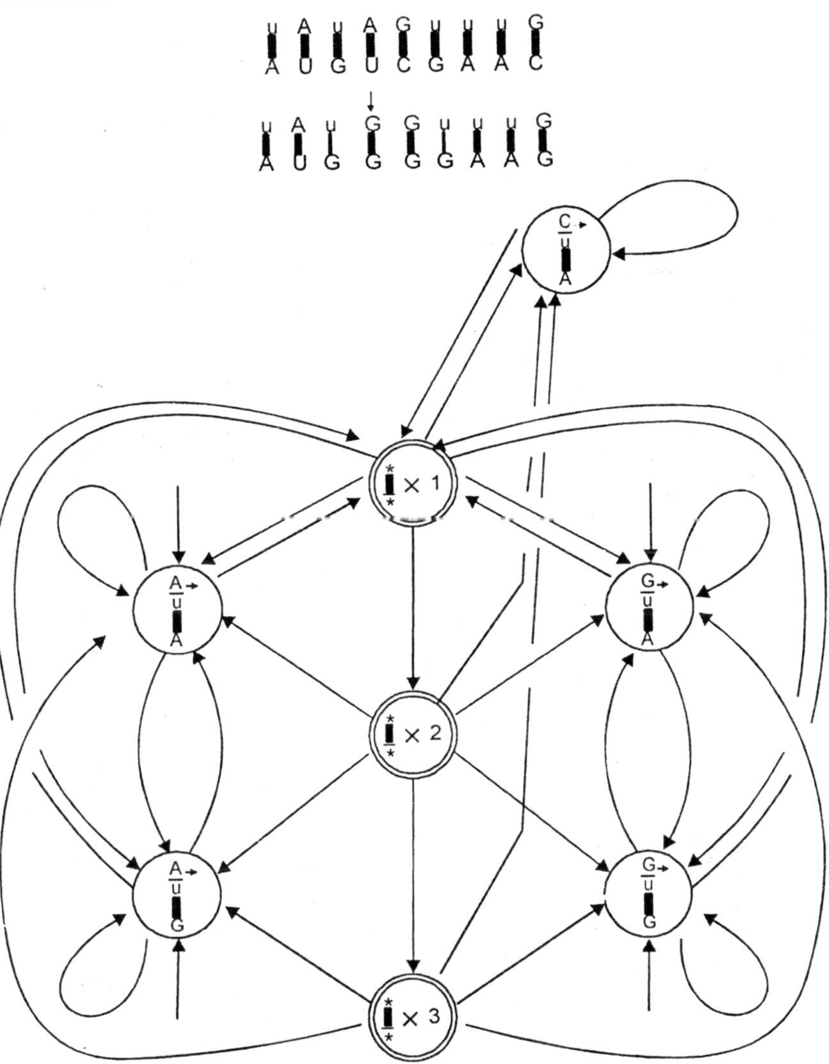

Fig. 8.11. Moore automaton for generating alignment of guide RNA and edited cryptogene for the guiding region.

The reason is that the first C in the above sequence (indicated by an arrow) does not satisfy the conditions we have stated above for edit events.

We continue by adding the probabilities to the Moore automaton thus generating the Markov chain. Recall that a Markov chain is a stochastic process producing a sequence of states $q_0\, q_1 \ldots q_t\, q_{t+1} \ldots$, where q_t is the state at time t. For the initial state q_0, we know that it must be an edit state (since the guiding region starts with an edit state). Consider for example the probability.

$$Pr\left[q_0 = \begin{matrix} A\rightarrow \\ u \\ | \\ A \end{matrix} \right]$$

This is the probability that we see a symbol A in the cryptogene (albeit it appears only later in the alignment), and a symbol A in the guide RNA, *i.e.*, $P_A\, q_A$. Similarly, we get

$$Pr\left[q_0 = \begin{matrix}\underline{A\rightarrow}\\ U\\ \blacksquare\\ G\end{matrix}\right] = p_A q_G, \quad Pr\left[q_0 = \begin{matrix}\underline{G\rightarrow}\\ U\\ \blacksquare\\ A\end{matrix}\right] = p_G q_A, \quad Pr\left[q_0 = \begin{matrix}\underline{G\rightarrow}\\ U\\ \blacksquare\\ G\end{matrix}\right] = p_G q_G,$$

and

$$Pr\left[q_0 = \begin{matrix}\underline{C\rightarrow}\\ U\\ \blacksquare\\ A\end{matrix}\right] = p_C q_A.$$

Since a guiding region does not start with a match, we get

$$Pr\left[q_0 = \begin{matrix}*\\ \blacksquare\\ *\end{matrix}\times 1\right] = 0, \quad Pr\left[q_0 = \begin{matrix}*\\ \blacksquare\\ *\end{matrix}\times 2\right] = 0, \quad Pr\left[q_0 = \begin{matrix}*\\ \blacksquare\\ *\end{matrix}\times 3\right] = 0.$$

By π, we denote the distribution for the initial state, *i.e.*,

$$\pi = (p_A\, q_A,\, p_A\, q_G,\, p_G\, q_A,\, p_G\, q_G,\, p_C\, q_A,\, 0, 0, 0).$$

Next, we have to calculate the transition probabilities. We want to describe two kinds of transitions in greater detail. The first are transitions from a match state into an edit state. For instance, consider

$$Pr\left[q_{t+1} = \begin{matrix}\underline{A\rightarrow}\\ U\\ |\\ G\end{matrix} \middle| q_t = \begin{matrix}*\\ \blacksquare\\ *\end{matrix}\times 1\right]$$

then this is again the probability that we see an A in the cryptogene, and an G in the guide RNA, *i.e.*, $p_A\, q_G$.

The second kind of transition is a transition from an edit state to a match state, *e.g.*,

$$Pr\left[q_{t+1} = \begin{matrix}*\\ \blacksquare\\ *\end{matrix}\times 1 \middle| q_t = \begin{matrix}\underline{A\rightarrow}\\ U\\ |\\ G\end{matrix}\right].$$

In order to investigate this probability, we have to consider all alignments that are allowed for any state sequence of the form

$$\cdots \begin{matrix}\underline{A\rightarrow}\\ U\\ |\\ G\end{matrix}\begin{matrix}*\\ \blacksquare\\ *\end{matrix}\times 1 \cdots$$

Since we know already that the next letter in the cryptogene must be A (by the definition of the edit state), and since we know that a match must follow, we know that the allowed alignments are of the form

$$\cdots \begin{matrix}U\\ |\\ G\end{matrix}\;\begin{matrix}A\\ \blacksquare\\ U\end{matrix} \cdots$$

Since the only new character we determine is the U in the guiding region, we get that the transition probability given in is q_U.

The transition probabilities are summarized in matrix M given in Table (8.1). Note that the matrix M is *substochastic*. Given this matrix M, we know that the probability for a guiding region of length k is given by the probability for a sequence of states. Edit states correspond to rows 1 ...5, and match states to row 6, 7 and 8. Thus, we get

$$Pr[L_e = k] = \sum_{i=1}^{5}\sum_{j=6}^{8} \pi_i M_{i,j}^{k}\left(1 - \sum_{s=1}^{8} M_{j,s}\right),$$

where the term $\Sigma_{s=1}^{8}(1 - M_{j,s})$ accounts for the probability for terminating the Markov chain with the jth state. Since M^k is positive for $k \geq 4$ (*i.e.* all entries are positive), the Perron-Frobenius theory can be applied to approximate M^k and therefore $Pr[L_e = k]$.

GenBank

GenBank (Genetic Sequence Data Bank) is a rapidly growing international repository of known genetic sequences from a variety of organisms. Its use is central to modern biology and to bioinformatics. This chapter shows you how to write Perl programs to extract information from GenBank files and libraries. Exercises include looking for patterns; creating special libraries; and parsing the flat-file format to extract the DNA, annotation, and features. You will learn how to make a DBM database to create your own rapid-access lookups on selected data in a GenBank library. *Perl* is a great tool for dealing with GenBank files. It enables you to extract and use any of the detailed data in the sequence and in the annotation, such as in the FEATURES table and elsewhere. When I first started using Perl, I wrote a program that searched GenBank for all sequence records annotated as being located on human chromosome 22. I found many genes where that information was so deeply buried within the annotation, that the major gene mapping database, Genome Database (GDB), hadn't included them in their chromosome map. I think you'll discover the same feeling of power over the information when your start applying Perl to GenBank files.

Most biologists are familiar with GenBank. Researchers can perform a search, *e.g.*, a BLAST search on some query sequence, and collect a set of GenBank files of related sequences as a result. Because the GenBank records are maintained by the individual scientists who discovered the sequences, if you find some new sequence of interest, you can publish it in GenBank. GenBank files have a great deal of information in them in addition to sequence data, including identifiers such as accession numbers and gene names, phylogenetic classification, and references to published literature.

A GenBank file may also include a detailed FEATURES table that summarizes facts about the sequence, such as the location of the regulatory regions, the protein translation, and exons and introns. GenBank is sometimes referred to as a databank or data store, which is different from a *database*. Databases typically have a relational structure imposed upon the data, including associated indices and links and a query language. GenBank in comparison is a flat file, that is, an ASCII text file that is easily readable by humans.[1]

From its humble beginnings GenBank has rapidly grown, and the flat-file format has seen signs of strain during the growth. With a quickly advancing body of knowledge, especially one that's growing as quickly as genetic data, it's difficult for the design of a databank to keep up. Several reworkings of GenBank have been done, but the flat-file format—in all its frustrating

[1] GenBank is also distributed in ASN. 1 format, for which you need specialized tools, provided by NCBI.

glory—still remains. Due to a certain flexibility in the content of some sections of a GenBank record, extracting the information you're looking for can be tricky. This flexibility is good, in that it allows you to put what you think is most important into the data's annotation. It's bad, because that same flexibility makes it harder to write programs that to find and extract the desired annotations.

As a result, the trend has been towards more structure in the annotations. Since Perl's data structures and its use of regular expressions make it a good tool for manipulating flat files, Perl is especially well-suited to deal with GenBank data. Using these features in Perl and building on the skills you've developed from previous chapters, you can write programs to access the accumulated genetic knowledge of the scientific community in GenBank. Since this is a beginning book that requires no programming experience, you should not expect to find the most finished, multipurpose software here. Instead you'll find a solid introduction to parsing and building fast lookup tables for GenBank files. If you've never done so. I strongly recommned you explore the nationl centro for Biotechnology informetion (NCBF) at the national institute of health (NIH). While you're at it, stop by the European Bioinformatics Institute (EBI) and the bioinformatics arm of the European Molecular Biology Laboratory (EMBL). These are large, heavily funded governmental bioinformatics powerhouses, and they have (and distribute) a great deal of state-of-the-art bioinformatics software.

GENBANK FILES

The primary repositories for genetic information are the NCBI GenBank, EMBL in Europe, and the DNA Data Bank of Japan (DDBJ). All have almost identical information due to international cooperative agreements. Each entry or record in GenBank or its mirror sites may contain identifying, descriptive, and genetic information in ASCII-format files. Each record is written in a specific standard format, organized so that both humans and computer programs can extract the desired information with reasonable ease.

Let's look at a relatively short GenBank record and at how the fields are defined, before writing any code. I'll save this information in a file called ***record, gb,*** for use in later programs.

```
LOCUS           AB031069 2487 bp mRNA PRI
DEFINITION      Homo sapiens PCCX1 mRNA for protein containing
CXXC domain 1,
                complete cds.
ACCESSION       AB031069
VERSION         AB031069.1 GI:8100074
EYWORDS
SOURCE          Homo sapiens embryo male lung fibroblast
cell_line:HuS-L12 cDNA to
                mRNA.
ORGANISM        Homo sapiens
                Eukaryota; Metazoa; Chordata;  Craniata;
Vertebrata;     Euteleostomi;
Mammalia;       Eutheria; Primates; Catarrhini;
Hominidae;      Homo.
REFERENCE       1 (sites)
```

```
AUTHORS          Fujino,T., Hasegawa,M., Shibata,S.,
Kishimoto,T., Imai,Si. and
                 Takano,T.
TITLE            PCCX1, a novel DNA-binding protein with PHD
finger and CXXC domain,
                 is regulated by proteolysis
JOURNAL          Biochem. Biophys. Res. Commun. 271 (2), 305-310
(2000)
MEDLINE          20261256
REFERENCE        2 (bases 1 to 2487)
AUTHORS          Fujino,T., Hasegawa,M., Shibata,S.,
Kishimoto,T., Imai,S. and
                 Takano,T.
TITLE            Direct Submission
JOURNAL          Submitted (15-AUG-1999) to the
DDBJ/EMBL/GenBank databases.
                 Tadahiro Fujino, Keio University School of
Medicine, Department of
                 Microbiology; Shinanomachi 35, Shinjuku-ku,
Tokyo 160-8582, Japan
                 (E-mail:fujino@microb.med.keio.ac.jp,
                 Tel:+81-3-3353-1211(ex.62692) , Fax:+81-3-5360-1508)
FEATURES         Location/Qualifiers
   Source        1..2487
                   /organism="Homo sapiens"
                   /db_xref="taxon:9606"
                   /sex="male"
                   /cell_line="HuS-L12"
                   /cell_type="lung fibroblast"
                   /dev_stage="embryo"
   gene            229..2199
                   /gene="PCCX1"
   CDS             229..2199
                   /gene="PCCX1"
                   /note="a nuclear protein carrying a
   PHD finger and CXXC
```

```
                     domain"
                     /codon_start=1
                     /product="protein containing CXXC
         domain 1"
                     /protein_id="BAA96307.1"
                     /db_xref="GI:8100075"
/translation="MEGDGSDPEPPDAGEDSKSENGENAPIYCICRKPDINCFMIGCD
NCNEWFHGDCIRITEKMAKAIREWYCRECREKDPKLEIRYRHKKSRERDGNERDSSEP
RDEGGGRKRPVPDPDLQRRAGSGTGVGAMLARGSASPHKSSPQPLVATPSQHHQQQQQ
QIKRSARMCGECEACRRTEDCGHCDFCRDMKKFGGPNKIRQKCRLRQCQLRARESYKY
FPSSLSPVTPSESLPRPRRPLPTQQQPQPSQKLGRIREDEGAVASSTVKEPPEATATP
EPLSDEDLPLDPDLYQDFCAGAFDDHGLPWMSDTEESPFLDPALRKRAVKVKHVKRRE
KKSEKKKEERYKRHRQKQKHKDKWKHPERADAKDPASLPQCLGPGCVRPAQPSSKYCS
DDCGMKLAANRIYEILPQRIQQWQQSPCIAEEHGKKLLERIRREQQSARTRLQEMERR
FHELEAIILRAKQQAVREDEESNEGDSDDTDLQIFCVSCGHPINPRVALRHMERCYAK
YESQTSFGSMYPTRIEGATRLFCDVYNPQSKTYCKRLQVLCPEHSRDPKVPADEVCGC
PLVRDVFELTGDFCRLPKRQCNRHYCWEKLRRAEVDLERVRVWYKLDELFEQERNVRT
                                AMTNRAGLLALMLHQTIQHDPLTTDLRSSADR"
BASE COUNT          564 a    715 c    768 g      440 t
ORIGIN
           1 agatggcggc gctgaggggt cttgggggct ctaggccggc
cacctactgg tttgcagcgg
          61 agacgacgca tggggcctgc gcaataggag tacgctgcct
gggaggcgtg actagaagcg
         121 gaagtagttg tgggcgcctt tgcaaccgcc tgggacgccg
ccgagtggtc tgtgcaggtt
         181 cgcgggtcgc tggcgggggt cgtgagggag tgcgccggga
gcggagatat ggagggagat
         241 ggttcagacc cagagcctcc agatgccggg gaggacagca
agtccgagaa tggggagaat
         301 gcgcccatct actgcatctg ccgcaaaccg gacatcaact gcttcatgat
         cgggtgtgac
         361 aactgcaatg agtggttcca tggggactgc atccggatca
ctgagaagat ggccaaggcc
         421 atccgggagt ggtactgtcg ggagtgcaga gagaaagacc
ccaagctaga gattcgctat
         481 cggcacaaga agtcacggga gcgggatggc aatgagcggg
acagcagtga gccccgggat
         541 gagggtggag ggcgcaagag gcctgtccct gatccagacc
```

tgcagcgccg ggcagggtca
601 gggacagggg ttggggccat gcttgctcgg ggctctgctt
cgccccacaa atcctctccg
661 cagcccttgg tggccacacc cagccagcat caccagcagc
agcagcagca gatcaaacgg
721 tcagcccgca tgtgtggtga gtgtgaggca tgtcggcgca
ctgaggactg tggtcactgt
781 gatttctgtc gggacatgaa gaagttcggg ggccccaaca
agatccggca gaagtgccgg
841 ctgcgccagt gccagctgcg ggcccgggaa tcgtacaagt
acttcccttc ctcgctctca
901 ccagtgacgc cctcagagtc cctgccaagg ccccgccggc
cactgcccac ccaacagcag
961 ccacagccat cacagaagtt agggcgcatc cgtgaagatg
agggggcagt ggcgtcatca
1021 acagtcaagg agcctcctga ggctacagcc acacctgagc
cactctcaga tgaggaccta
1081 cctctggatc ctgacctgta tcaggacttc tgtgcagggg
cctttgatga ccatggcctg
1141 ccctggatga gcgacacaga agagtcccca ttcctggacc
ccgcgctgcg gaagagggca
1201 gtgaaagtga agcatgtgaa gcgtcgggag aagaagtctg
agaagaagaa ggaggagcga
1261 tacaagcggc atcggcagaa gcagaagcac aaggataaat
ggaaacaccc agagagggct
1321 gatgccaagg accctgcgtc actgccccag tgcctggggc
ccggctgtgt gcgccccgcc
1381 cagcccagct ccaagtattg ctcagatgac tgtggcatga
agctggcagc caaccgcatc
1441 tacgagatcc tcccccagcg catccagcag tggcagcaga
gcccttgcat tgctgaagag
1501 cacggcaaga agctgctcga acgcattcgc cgagagcagc
agagtgcccg cactcgcctt
1561 caggaaatgg aacgccgatt ccatgagctt gaggccatca
ttctacgtgc caagcagcag
1621 gctgtgcgcg aggatgagga gagcaacgag ggtgacagtg

```
atgacacaga cctgcagatc
          1681 ttctgtgttt cctgtgggca ccccatcaac ccacgtgttg
ccttgcgcca catggagcgc
          1741 tgctacgcca agtatgagag ccagacgtcc tttgggtcca
tgtaccccac acgcattgaa
          1801 ggggccacac gactcttctg tgatgtgtat aatcctcaga
gcaaaacata ctgtaagcgg
          1861 ctccaggtgc tgtgccccga gcactcacgg gaccccaaag
tgccagctga cgaggtatgc
          1921 gggtgccccc ttgtacgtga tgtctttgag ctcacgggtg
acttctgccg cctgcccaag
          1981 cgccagtgca atcgccatta ctgctgggag aagctgcggc
gtgcggaagt ggacttggag
          2041 cgcgtgcgtg tgtggtacaa gctggacgag ctgtttgagc
aggagcgcaa tgtgcgcaca
          2101 gccatgacaa accgcgcggg attgctggcc ctgatgctgc
accagacgat ccagcacgat
          2161 cccctcacta ccgacctgcg ctccagtgcc gaccgctgag
cctcctggcc cggacccctt
          2221 acaccctgca ttccagatgg gggagccgcc cggtgcccgt
gtgtccgttc ctccactcat
          2281 ctgtttctcc ggttctccct gtgcccatcc accggttgac
cgcccatctg cctttatcag
          2341 agggactgtc cccgtcgaca tgttcagtgc ctggtggggc
tgcggagtcc actcatcctt
          2401 gcctcctctc cctgggtttt gttaataaaa ttttgaagaa
accaaaaaaa aaaaaaaaaa
          2461 aaaaaaaaaa aaaaaaaaaa aaaaaaa
//
```

Even if you're used to seeing GenBank files, it's worth taking the time to look one over, while considering how you would write a program to extract various parts of the data. For instance, how would you extract the sequence data? What's the format of the FEATURES table and its various subfields? There's a lot of information packed into a typical GenBank entry, and it's important to be able to separate the different parts. For instance, if you can extract the sequence, you can search for motifs, calculate statistics on the sequence, look for similarity with other sequences, and so forth.

Similarly, you'll want to separate out—or parse—the various parts of the data annotation. In GenBank, this includes ID numbers, gene names, genus and species, publications, etc. The FEATURES table part of the annotation can include specific information about the DNA, such as the locations of exons, regulatory regions, important mutations, and so on. The format specification of GenBank files and a great deal of other information about GenBank can be found in the GenBank release notes, ***gbrel.txt,*** on the GenBank web site at ***ftp://ncbi.nlm.nlm.nih.gov/genbank/gbrel.txt. gbrel.txt*** gives complete detail about the structure of GenBank files to help programmers, so you may want to refer to it as your searches become more complex. As a Perl programmer, you won't need all of the detail because you can parse data using regular expressions or the *split* function. You need to get the data out and make it available to your programs. The code that accomplishes this task can be fairly simple.

GEN BANK LIBRARIES

GenBank is distributed as a set of libraries—flat files containing many records in succession[2]. As of GenBank release 125.0, August 2001, there are 243 files, most of which are over 200 MB in size. Altogether, GenBank contains 12,813516 loci and 13,543,364,296 bases from 12,813,516 reported sequences. The libraries are usually distributed compressed, which means you can download somewhat smaller files, but you need to uncompress them after you received them. Uncompressed, this amounts to about 50 GB of data. Since 1982, the number of sequences in GenBank has doubled about every 14 months.

GenBank libraries are further organized into divisions by the classification of the sequences they contain, either phylogenetically or by sequencing technology. Here are the divisions:

PRI: primate sequences

ROD: rodent sequences

MAM: other mammalian sequences

VRT: other vertebrate sequences

INV: invertebrate sequences

PLN: plant, fungal, and algal sequences

BCT: bacterial sequences

VRL: viral sequences

PHG: bacteriophage sequences

SYN: synthetic and chimeric sequences

UNA: unannotated sequences

EST: EST sequences (expressed sequence tags)

PAT: patent sequences

STS: STS sequences (sequence tagged sites)

GSS: GSS sequences (genome survey sequences)

HTG: HTGS sequences (high throughput genomic sequencing data)

HTC: HTC sequences (high throughput cDNA sequencing data)

Some divisions are very large: the largest, the EST, or expressed sequence tag division, is comprised of 123 library files! A portion of human DNA is stored in the PRI division, which contains (as of this writing) 13 library files, for a total of almost 3.5 GB of data.

---

[2] The data is also distributed in the ASN. 1 format.

Human data is also stored in the STS, GSS, HTGS, and HTC divisions. Human data alone in GenBank makes up almost 5 million record entries with over 8 trillion bases of sequence. The public database servers such as Entrez or BLAST at *http://www.ncbi.nlm.nih.gov/* give you access to well-maintained and updated sequence data and programs, but many researchers find that they need to write their own programs to manipulate and analyze the data. The problem is, there's so much data. For many purposes, you can download a selected set of records from NCBI or other locations, but sometimes you need the whole dataset. It's possible to set up a desktop workstation (Windows, Mac, Unix, or Linux) that contains all of GenBank; just be sure to buy a very large hard disk! Getting all that data onto your hard drive, however, is more difficult.

A Perl program called *mirror.pl* helps to address this need. Downloading it, even with a university-standard, high-speed Internet connection can be time-consuming; downloading an entire dataset with a modem can be an exercise in frustration. The best solution is to download only the files you need, in compressed form. The EST data, for example, is about half the entire database; don't download it unless you really need to. Since you're learning to program, it makes more sense to practice on a tiny, five-record library file, but the programs you'll write will work just fine on the real files.

SEPARATING SEQUENCE AND ANNOTATION

In previous chapters you saw how to examine the lines of a file using Perl's array operations. Usually, you do this by saving the data in an array with each appearing as an element of the array. Let's look at two methods to extract the annotation and the DNA from a GenBank file. In the first method, you'll slurp the file into an array and look through the lines, as in previous programs. In the second, you'll put the whole GenBank record into a scalar variable and use regular expressions to parse the information. Is one approach better than the other? Not necessarily: it depends on the data. There are advantages and disadvantages to each, but both get the job done.

Using Arrays

I have put five GenBank records in a file called ***library.gb***. As before, you can download the file from this book's web site. You'll use this datafile and the file ***record.gb*** in the next few examples.

Example shows the first method, which operates on an array containing the lines of the GenBank record. The main program is followed by a subroutine that does the real work.

Example 9.1. Extract annotation and sequence from GenBank file

```
#!/usr/bin/perl
# Extract annotation and sequence from GenBank file

use strict;

use warnings;

use BeginPerlBioinfo;       # see Chapter 6 about this module

# declare and initialize variables
my @annotation = ( );
```

```
my $sequence = '';
my $filename = 'record.gb';

parsel(\@annotation, \$sequence, $filename);

# Print the annotation, and then
# print the DNA in new format just to check if we got it
okay.
print @annotation;
**.
print_sequence($sequence, 50); exit;

##########################################################1
########«##########
# Subroutine
##########################################t################f#
#####################
# parsel
#
# --parse annotation and sequence from GenBank record
sub parsel {
    my($annotation, $dna, $filename) = @_;
    # $annotation-reference to array
    # $dna -reference to scalar
    # $filename -scalar
    # declare and initialize variables
    my $in_sequence = 0;
    my SGenBankFile = ( ) ;
    # Get the GenBank data into an array from a file
    @GenBankFile = get_f ile_data ($f ilename) ;
    # Extract all the sequence lines
    foreach my $line (SGenBankFile) {
if ( $line =~ /^\/\/\n/ ) { # If $line is end-of-record line //\n,
   last; tbreak out of the foreach loop.
} elsif ( $in_sequence) { # If we know we're in a sequence,
        $$dna .= $line; # add the current line to $$dna.
   } elsif ( $line =~ /^ORIGIN/ ) { # If $line begins a sequence,
        $in_sequence = 1; # set the $in_sequence flag.
   } else{ # Otherwise
```

```
            push (@$annotation, $line) ; # add the current line to
@annotation.


    # remove whitespace and line numbers from DNA sequence
    $$dna =~ s/ [\s0-9] //g;
```

Here's the beginning and end of Example 9–1's output of the sequence data:

```
agatggcggcgctgaggggtcttgggggctctaggccggccacctactgg
tttgcagcggagacgacgcatggggcctgcgcaataggagtacgctgcct
gggaggcgtgactagaagcggaagtagttgtgggcgcctttgcaaccgcc
tgggacgccgccgagtggtctgtgcaggttcgcgggtcgctggcgggggt
cgtgagggagtgcgccgggagcggagatatggagggagatggttcagacc
. . .
cggtgcccgtgtgtccgttcctccactcatctgtttctccggttctccct
gtgcccatccaccggttgaccgcccatctgcctttatcagagggactgtc
cccgtcgacatgttcagtgcctggtggggctgcggagtccactcatcctt
gcctcctctccctgggttttgttaataaaattttgaagaaaccaaaaaaa
aaaaaaaaaaaaaaaaaaaaaaaaaaaaaaaaaaa
```

The foreach loop in subroutine `parsel` Example 9–1's moves one by one through the lines from the GenBank file stored. It takes advantage of the structure of a GenBank file, which begins with annotation and runs until the line:

```
ORIGIN
```

is found, after which sequence appears until the end-of-record line // is reached. The loop uses a flag variable `$in_sequence` to remember that it has found the `ORIGIN` line and is now reading sequence lines.

The `foreach` loop has a new feature: the Perl built-in function `last`, which breaks out of the nearest enclosing loop. It's triggered by the end-of-record line //, which is reached when the entire record has been seen.

A regular expression is used to find the end-of-record line. To correctly match the end-of-record (forward) slashes, you must escape them by putting a backslash in front of each one, so that Perl doesn't interpret them as prematurely ending the pattern. The regular expression also ends with a newline `\/ \ / \n`, which is then placed inside the usual delimiters: `/ \ / \ / \ n /`. (When you have a lot of forward slashes in a regular expression, you can use another delimiter around the regular expression and precede it with an m, thus avoiding having to backslash the forward slashes. It's done like so: `m! //\ n!`). An interesting point about subroutine `parsel` is the order of the tests in the foreach loop that goes through the lines of the GenBank record.

As you read through the lines of the record, you want to first gather the annotation lines, set a flag when the `ORIGIN` start-of-sequence line is found, and then collect the lines until the end-of-record / / line is found. Notice that the order of the tests is exactly the opposite. First, you test for the end-of-record line, collect the sequence if the `$ in sequence` flag is set, and then test for the start-of-sequence `ORIGIN` line. Finally, you collect the annotation. The technique of reading lines one by one and using flag variables to mark what section of the file you're in, is a common programming technique. So, take a moment to think about how the loop would behave if you changed the order of the tests.

If you collected sequence lines before testing for the end-of-record, you'd never get to the end-of-record test! Other methods of collecting annotation and sequence lines are possible, especially if you go through the lines of the array more than once. You can scan through the array, keeping track of the start-of-sequence and end-of-record line numbers, and then go back and extract the annotation and sequence using an array splice (which was described in the `parseREBASE` subroutine. Here's an example:

```
# find line numbers of ORIGIN and // in the GenBank record
$linenumber = 0;
foreach my $line (@GenBankFile) {
if      ($line =~ /^\/\/\n/) {#end-of-record//line
        $end = $linenumber;
        last;
} elsif ( $line =~ /^ORIGIN/ ) { # end annotation, begin sequence
        $origin = $linenumber;
        }
        $linenumber++;
}
# extract annotation and sequence with "array splice"
@annotation = @GenBankFile [0.. ($origin-1) ] ;
@sequence = @GenBankFile [ ($origin+1). . ($end-1) ];
</programlisting>
```

USING SCALARS

A second way to separate annotations from sequences in GenBank records is to read the entire record into a scalar variable and operate on it with regular expressions. For some kinds of data, this can be a more convenient way to parse the input (compared to scanning through an array.

Usually string data is stored one line per scalar variable with its newlines, if any, at the end of the string. Sometimes, however, you store several lines concatenated together in one string that is, in turn, stored in a single scalar variable. These multiline strings aren't uncommon; you used them to gather the sequence from a FASTA file in Examples Example 9–2 and Example 9–3. Regular expressions have pattern modifiers that can be used to make multiline string with their embedded newlines easy to use.

Pattern Modifiers

The pattern modifiers we've used so far are `/g`, for global matching, and `/i`, for case-insensitive matching. Let's take a look at two more that affect the way regular expressions interact with the newlines in scalars.

Recall that previous regular expressions have used the caret (^), dot (.), and dollar sign ($) metacharacters. The ^ anchors a regular expression to the beginning of a string, by default, so that `/^THE BEGUINE/` matches a string that begins with "THE BEGUINE". Similarly, $ anchors an expression to the end of the string, and the dot (.) matches any character except a newline.

The following pattern modifiers affect these three metacharacters:

The `/s` modifier assumes you want to treat the whole string as a single line, even with embedded newlines, so it makes the dot metacharacter match any character *including* newlines.

The `/m` modifier assumes you want to treat the whole string as a multiline, with embedded newlines, so it extends the ^ and the $ to match after, or before, a newline, embedded in the string.

Examples of Pattern Modifiers : Here's an example of the default behavior of caret (^), dot (.), and dollar sign ($):

```
use warnings;
"AAC\nGTT" =-/^.*$/;
print $&, "\n";
```

This demonstrates the default behavior without the `/m` or `/s` modifiers and prints the warning:

```
Use of uninitialized value in print statement at line 3.
```

The print statement tries to print $& , a special variable that is always set to the last successful pattern match. This time, since the pattern doesn't match, the variable $ & isn't set, and you get a warning message for attempting to print an uninitialized value.

Why doesn't the match succeed? First, let's examine the `^.*$` pattern. It begins with a ^, which means it must match from the beginning of the string. It ends with a $, which means it must also match at the end of the string (the end of the string may contain a single newline, but no other newlines are allowed). The `.*` means it must match zero or more (*) of any characters (.) except the newline. So, in other words, the pattern `^.*$` matches any string that doesn't contain a newline except for a possible single newline as the last character. But since the string in question, `"ACC\nGTT"` does contain an embedded newline \n that isn't the last character, the pattern match fails.

In the next examples, the pattern modifiers/m and /B change the default behaviors for the metacharacters ^, and *$, and* the dot:

```
"AAC\nGTT" =~ /^.*$/m;
print $&, "\n";
```

This snippet prints out `AAC` and demonstrates the `/m` modifier. The `/m` extends the meaning of the ^ and the $ so they also match around embedded newlines. Here, the pattern matches from the beginning of the string up to the first embedded newline.

The next snippet of code demonstrates the `/s` modifier:

```
"AACXnGTT" =~ /A.*$/s;
print $&, "\n";
```

which produces the output:

```
AAC
GTT
```

The /s modifier changes the meaning of the dot metacharacter so that it matches any character including newlines. With the /s modifier, the pattern matches everything from the beginning of the string to the end of the string, including the newline. Notice when it prints, it prints the embedded newline.

Separating Annotations from Sequence

Now that you've met the pattern-matching modifiers and regular expressions that will be your main tools for parsing a GenBank file as a scalar, let's try separating the annotations from the sequence.

The first step is to get the GenBank record stored as a scalar variable. Recall that a GenBank record starts with a line beginning with the word "LOCUS" and ends with the end-of-record separator: a line containing two forward slashes.

First you want to read a GenBank record and store it in a scalar variable. There's a device called an input record separator denoted by the special variable `$/` that lets you do exactly that. The input record separator is usually set to a newline, so each call to read a scalar from a filehandle gets one line. Set it to the GenBank end-of-record separator like so:

```
$/ = "// /n";
```

A call to read a scalar from a filehandle takes all the data up to the GenJrJank end-of-record separator. So the line `$ record = <GBFILE>` in Example 9.2 stores the multiline GenBank record into the scalar variable `$record`. Later, your'll see that you can keep repeating this call in order to read in successive GenBank records from a GenBank library file.

After reading in the record, you'll parse it into the annotation and sequence parts making use of /s and /m pattern modifiers. Extracting the annotation and sequence is the easy part; parsing the annotation will occupy most of the remainder of the chapter.

Example 9.2. Extract annotation and sequence from Genbank record

```
#! /usr/bin/perl
# Extract the annotation and sequence sections from the first
# record of a GenBank library

use strict;
use warnings;
use BeginPerlBioinfo;        # see Chapter 6 about this module

# Declare and initialize variables
my $annotation = '';
my $dna = '';
my $record = '';
```

```
my $filename = 'record.gb';
my $save_input_separator = $/;

# Open GenBank library file
unless (open(GBFILE, $filename)) {
   print "Cannot open GenBank file \"$filename\"\n\n";
   exit;
}

# Set input separator to "//\n" and read in a record to a
scalar
$/ = "/An";

$record = <GBFILE>;

# reset input separator
$/ = $save_input_separator;

# Now separate the annotation from the sequence data
($annotation, $dna) = ($record =~
/^ (LOCUS. *ORIGIN\s*\n) (.*) \/\/\n/s) ;

# Print the two pieces, which should give us the same as the
# original GenBank file, minus the // at the end
print $annotation, $dna;

exit;
```

The output from this program is the same as the GenBank file listed previously, minus the last line, which is the end-of-record separator //.

Let's focus on the regular expression that parses the annotation and sequence out of the `$ record` variable. This is the most complicated regular expression so far:

```
$record = /^ (LOCUS . *ORIGIN\s*\n) (.*) \/\/\n/s .
```

There are two pairs of parentheses in the regular expression: `(LOCUS.*ORIGIN\s*\n )` and `( . * )`. The parentheses are metacharacters whose purpose is to remember the parts of the data that match the pattern within the parentheses, namely, the annotation and the sequence. Also note that the pattern match returns an array whose elements are the matched parenthetical patterns. After you match the annotation and the sequence within the pairs of parentheses in the regular expression, you simply assign the matched patterns to the two variables `$annotation` and `$dna`, like so:

```
($annotation, $dna) = ($record =~
/^ (LOCUS. *ORIGIN\s*\n) (.*) \/\/\n/s);
```

Notice that at the end of the pattern, we've added the `/s` pattern matching modifier, which, as you've seen earlier, allows a dot to match any character including an embedded newline. (Of course, since we've got a whole GenBank record in the `$record` scalar, there are a lot of embedded newlines.)

Next, look at the first pair of parentheses:

```
(LOCUS . *ORIGIN\s*\n)
```

This whole expression is anchored at the beginning of the string by preceding it with a ^ metacharacter, (/s doesn't change the meaning of the ^ character in a regular expression.)

Inside the parentheses, you match from where the string LOCUS appears at the beginning of the GenBank record, followed by any number of characters including newlines with .*, followed by the string ORIGIN, followed by possibly some whitespace with \s*, followed by a newline \n. This matches the annotation part of the GenBank record.

Now, look at the second parentheses and the remainder:

```
(.*)\/\/\n
```

This is easier. The .* matches any character, including newlines because of the /s pattern modifier at the end of the pattern match. The parentheses are followed by the end-of-record line, //, including the newline at the end, with the slashes preceded by backslashes to show that you want to match them exactly. They're not delimiters of the pattern matching operator. The end result is the GenBank record with the annotation and the sequence separated into the variables $annotation and $sequence. Although the regular expression I used requires a bit of explanation, the attractive thing about this approach is that it took only one line of Perl code to extract both annotation and sequence.

PARSING ANNOTATIONS

Now that you've successfully extracted the sequence, let's look at parsing the annotations of a GenBank file.

Looking at a GenBank record, it's interesting to think about how to extract the useful information. The FEATURES table is certainly a key part of the story. It has considerable structure: what should be preserved, and what is unnecessary? For instance, sometimes you just want to see if a word such as "endonuclease" appears anywhere in the record. For this, you just need a subroutine that searches for any regular expression in the annotation. Sometimes this is enough, but when detailed surgery is necessary, Perl has the necessary tools to make the operation successful.

Using Arrays

Example 9.3 parses a few pieces of information from the annotations in a GenBank file. It does this using the data in the form of an array.

Example 9.3. Parsing GenBank annotations using arrays

```
# !/usr/bin/perl
# Parsing GenBank annotations using arrays

use strict;
use warnings;
use BeginPerlBioinfo; # see Chapter 6 about this module

# Declare and initialize variables
my Sgenbank = ( ) ;
```

```
my $ locus = ' ' ;
my $accession = ' ' ;
my $organism = ' ' ;

# Get GenBank file data
@genbank = get_f ile_data ('record. gb');

# Let's start with something simple. Let's get some of the identifying
# information, let's say the locus and accession number
  (here the same
# thing) and the definition and the organism.

for my $line (@genbank) {
   if($line =~ /^LOCUS/) {
     $line =~ s/^LOCUS\s*//;
     $locus = $line;
   }elsif($line =~ /"ACCESSION/) {
     $line =~ s/^ACCESSION\s*//;
     $accession = $line;
   }elsif($line =~ /^ ORGANISM/) {
     $line =~ s/^\s*ORGANISM\s*//;
     $organism = $line;
   }
}
print "*** LOCUS ***\n";
print $locus;
print "*** ACCESSION ***\n";
print $accession;
print "*** ORGANISM ***\n";
print $organism;

exit;
```

Here's the output from Example 9.3

```
*** LOCUS ***
AB031069 2487 bp mRNA PRI 27-MAY-2000
*** ACCESSION ***
AB031069
*** ORGANISM ***
Homo sapiens
```

Now let's slightly extend that program to handle the *DEFINITION* field. Notice that the *DEFINITION* field can extend over more-than one line. To collect that field, use a trick you've already seen in Example 9.1 set a flag when you're in the "state" of collecting a definition. The flag variable is called, unsurprisingly, `$flag`.

Example 9–4 . Parsing GenBank annotations using arrays, take 2

```
#!/usr/bin/perl
# Parsing GenBank annotations using arrays, take 2

use strict;
use warnings;
use BeginPerlBioinfo;       #see Chapter 6 about this module

# Declare and initialize variables
my @genbank = ( );
my $locus = '';
my $accession = '';
my $organism = '';
my $definition = '';
my $flag = 0;

# Get GenBank file data
@genbank = get_file_data('record.gb');

# Let's start with something simple. Let's get some of the identifying
# information, let's say the locus and accession number (here the same
# thing) and the definition and the organism.

for my $line (Qgenbank) {
    if($line =~ /^LOCUS/) {
        $line =~ s/^LOCUS\s*//;
        $locus = $line;
    }elsif($line =~ /^DEFINITION/) {
        $line =~ s/^DEFINITION\s*//;
        $definition = $line;
        $flag = 1;
    }elsif($line =~ /^ACCESSION/) {
        $line =~ s/^ACCESSION\s*//;
        $accession = $line;
        $flag = 0;
    }elsif ($flag) {
        chomp($definition) ;
        $definition .= $line;
```

```
        }elsif($line =~ /^ ORGANISM/) {
            $line =~ s/^\s*ORGANISM\s*//;
            $organism = $line;
        }
    }
    print "*** LOCUS ***\n";
    print $locus;
    print "*** DEFINITION ***\n";
    print $def inition;
    print "*** ACCESSION ***\n";
    print $accession;
    print "*** ORGANISM ***\n";
    print $organism;

    exit;
```

Example 9.4 outputs:

```
*** LOCUS ***
AB031069 2487 bp mRNA PRI 27-MAY-2000
*** DEFINITION ***
Homo sapiens PCCX1 mRNA for protein containing CXXC domain
1, complete cds .
*** ACCESSION *** AB031069
*** ORGANISM *** Homo sapiens
```

This use of flags to remember which part of the file you're in, from one iteration of a loop to the next, is a common technique when extracting information from files that have multiline sections. As the files and their fields get more complex, the code must keep track of many flags at a time to remember which part of the file it's in and what information needs to be extracted. It works, but as the files become more complex, so does the code. It becomes hard to read and hard to modify. So let's look at regular expressions as a vehicle for parsing annotations.

When to Use Regular Expressions

We've used two methods to parse GenBank files: regular expressions and looping through arrays of lines and setting flags. We used both methods to separate the annotation from the sequence in a previous section of this chapter. Both methods were equally well suited, since in GenBank files, the annotation is followed by the sequence, clearly delimited by an `ORIGIN` line: a simple structure.

However, parsing the annotations seems a bit more complicated; therefore, let's try to use regular expressions to accomplish the task. To begin, let's wrap the code we've been working on into some convenient subroutines to focus on parsing the annotations. You'll want to fetch GenBank records one at a time from a library (a file containing one or more GenBank records), extract the annotations and the sequence, and then if desired parse the annotations.

This would be useful if, say, you were looking for some motif in a GenBank library. Then you can search for the motif, and, if found, you can parse the annotations to look for additional information about the sequence. As mentioned previously, we'll use the file `library, gb,` which you can download from this book's web site. Since dealing with annotation data is somewhat complex, let's take a minute to break our tasks into convenient subroutines. Here's the pseudocode:

```
sub open_file
    given the filename, return the filehandle

sub get_next_record
    given the filehandle, get the record
    (we can get the offset by first calling "tell")

sub get_annotation_and_dna
    given a record, split it into annotation and cleaned-up sequence

sub search_sequence
    given a sequence and a regular expression,
    return array of locations of hits

sub search_annotation
    given a GenBank annotation and a regular expression,
        return array of locations of hits

sub parse_annotation
    separate out the fields of the annotation in a
convenient form

sub parse_features
    given the features field, separate out the components
```

The idea is to make a subroutine for each important task you want to accomplish and then combine them into useful programs. Some of these can be combined into other subroutines: for instance, perhaps you want to open a file and get the record from it, all in one subroutine call.

You're designing these subroutines to work with library files, that is, files with multiple GenBank records. You pass the filehandle into the subroutines as an argument, so that your subroutines can access open library files as represented by the filehandles. Doing so enables you to have a `get_next_record` function, which is handy in a loop. Using the Perl function *tell* also allows you to save the byte offset of any record of interest, and then return later and extract the record at that byte offset very quickly. (A byte offset is just the number of characters into the file where the information of interest lies.)

The operating system supports Perl in letting you go immediately to any byte offset location in even huge files, thus bypassing the usual way of opening the file and reading from the beginning until you get where you want to be. Using a byte offset is important when you're dealing with large files. Perl gives you built-in functions such as seek that allow you, on an open file, to go immediately to any location in the file. The idea is that when you find something in a file, you can save the byte offset using the Perl function *tell*. Then, when you want to return to that point in the file, you can just call the Perl function *seek* with the byte offset as an argument.

You'll see this later in this chapter when you build a DBM file to look up records based on their accession numbers. But the main point is that with a 250-MB file, it takes too long to find something by searching from the beginning, and there are ways of getting around it. The parsing of the data is done in three steps, according to the design: You'll separate out the annotation and the sequence (which you'll clean up by removing whitespace, etc., and making it a simple string of sequence). Even at this step, you can search for motifs in the sequence, as well as look for text in the annotation.

Extract out the fields.

Parse the features table.

These steps seem natural, and, depending on what you want to do, allow you to parse to whatever depth is needed.

Here's a main program in pseudocode that shows how to use those subroutines:

```
open_file
while (get_next_record)
   get_annotation_and_dna
   if ( search_sequence for a motif AND
        search_annotation for chromosome 22 )
         parse_annotation
         parse_features to get sizes of exons, look for small sizes
   }
}
return accession numbers of records meeting the criteria
```

This example shows how to use subroutines to answer a question such as: what are the genes on chromosome 22 that contain a given motif and have small exons?

Main Program

Let's test these subroutines with Example 9.5. which has some subroutine definitions that will be added to the *BeginPerlBioinfo.pm* module:

Example 9.5. GenBank library subroutines

```
# !/usr/bin/perl
# - test program of GenBank library subroutines

use strict;
use warnings;
# Don't use BeginPerlBioinfo
# Since all subroutines defined in this file
# use BeginPerlBioinfo;     # see Chapter 6 about this module

# Declare and initialize variables
my $fh; f variable to store filehandle
my $record;
my $dna;
```

```
my $annotation;
my $offset;
my $library = 'library.gb';

# Perform some standard subroutines for test
$fh = open_file ($library)/

$offset = tell ($fh);
while ($record = get_next_record ($fh) ) {
    ($annotation, $dna) = get_annotation_and_dna ($record) ;
    if(search_sequence ($dna, 'AAAfCG].')) {
        print "Sequence found in record at offset
    $offset\n";
    }
    if(search_annotation ($annotation, 'homo sapiens')) {
        print "Annotation found in record at offset
    $offset\n";
        }
        $offset = tell ($fh) ;
    }
    exit;
############################################################
#####################
# Subroutines
############################################################
#####################
# open_file
#
# - given filename, set filehandle
sub open_file {

    my ($f ilename) = @_;
    my $fh;

    unless (open ($fh, $f ilename) ) {
        print "Cannot open file $f ilename\n";
        exit;
    }
    return $fh;
}
# get_next_record
#
# - given GenBank record, get annotation and DNA
```

```
sub get_next_record {
    my($fh) = @_;

    my ($offset) ;
    my ($record) = '' ;
    my ($save_input_separator) = $/;

    $/ = "//\n";
    $record = <$fh>;
    $/ = $save_input_separator;
    return $record;
}
# get_annotation_and_dna
#
# - given GenBank record, get annotation and DNA

sub get_annotation_and_dna
    {my($record) = @_;

    my ($annontation) = ' ';
    my ($dna) = '' ;

# Now separate the annotation from the sequence data
    ($annotation, $dna) = ($record =~
/^ (LOCUS.*ORIGIN\s*\n) (.*) \/\/\n/s);
    # clean the sequence of any whitespace or / characters
    # (the / has to be written \/ in the character class, because
    # / is a metacharacter, so it must be "escaped" with \)
    $dna =~ s/[\s\/]//g;
    return ($annotation, $dna)
}
# search_sequence
#
# - search sequence with regular expression
sub search_sequence {
    my ($sequence, $regularexpression) = @_;
    my (Olocations) = ( );
    while ( $sequence =~ /$regularexpression/ig ) {
            push ( @locations, pos );
    }
    return (@locations);
```

```
}
# search_annotation
#
# - search annotation with regular expression

sub search_annotation {
   my ($annotation, $regularexpressio.n) = @_;
   my(Slocations) = ( );
   # note the /s modifier--, matches any character including newline
   while ( $annotation =~ /$regularexpression/isg ) {
         push ( @locations, pos );
   }
   return (@locations);
}
```

Example 9.5 generates the following output on our little GenBank library:

```
Sequence found in record at offset 0
Annotation found in record at offset 0
Sequence found in record at offset 6256
Annotation found in record at offset 6256
Sequence found in record at offset 12.366
Annotation found in record at offset 12366
Sequence found in record at offset 17730
Annotation found in record at offset 17730
Sequence found in record at offset 22340
Annotation found in record at offset 22340
```

The `tell` function reports the byte offset of the file up to the point where it's been read; so you want to first call `tell` and then read the record to get the proper offset associated with the beginning of the record.

Parsing Annotations at the Top Level

Now let's parse the annotations.

There is a document from NCBI we mentioned earlier that gives the details of the structure of a GenBank record. This file is *gbrel. txt* and is part of the GenBank release, available at the NCBI web site and their FTP site. It's updated with every release (every two months at present), and it includes notices of changes to the format. If you program with GenBank records, you should read this document and keep a copy around for reference use, and check periodically for announced changes in the GenBank record format.

If you look back at the complete GenBank record earlier in this chapter, you'll see that the annotations have a certain structure. You have some fields, such as LOCUS, DEFINITION, ACCESSION, VERSION, KEYWORDS, SOURCE, REFERENCE, FEATURES, and BASE COUNT that start at the beginning of a line. Some of these fields have subfields, especially the FEATURES field, which has a fairly complex structure.

But for now, let's just extract the top-level fields. You will need a regular expression that matches everything from a word at the beginning of a line to a newline that just precedes another word at the beginning of a line.

Here's a regular expression that matches our definition of a field:

```
/^[A-Z] .*\n(^\s.*\n) */m
```

What does this regular expression say? First of all, it has the `/m` pattern matching modifier, which means the caret ^ and the dollar sign $ also match around embedded newlines (not just at the beginning and end of the entire string, which is the default behavior).

The first part of the regular expression:

```
^[A-Z].*\n
```

matches a capital letter at the beginning of a line, followed by any number of characters (except newlines), followed by a newline. That's a good description of the first lines of the fields you're trying to match.

The second part of the regular expression:

```
r\s.*\n)*
```

matches a space or tab `\s` at the beginning of a line, followed by any number of characters (except newlines), followed by a newline. This is surrounded by parentheses and followed by a *, which means 0 or more such lines. This matches succeeding lines in a field, lines that start with whitespace. A field may have no extra lines of this sort or several such lines.

So, the two parts of the regular expression combined match the fields with their optional additional lines.

Example 9.6 shows a subroutine that, given the annotations section of a GenBank record storedma scalar variable, returns a hash with keys equal to the names of the top-level fields and values equal to the contents of those fields.

Example 9.6. Parsing Genbank annotation

```
#!/usr/bin/perl
# - test program for parse_annotation subroutine
use strict;
use warnings;
use BeginPerlBioinfo;      # see Chapter 6 about this module

# Declare and initialize variables
my $fh;
my $record;

my $dna;
my $annotation;
my %fields;
my $library = 'library.gb';

# Open library and read a record
```

```
$fh = open_file($library) ;

$record = get_next_record($fh);

# Parse the sequence and annotation
($annotation, $dna) = get_annotation_and_dna($record)

# Extract the fields of the annotation
%fields = parse_annotation($annotation);

# Print the fields
foreach my $key (keys %fields) {
    print "******** $key *********\n";
    print $fields{$key};
}
exit;
##########################################################
#####################
# Subroutine
##########################################################
#####################
# parse_annotation
#
# given a GenBank annotation, returns a hash with
# keys: the field names
# values: the fields

sub parse_annotation {

my($annotation) = @_;
my(%results) = ( );

while ( $annotation =~ /^[A-Z] .*\n(^\s.*\n)*/gm ) {
    my $value = $&;
    (my $key = $value) =~ s/^([A-Z]+).*/$1/s;
    $results{$key} = $value;
}
return %results;
}
```

In the subroutine *parse_annotation*, note how the variables `$key` and `$value` are scoped within the `while` block. One benefit of this is that you don't have to reinitialize the variables each time through the loop. Also note that the key is the name of the field, and the value is the whole field.

You should take the time to understand the regular expression that extracts the field name for the key:

```
(my $key = $value) =~ s/^([A-Z]+).*/$1/s;
```

This first assigns `$key` the value `$value`. It then replaces everything in `$key` (note the /s modifier for embedded newlines) with `$1`, which is a special variable pattern between the first pair of parentheses (`[A-Z]+`). This pattern is one or more capital letters (anchored to the beginning of the string, *i.e.*, the field name), so it sets `$key` to the value of the first word in the field name.

You get the following output from Example 9.6 the test just fetches the first record in the GenBank library):

```
******** SOURCE *********
SOURCE          Homo sapiens embryo male lung fibroblast
cell_line:HuS-L12 cDNA to
                mRNA.
  ORGANISM      Homo sapiens
                Eukaryota; Metazoa; Chordata; Craniata;
Vertebrata;     Euteleostomi;
                Mammalia; Eutheria; Primates; Catarrhini;
Hominidae;      Homo.
******** DEFINITION *********
DEFINITION Homo sapiens PCCX1 mRNA for protein containing CXXC domain 1,
                complete cds.
******** KEYWORDS *********
KEYWORDS
******** VERSION ********* AB031069.1 61:8100074
******** FEATURES *********
FEATURES        Location/Qualifiers
   source       1..2487
                /organism="Homo sapiens"
                /db_xref="taxon:9606"
                /sex="male"
                /cell_line="HuS-L12"
                /cell_type="lung fibroblast"
                /dev stage="embryo"

gene            229..2199
                /gene="PCCX1"

CDS             229..2199
                /gene="PCCX1"
                /note="a nuclear protein carrying a
```

```
PHD finger and a CXXC
             domain"
             /codon_start=1
             /product="protein containing CXXC
domain 1"    /protein_id="BAA96307.1"
             /db_xref="GI:8100075"
/translation="MEGDGSDPEPPDAGEDSKSENGENAPIYCICRKPDINCFMIGCD
NCNEWFHGDCIRITEKMAKAIREWYCRECREKDPKLEIRYRHKKSRERDGNERDSSEP
RDEGGGRKRPVPDPDLQRRAGSGTGVGAMLARGS?SPHKSSPQPLVATPSQHHQQQQQ
QIKRSARMCGECEACRRTEDCGHCDFCRDMKKFGGPNKIRQKCRLRQCQLRARESYKY
FPSSLSPVTPSESLPRPRRPLPTQQQPQPSQKLGRIREDEGAVASSTVKEPPEATATP
EPLSDEDLPLDPDLYQDFCAGAFDDHGLPWMSDTEESPFLDPALRKRAVKVKHVKRRE
KKSEKKKEERYKRHRQKQKHKDKWKHPERADAKDPASLPQCLGPGCVRPAQPSSKYCS
DDCGMKLAANRIYEILPQRIQQWQQSPCIAEEHGKKLLERIRREQQSARTRLQEMERR
FHELEAIILRAKQQAVREDEESNEGDSDDTDLQIFCVSCGHPINPRVALRHMERCYAK
YESQTSFGSMYPTRIEGATRLFCDVYNPQSKTYCKRLQVLCPEHSRDPKVPADEVCGC
PLVRDVFELTGDFCRLPKRQCNRHYCWEKLRRAEVDLERVRVWYKLDELFEQERNVRT
                           AMTNRAGLLALMLHQTIQHDPLTTDLRSSADR"
******** REFERENCE *********
REFERENCE    2 (bases 1 to 2487)
AUTHORS      Fujino,T., Hasegawa,M., Shibata,S.,
Kishimoto,T., Imai,S. and
             Takano,T.
TITLE        Direct Submission
JOURNAL      Submitted (15-AUG-1999) to the
DDBJ/EMBL/GenBank databases.
             Tadahiro Fujino, Keio University School of
Medicine, Department of
             Microbiology; Shinanomachi 35, Shinjuku-ku,
Tokyo 160-8582, Japan
             (E-mail:fujino@microb.med. keio.ac.jp,
             Tel: +81 -3-3353-12 11 (ex.62692) , Fax:+81-3-5360-1508)
******** ACCESSION *********
ACCESSION AB031069
******** LOCUS *********
LOCUS AB031069 2487 bp mRNA PRI
27-MAY-2000
******** ORIGIN *********
```

```
ORIGIN
******** BASE *********
BASE COUNT 564 a 715 c 768 g 440 t
```

As you see, the method is working, and apart from the difficulty of reading the regular expressions (which will become easier with practice), the code is very straightforward, just a few short subroutines.

Parsing the Features Table

Let's take this one step further and parse the features table to its next level, composed of the `source` , `gene`, and `CDS` features keys. (See later in this section for a more complete list of these features keys.) In the exercises at the end of the chapter, you'll be challenged to descend further into the FEATURES table.

To study the FEATURES table, you should first liik over the NCBI *gbrel.txt* document mentioned previously. Then you should study the most complete documentation for the FEATURES table available at http://www.ncbi.nlm.nih.gov/collab/FT/index.html.

Features : Although our GenBank entry is fairly simple and includes only three features, there are actually quite a few of them. Notice that the parsing code will find all of them, because it's just looking at the structure of the document, not for specific features.

The following is a list of the features defined for GenBank records. Although lengthy, I think it's important to read through it to get an idea of the range of information that may be present in a GenBank record.

allele

Obsolete; see `variation` feature key

attenuator

Sequence related to transcription termination

C_region

Span of the C immunological feature

CAAT_signal

CAAT box in eukaryotic promoters

CDS

Sequence coding for amino acids in protein (includes stop codon)

conflict

Independent sequence determinations differ

D-loop

Displacement loop

D_segment

Span of the D immunological feature

enhancer

Cis-acting enhancer of promoter function

[illegible]

Region that codes for part of spliced mRNA

gene

Region that defines a functional gene, possibly including upstream (promoter, enhancer, etc.) and downstream control elements, and for which a name has been assigned

GC_signal

GC box in eukaryotic promoters

iDNA

Intervening DNA eliminated by recombination

intron

Transcribed region excised by mRNA splicing

J_region

Span of the J immunological feature

LTR

Long terminal repeat

mat_peptide

Mature peptide coding region (doesn't include stop codon)

misc_binding

Miscellaneous binding site

misc_difference

Miscellaneous difference feature

misc_feature

Region of biological significance that can't be described by any other feature

misc_recomb

Miscellaneous recombination feature

misc_RNA

Miscellaneous transcript feature not defined by other RNA keys

misc_signal

Miscellaneous signal

misc_structure

Miscellaneous DNA or RNA structure

modified_base

The indicated base is a modified nucleotide

mRNA

Messenger RNA

mutation

Obsolete: see variation feature key

N_region

Span of the N immunological feature

old_sequence

Presented sequence revises a previous version

polyA_signal

Signal for cleavage and polyadenylation

polyA_site

Site at which polyadenine is added to mRNA

precursor_RNA

Any RNA species that isn't yet the mature RNA product

prim_transcript

Primary (unprocessed) transcript

primer

Primer binding region used with PCR

primer_bind

Noncovalent primer binding site

promoter

A region involved in transcription initiation

protein_bind

Noncovalent protein binding site on DNA or RNA

RBS

Ribosome binding site

rep_origin

Replication origin for duplex DNA

repeat_region

Sequence containing repeated subsequences

repeat_unit

One repeated unit of a repeat_region

rRNA

Ribosomal RNA

S_region

Span of the S immunological feature

satellite

Satellite repeated sequence

SCRNA

Small cytoplasmic RNA

sig_peptide

Signal peptide coding region

snRNA

Small nuclear RNA

source

Biological source of the sequence data represented by a GenBank record; mandatory feature, one or more per record; for organisms that have been incorporated within the NCBI taxonomy database, an associated `/db_xref="taxon:NNNN"` qualifier will be present (where `NNNNN` is the numeric identifier assigned to the organism within the NCBI taxonomy database) *stem__loop*

Hairpin loop structure in DNA or RNA

STS

Sequence Tagged Site: operationally unique sequence that identifies the combination of primer spans used in a PCR assay

TATA_signal

TATA box in eukaryotic promoters

terminator

Sequence causing transcription termination

transit_peptide

Transit peptide coding region

transposon

Transposable element (TN)

tRNA

Transfer RNA

unsure

Authors are unsure about the sequence in this region

V_region

Span of the V immunological feature

variation

A related population contains stable mutation

—

Placeholder (hyphen)

-10_signal

ribnow box in prokaryotic promoters

-35_signal

-35 box in prokaryotic promoters

3'Clip

3′-most region of a precursor transcript removed in processing

3'UTR

3′ untranslated region (trailer)

5'clip

5′-most region of a precursor transcript removed in processing

5'UTR

5′ untranslated region (leader)

These feature keys can have their own additional features, which you'll see here and in the exercises.

Example 9.7 finds whatever features are present and returns an array populated with them. It doesn't look for the complete list of features as presented in the last section; it finds just the features that are actually present in the GenBank record and returns them for further use.

It's often the case that there are multiple instances of the same feature in a record. For instance, there may be several exons specified in the FEATURES table of a GenBank record. For this reason we'll store the features as elements in an array, rather than in a hash keyed on the feature name (as this allows you to store, for instance, only one instance of an exon).

Example 9.7. Testing subroutine parse_features

```
# ! /usr/bin/perl
# - main program to test parse_features

use strict;
use warnings;
use BeginPerlBioinf o;      # see Chapter 6 about this module

# Declare and initialize variables
my $fh;
my $ re cord;
my $dna;
my $annotation;
my %fields;
my @ features;
my $library = ' library . gb ';

# Get the fields from the first GenBank record in a library
$fh = open_f ile ($library) ;

$record = get_next_record ($fh) ;
($annotation, $dna) = get_annotation_and_dna ($record) ;
%fields = parse_annotation ($annotation) ;
# Extract the features from the FEATURES table
@features = parse features ($fields {' FEATURES '});

# Print out the features
foreach my $feature (@features) {

# extract the name of the feature (or "feature key")
my($featurename) = ($feature =~ /^ (5

print "******** $featurename ********\n";
print $feature;
}
exit;
################################################################
#################
# Subroutine
```

```
#################################################################
#################
# parse_features
#
# extract the features from the FEATURES field of a GenBank record
sub parse_features {
   my ($features) = @_; # entire FEATURES field in a scalar variable
# Declare and initialize variables
   my (@features) = (); # used to store the individual features
# Extract the features
while ( $features =~ /^ {5}\S.*\n(^ {21 } \S . *\n) */gm ) {
     my $ feature = $&;
push (@features, $ feature);
}
return @features;
}
```

Example 9.7 gives the output:

```
********** source ***********
    source              1..2487
                        /organism="Homo sapiens"
                        /db_xref="taxon: 9606"
                        /sex="male"
                        /cell_line="HuS-L12"
                        /cell_type="lung fibroblast"
                        /dev_stage="embryo"
******** gene *********
gene                    229..2199
                        /gene="PCCX1"
******** CDS *********
CDS                     229..2199
                        /gene="PCCX1"
                        /note="a nuclear protein carrying a PHD finger and
                        a CXXC
                        domain"
                        /codon_start=1
                        /product="protein containing CXXC
domain 1"
                        /protein_id="BAA96307.1"
                        /db_xref="GI .-8100075"
/translation="MEGDGSDPEPPDAGEDSKSENGENAPIYCICRKPDINCFMIGCD
```

```
NCNEWFHGDCIRITEKMAKAIREWYCRECREKDPKLEIRYRHKKSRERDGNERDSSEP
RDEGGGRKRPVPDPDLQRRAGSGTGVGAMLARGSASPHKSSPQPLVATPSQHHQQQQQ
QIKRSARMCGECEACRRTEDCGHCDFCRDMKKFGGPNKIRQKCRLRQCQLRARESYKY
FPSSLSPVTPSESLPRPRRPLPTQQQPQPSQKLGRIREDEGAVASSTVKEPPEATATP
EPLSDEDLPLDPDLYQDFCAGAFDDHGLPWMSDTEESPFLDPALRKRAVKVKHVKRRE
KKSEKKKEERYKRHRQKQKHKDKWKHPERADAKDPASLPQCLGPGCVRPAQPSSKYCS
DDCGMKLAANRIYEILPQRIQQWQQSPCIAEEHGKKLLERIRREQQSARTRLQEMERR
FHELEAIILRAKQQAVREDEESNEGDSDDTDLQIFCVSCGHPINPRVALRHMERCYAK
YESQTSFGSMYPTRIEGATRLFCDVYNPQSKTYCKRLQVLCPEHSRDPKVPADEVCGC
PLVRDVFELTGDFCRLPKRQCNRHYCWEKLRRAEVDLERVRVWYKLDELFEQERNVRT
                                  AMTNRAGLLALMLH^r^QHDPLTTDLRSSADR
```

In subroutine `parse_features` of Example 9.7 the regular expression that extracts the features is much like the regular expression used in Example 9.7 to parse the top level of the annotations. Let's look at the essential parsing code of Example 9.7.

```
while( $features =~ /^ {5}\S.*\n(^ {21}\S.*\n)*/gm ) {
```

On the whole, and in brief, this regular expression finds features formatted with the first lines beginning with 5 spaces, and optional continuation lines beginning with 21 spaces.

First, note that the pattern modifier `/m` enables the ^ metacharacter to match after embedded newlines. Also, the `{5}` and `{21}` are quantifiers that specify there should be exactly 5, or 21, of the preceding item, which in both cases is a space.

The regular expression is in two parts, corresponding to the first line and optional continuation lines of the feature. The first part `^ {5}\S.*\n` means that the beginning of a line (`^`) has 5 spaces (`{5}`), followed by a non-whitespace character (`\S`) followed by any number of non-newlines (`.*`) followed by a newline (\n). The second part of the regular expression, `(^ {21}\S.*\n)*` means the beginning of a line (`^`) has 21 spaces (`{21}`) followed by a non-whitespace character (`\S`) followed by any number of non-newlines (. *) followed by a newline (`\n`); and there may be `0` or more such lines, indicated by the `()*` around the whole expression.

The main program has a short regular expression along similar lines to extract the feature name (also called the feature key) from the feature. So, again, success. The FEATURES table is now decomposed or "parsed" in some detail, down to the level of separating the individual features. The next stage in parsing the FEATURES table is to extract the detailed information for each feature. This includes the location (given on the same line as the feature name, and possibly on additional lines); and the qualifiers indicated by a slash, a qualifier name, and if applicable, an equals sign and additional information of various kinds, possibly continued on additional lines.

I'll leave this final step for the exercises. It's a fairly straightforward extension of the approach we've been using to parse the features. You will want to consult the documentation from the NCBI web site for complete details about the structure of the FEATURES table before trying to parse the location and qualifiers from a features.

You will want to consult the documentation from the NCBI web site for complete details about the structure of the FEATURES table before trying to parse the location and qualifiers from a feature. However, sometimes you just want to see if some word such as "endonulease" appears anywhere in the record. Fofcfckis, recall that you created a `search_annotation` subroutine in Example 9.5 that searches for any regular expression in the entire annotation; very often, this is mt-you really need. As you've now seen, however, when you really need to dig into the FEATURES table, Perl has its own features that make the job possible and even fairly easy.

INDEXING GENBANK WITH DBM

DBM stands for Database Management. Perl provides a set of built-in functions that give Perl programmers access to DBM files.

DBM Essentials

When you open a DBM file, you access it like a hash: you give it keys and it returns values, and you can add and delete key-value pairs. What's useful about DBM is that it saves the key-value data in a permanent disk file on your computer. It can thus save information between the times you run your program; it can also serve as a way to share information between different programs that need the same data. A DBM file can get very big without killing the main memory on your computer and making your program — and everything else — slow to a crawl. There are two functions, *dbmopen* and *dbmdose,* that "tie" a hash to a DBM file; then you just use the hash. As you've seen, with a hash, lookups are easy, as are definitions.

You can get a list of all the keys from a hash called `%my_hash` by typing `keys %my_hash`. You then can get a list of all values by typing values `%my_hash`. For large DBM files, you may not want to do this; the Perl function `each` allows you to read key-value pairs one at a time, thus saving the memory of your running program. There is also a *delete* function to remove the definitions of keys:

```
delete $my_hash{'DNA'}
```

entirely removes that key from the hash.

DBM files are a very simple database. They don't have the power of a relational database such as *MySQL, Oracle,* or *PostgreSQL*; however, it's remarkable how often a simple database is all that a problem really needs. When you have a set of key-value data (or several such sets), consider using DBM. It's certainly easy to use with Perl.

The main wrinkle to using DBM is that there are several, slightly different DBM implementations—NDBM, GDBM, SDBM, and Berkeley DB. The differences are small but real; but for most purposes, the implementations are interchangeable. Newer versions of Perl give you Berkeley DB by default, and it's easy to get it and install it for your Perl if you want. If you don't have really long keys or values, it's not a problem. Some older DBMs require you to add null bytes to keys and delete them from values:

```
$ value = $my_hash{"$key\0"};
chop $value;
```

Chances are good that you won't have to do that. Berkeley DB handles long strings well (some of the other DBM implementations have limits), and because you have some potentially long strings in biology, I recommend installing Berkeley DB if you don't have it.

A DBM Database for GenBank

You've seen how to extract information from a GenBank record or from a library of GenBank records. You've just seen how DBM files can save your hash data on your hard disk between program runs. You've also seen the use of `tell` and `seek` to quickly access a location in a file.

Now let's combine the three ideas and use DBM to build a database of information about a GenBank library. It'll be something simple: you'll extract the accession numbers for the keys and store the byte offsets in the GenBank library of records for the values. You'll add some code that, given a library and an offset, returns the record at that offset, and write a main program that allows the user to interactively request GenBank records by accession number. When complete, your program should very quickly return a GenBank record if given its accession number.

This general idea is extended in the exercises at the end of the chapter to a considerable extent; you may want to glance ahead at them now to get an idea of the potential power of the technique I'm about to present.

With just the appropriate amount of further ado, here is a code fragment that opens (creating if necessary) a DBM file:

```
unless (dbmopen (%my_hash, 'DBNAME', 0644)) {

      print "Cannot open DBM file DBNAME with mode 0644\n";
      exit;
}
```

`%my_hash` is like any other hash in Perl, but it will be tied to the DBM file with this statement. DBNAME is the basename of the actual DBM files that will be created. Some DBM versions create one file of exactly that name; others create two files with file extensions *.dir and .pag.*

Another parameter is called the mode. Unix or Linux users will be familiar with file permissions in this form. Many possibilities exist; here are the most common ones:

0644

You can read and write; others can just read.

0600

Only you can read or write.

0666

Anyone can read or write.

0444

Anyone can read (nobody can write).

0400

Only you can read (nobody else can do anything).

The *dbmopen* call fails if you try to open a file with a mode that assumes there are more permissions than were conferred on the DBM file when it was created. Usually, the mode 0644 is declared by the owner if only the owner should be allowed to write, and 0444 is declared by readers. Mode 0666 is declared by the owner and others if the file is meant to be read or written by anyone.

That's pretty much it; DBM files are that simple. Example 9.8 displays a DBM file that stores key-value pairs of accession numbers of GenBank records for keys, and byte offsets of the records as values.

Example 9.8. A DBM index of a GenBank library

```
#! /usr/bin/perl
#    -make a DBM index of a GenBank library,
#         and demonstrate its use interactively

use strict;
use warnings;
use BeginPerlBioinfo;       # see Chapter 6 about this module

# Declare and initialize variables
my $fh;
my $record;
my $dna;
my $annotation;
my %fields;
my %dbm;
my $answer;
my $offset;
my $library = 'library. gb';

# open DBM file, creating if necessary
unless (dbmopen (%dbm, 'GB', 0644)) {
    print "Cannot open DBM file GB with mode 0644\n";
    exit;
}

# Parse GenBank library, saving accession number and offset
in DBM file
$fh = open_file ($library) ;

$offset = tell ($fh);

while ($record = get_next_record ($fh) ) {
      # Get accession field for this record.
      ($annotation, $dna) = get_annotation_and_dna ($record);
%fields = parse_annotation ($annotation);
my $accession = $fields{ 'ACCESSION' };

# extract just the accession number from the accession field
# --remove any trailing spaces
$accession =~ s/^ACCESSION\s*//;

$accession =~ s/\s*$///;

# store the key/value of accession/offset
```

```
    $dbm{ $accession} = $offset;

    # get offset for next record
    $offset = tell($fh);
}
# Now interactively query the DBM database with accession numbers
# to see associated records
print "Here are the available accession numbers :"\n";
print join ("\n", keys %dbm), "\n";
print "Enter accession number (or quit) : ";
    while ( $answer = <STDIN> ) {
            chomp $answer;
            if ($answer =~ /^\s*q/) {
                    last;
            }
            $offset = $dbm{$answer};

    if ($offset) {
            seek($fh, $offset, 0) /
            $record = get_next_record ($fh) ;
            print $record;

    }else{
            print "Do not have an entry for accession number $answer\n";
    }
            print "\nEnter accession number (or quit) : ";
    }
    dbmclose (%dbm);
    close ($fh) ;
    exit;
```

Here's the truncated output of *Example 9.8:*

```
Here are the available accession numbers:
XM_006271
NM_021964
XM_009873
AB031069
XM_006269
Enter accession number (or quit) : NM_021964
LOCUS       NM_021964     3032 bp    mRNA      PRI
14-MAR-2001
DEFINITION Homo sapiens zinc finger protein 148 (pHZ-52)
(ZNF148), mRNA.
...
//
Enter accession number (or quit) : q
```

9.6 Exercise

Exercise 9.1

Go to the NCBI, EMBL, and EBI web sites and become familiar with their use.

Exercise 9.2

Read the GenBank format documentation, *gbrel. txt.*

Exercise 9.3

Write a subroutine that passes a hash by value. Now rewrite it to pass the hash by reference.

Exercise 9.4

Design a module of subroutines to handle the following kinds of data: a flat file containing records consisting of gene names on a line and extra information of any sort on succeeding lines, followed by a blank line. Your subroutines should be able to read in the data and then do a fast lookup on the information associated with a gene name. You should also be able to add new records to the flat file. Now reuse this module to build an address book program.

Exercise 9.5

Descend further into the *FEATURES* table. Parse the features in the table into their next level by parsing the feature names, locations, and qualifiers. Check the document *gbrel. txt* for definitions of the structures of the fields.

Exercise 9.6

Write a program that takes a long DNA sequence as input and outputs the counts of all four-base subsequences (256 of them in all), sorted by frequency.

PROTEIN STRUCTURE PREDICTION BY PROTEIN THREADING AND PARTIAL EXPERIMENTAL DATA

The Human Genome Project and other genome sequencing efforts are producing DNA sequence at a prodigious rate, yielding thousands to millions of new genes and proteins. A major challenge facing the entire biological community is to understand the functions of these newly identified proteins in biological systems and how the functions are performed at the molecular level. The three-dimensional (3D or tertiary) structures provide the essential information for defining the proteins' biological functions. In addition, the 3D structures provide a key to engineering proteins and designing drugs targeted at proteins related to diseases. As we move toward the post-genome era, we expect that the demand for rapid protein structure determination will grow drastically.

Traditional experimental methods for protein structure determination alone, such as X-ray crystallography and NMR (nuclear magnetic resonance), will probably not be able to keep up with the pace at which protein sequences are being generated. Computational methods could play a significant role, in conjunction with experimental methods, in protein structure determination on a genome-scale. The earliest work on prediction of protein tertiary structures can be traced back to 1970s. Significant advancement has been made in the capability of protein structure prediction since then. In the past several years, many nontrivial structure predictions prior to the experimental structures, turned out to be fairly accurate.

Most notably, the success of protein structure prediction has been demonstrated in the community-wide experiments on the "critical assessment of techniques for protein structure prediction" (CASP) which started about six years ago. In each of the past three CASP experiments and the ongoing CASP4, predictors are given a list of protein sequences whose structures have been solved experimentally (or are expected to be solved during the CASP prediction season) but unpublished. The prediction teams submit their predictions before a preset expiration date for each prediction target. Their prediction results are then evaluated against the experimental structures at the end of the prediction season. It has been shown that many predictions in CASP have good enough qualities to make some functional inferences.

Even in cases where the entire predicted structure is not accurate enough to make a direct functional analysis, the predictions, when in conjunction with supporting experimental data, can provide valuable insights into the structure and function of a protein. In particular, predicted structures can often give experimenters some leads for further studies. There are two main classes of tertiary structure prediction methods: ab initio structure prediction, and template-based structure prediction. Ab initio prediction methods attempt to predict protein structure based on physi-chemical principles directly. A comprehensive review of ab initio methods appears in chapter 16.

Template-based methods use known 3D structures in PDB (Protein Data Bank) as templates to derive the structure of a query protein. Homology modeling and fold recognition methods belong to this class. A *homology modeling* method (also known as *comparative modeling)* predicts the 3D coordinates of atoms of a query protein mainly based on an optimal sequence alignment between the query protein and a template protein with known structure. It is discussed in detail in chapter 17. *Fold recognition* methods attempt to recognize the "correct" templates from a structure library for the query protein and generate an alignment between the query and the recognized template protein, from which the backbone structure of the query protein can be predicted.

Recognition of the correct templates are typically achieved by either the sequence profile approach or the sequence-structure comparison (or threading) approach. *Sequence profil methods* detect remote homologs through recognizing similar sequence profiles exhibited by the templates and by the query protein and its homologs through multiple sequence alignments. Threading identifies a homolog or analog through aligning the query protein sequence onto template structures. The main difference between threading and sequence profile methods is that threading uses not only information related to sequence but also structure information such as residue-residue contact patterns in a structure. As researchers start to develop hybrid methods through combining information from different methods, the boundaries between these methods become more and more blurred. For example, one can use the alignment derived from fold recognition in homology modeling, or assemble partial structures predicted by threading in an ab initio prediction. Each prediction method has its own strength and limitations. In theory, ab initio methods are the most general prediction methods, but at the current stage, they are far from being practical due to their poor prediction reliability and long computing time, though a few isolated successful predictions have been reported.

Homology modeling, as a relatively mature method, has been widely used. It can provide relatively reliable atomic coordinates with a low root mean square deviation (rmsd) between a model and a high-quality experimental structure. However, homology modeling applies to only proteins having structure templates with high sequence similarity. Fold recognition, which does not require significant sequence similarity between a query and the template protein, has gained significant ground in recent years. It has substantially extended the applicability of template-based methods and often generates useful backbone structures.

When compared with ab initio methods, the search space for threading is intrinsically discrete rather than continuous, implying a much lower computational cost (time) needed for searching for an optimally folded state. Among these prediction methods, protein threading probably applies to the largest class of proteins, though it is limited to backbone predictions only. A new trend in structure prediction is to incorporate partial experimental data as constraints in the computation process, blurring the boundary between structure prediction and determination. Such data may include anything that is relatively easy to obtain and helpful for structure calculation, ranging from disulfide bonds, active sites, cross-linking data from mass spectrometry, or sparse NMR data, to X-ray or neutron scattering data.

The merge of sophisticated computational methods and partial experimental data could potentially lead to new and more efficient paradigms of protein structure determination. This chapter focuses on protein structure prediction based on protein threading. We will first introduce the fundamentals of protein threading and use our own program PROSPECT as an example to illustrate the basic ideas of threading methods. Then we will describe enhanced threading methods with partial experimental data. In the last part of this chapter, we will discuss challenging issues and future outlook of the threading method. A reader can find more information about threading from the cited literature and the Web pages listed in the appendix

FUNDAMENTALS OF PROTEIN THREADING

The idea of threading was originated from the observation that proteins with no apparent sequence similarity could have similar structural folds. Recent studies have further indicated that the total number of different structural folds in nature may be quite small, possibly in the range of a few thousand or even fewer, which is at least two orders of magnitude fewer than the number of known protein sequences. Having realized this, a structure prediction problem can be potentially reduced to a recognition problem—which fold(s) the query protein sequence will fold into—plus a modeling problem of side chains. The realization shifted the paradigm of protein structure prediction. An inverse folding problem was then formulated—searching for sequences that are compatible with a specified structural fold based on sequence-structure relationships.

It has long been known that different amino acids may prefer different structural environments: for example, a hydrophobic amino acid tends to be in the interior of a globular protein, and proline rarely occurs in an α-helix. In addition, certain amino-acid pairs (*e.g.*, salt bridges) in space are more favorable than other pairs. Using these sequence-structure relationships, one can differentiate which sequence-structure pairs are more favorable in order to predict which structure is most likely to be adopted 3 by a given sequence. This new paradigm for protein structure prediction, called*"threading,"* achieved its initial success by the pioneering work of Eisenberg and colleagues. Since then, threading has been developed into a widely used technology for structure prediction through the further efforts of many other research groups. The computational formulation of threading can be summarized as follows.

Given a query protein sequence s of unknown structure, threading searches a structure template library T to find the best sequence-structure pair $\boldsymbol{s}$-$\boldsymbol{t}$, $\boldsymbol{t} \in \boldsymbol{T}$, measured by the overall preference of individual residues to their structural environment and of residue-residue contacts. A threading method typically consists of four components: (1) a library T of representative 3D protein structures for use as templates; (2) an energy function for measuring the fitness between a query $\boldsymbol{s}$ and a template $\boldsymbol{t}$, where $\boldsymbol{t} \in \boldsymbol{T}$; (3) a threading algorithm for searching for the lowest energy among the possible alignments for a given ***s-t*** pair; and (4) a criterion for estimating the confidence level of the predicted structure.

Fold Templates

Protein threading rests on the premise that the 3D structures of proteins have been better conserved during evolution than their sequences. Proteins can be classified into groups, at different levels, according to their structural and evolutionary relationships. A widely used classification scheme consists of three levels of groups: family, super-family, and fold. A *family* consists of proteins that have high sequence identity among each other and share a common evolutionary ancestor. Proteins of different families sharing a common evolutionary origin (reflected by their common structural and functional features) are placed in the same *superfamily*. Different superfamilies are grouped into a *fold* family if their proteins have the same major secondary structures in the same arrangement and with the same topological connections.

The structural similarities among proteins of the same fold family (but not the same superfamily) may arise just from the protein energetics favoring certain packing arrangements instead of a common evolutionary origin. The structural difference generally gets larger among proteins of the same family, superfamily, and fold. Most template libraries of the existing threading programs are based on three widely used databases of protein structure classifications: CATH FSSP and SCOP. CATH is a hierarchical classification of protein domain structures. FSSP contains the similar information but is based on protein chains rather than domains; in addition, it contains sequence neighbours and multiple structure alignments. SCOP uses a manual procedure to classify protein domains into folds, superfamilies, and families. Hence, its classification is probably of higher quality than the other two. However, SCOP has not been updated as frequently as desired simply due to the amount of manual work involved, whereas FSSP and CATH have been following the Protein Data Bank (PDB) updates closely.

The classifications for folds by the three databases differ somewhat due to their different classification criteria (*e.g.,* classification on a whole chain or a structure domain) and structure-structure comparison methods. As a result, the number of folds (or *unique* folds) differ among the three databases. Based on the data available as of June 2000, CATH contains 580 folds FSSP has 556 folds and SCOP consists of 548 folds. In principle, one can use one representative from each fold family as the template library for threading. However, it is often helpful to include a representative for each superfamily or even each family in the template library to cover variations among structures of a fold family and to achieve better prediction accuracy. In designing the template library, one can use both protein chains as well as protein domains. The general applicability of protein threading is based on the widely held belief that there are only a small number of protein folds.

The argument used to support this is that the existing protein structures have evolved from a small number of *"primordial folds."* During evolution, although protein sequences have diverged significantly, their 3D structural folds have been more conserved and maintained a small number. Another line of argument for the small size of the fold universe is a stereo-chemical one—the vast majority of geometrically possible folds are simply not energetically favorable enough to exist due to stereo-chemical constraints and physical interactions. This suggestion is supported by a computer simulation of protein folding using a simple lattice model.

The simulation has demonstrated that only a handful of protein like folds can adopt many different sequences and keep the energy favorable, whereas other folds do not. The results suggest that protein structures have been selected in nature because they are readily designed and stable against mutations, and that such a selection simultaneously leads to thermodynamic stability.

This line of argument seems to be convincing and may possibly explain the assumption used in the first argument, that the number of primordial folds was small. The next question is how many unique folds for globular proteins exist in nature. Many suggest that this number is less than a thousand, so about half of all possible unique folds are already known. A recent estimate even suggests that the total number of unique folds may be as low as 650. These estimates are based on statistical analysis of the novel-fold occurrence in the PDB. However, there is a danger in deriving incorrect conclusions from such an analysis due to the possibility that the distribution of the PDB structures is biased. For example, some types of proteins are difficult to purify or are not soluble enough to have either crystal structures or NMR structures. It is quite possible that one thousand or fewer folds may be an underestimate. On the other hand, some have suggested that the number of unique folds could be as high as eight thousand.

The argument was that 3 per cent of sequence databases has yielded 80 folds at the time of estimation. If the sequences represent one-third of all superfamilies, the number of folds should be $80 \times 3/0.03 = 8{,}000$. The underlying assumption used here is that every fold is equally populated in protein sequences. This is evidently incorrect. For example, the TM barrel fold occurs much more frequently than other folds. Hence, the eight thousand folds may be an overestimate. Though it is still an open question about the number of folds in nature, it is clear that the percentage of proteins being submitted to PDB that have new folds has been decreasing.

As one can see from the statistics at the PDB Web site, http:// www.rcsb.org/pdb/ holdings.html, this percentage has dropped from about 30 per cent in the 1980s to about 10 percent during the past three years. These numbers may have a more practical meaning for threading: given a new sequence of globular protein, what is the probability that it has a template in PDB? The Live Bench program recently carried out a survey on 125 structures submitted to PDB between October 29, 1999, and April 6, 2000. Only five proteins had no structurally similar proteins in PDB; 96 per cent proteins submitted to PDB during this period are not unique folds. This percentage may be a bit too high for globular proteins genome-wide, given the biased distribution of proteins in PDB. However, it is safe to say that the percentage should be at least 50-70 per cent. This number is clearly growing as more and more structures are being solved and put into PDB.

Energy Function

The energy function describes how favorable an alignment between a query sequence and a template structure is. Threading generally uses knowledge-based energy functions rather than physical energies. The main consideration is that even though a query protein may share the same fold of the template, the exact 3D coordinates of their corresponding atoms may differ to a degree that the physical energy (including bond, van der Waals, electrostatics based on CHARMM potential of the correct query-template alignment may not be favorable. This is simply because these physical energies are too sensitive to small displacement of atomic coordinates, making them less suitable for threading.

Another consideration comes from the fact that the calculation of these physical energies is too time-consuming. A recent study has shown that knowledge-based threading energy by and large captures the detailed atomic potentials. Theoretically, one could build a probabilistic model of knowledge-based threading energy that accounts for different factors and the correlations between these factors. Such a model could directly give the probability of adopting a certain fold by a given sequence.

However, this type of model has not been successfully developed yet, due to our limited understanding of the complicated factors that determine a fold. Currently, most researchers use simplified energy functions without calculating probabilities or considering any correlation between different energy terms. A typical threading energy function has the following form:

$$E_{total} = E_{mutate} + E_{gap} + E_{single} + E_{pair} \quad ..(9.1)$$

The mutation energy E_{mutate} describes the compatibility of substituting one amino acid type by another; E_{gap} is the alignment gap penalty; the singleton energy E_{single} represents a residue's preference to its local secondary structures (α-helix, β-strand, and loop) and its preference to being in certain solvent environment (either exposed to solvent or in the interior of the protein); E_{pair} is the pairwise potential between spatially close residues that are not neighbours in the protein sequence. The different energy terms are weighted. The mutation energy and the alignment gap penalty are similar to the ones used in sequence alignments.

The original concept of threading did not include the mutation energy but assumed that sequence-structure alignment (*i.e.*, E_{sinlge} and E_{pair}) should be sufficient for fold recognition. Prediction practices in the past few years have clearly demonstrated the usefulness of the mutation energy, so, most existing threading programs include it. Several matrices have been developed based on mutation rates found in sequence databases; the most popular ones are the PAM and BLOSUM matrices. BLOSUM-62 is a widely used matrix for searching for close homologs, whereas PAM250 is more suitable for identifying remote homologs. It has been shown that PAM250 is one of the best substitution matrices available for threading.

The gap penalty is often a linear function of the gap size, with a penalty for opening a gap and a small penalty for each extension thereafter. Both E_{single} and E_{pair} are typically derived from Boltzmann statistics from a non-homologous protein database. The basic idea is that if an amino acid is frequently observed in the interior of protein structures, a favorable energy value will be rewarded when it is aligned to an interior position of a template. Though most of the threading programs follow a similar principle in implementing their energy functions, detailed variations may exist. For example, some use the degree of environmental polarity (the degree of burial by polar rather than apolar atoms [Bow'e et. al. 1991] and/or secondary structure fitness between the template and the query protein [Rost 1995], whereas others may not explicitly use the singleton term [Bryant and Lawrence 1993] as they believe that the information of singleton energy is covered in the pairwise term (*e.g.*, a favorable packing between hydrophobic residues reflects a similar information of solvent accessibilities). The E_{pair} term is based on a pairwise energy parameter $e_{pair}(i, j)$, which describes the preference for the combination of two spatially close residues of type i and j (say, up to 15 Å between the C_β atoms). Some threading programs further divide this range into subintervals and have a different preference value for each subinterval, whereas others simply treat the whole range as one interval.

Different programs may measure the distance between two residues somewhat differently. Some measure the distance between the C_β atoms whereas others may measure between the backbone atoms or even between projected centroids of side chains. There are also more sophisticated methods for measuring residue-residue contacts, such as *"visible volume"* or Voronoi contacts. We now use our own threading program PROSPECT as an example to illustrate how the E_{single} and E_{pair} terms are calculated. We calculate Esingle based on a singleton energy parameter $e_{single}(i,ss,sol)$ that is, $E_{Single} = \Sigma_i\, e_{single}(i,ss_i,sol_i)$. $e_{single}(I,ss,sol)$ describes the energy or preference for a particular combination of amino acid type *i*, secondary structure type *ss*, and solvent accessibility type *sol*, that is

$$e_{single}(i,\ ss,\ sol) \;=\; -\log\frac{N(i,ss,sol)}{N_E(i,ss,sol)} \qquad ...(9.2)$$

where log denotes the natural logarithm; $N(i,ss,sol)$ is the number of amino acids of type *i* in *ss* with *sol*, counted from the database; $N_E(i,ss,sol)$ is the average of the reference state—the estimated number of amino acids of type *i* in *ss* and *sol* assuming *i*, *ss*, and *sol* are independent. $N_E(i,ss,sol)$ is calculated by

$$N_E(i,ss,sol) \;=\; \frac{N(i)N(ss)N(sol)}{N^2} \qquad ..(9.3)$$

where $N(i)$ is the number of amino acids of type *i*, $N(ss)$ is the number of residues in secondary structures of type *ss*, $N(sol)$ is the number of residues with solvent accessibility type *sol*, and *N* is the total number of residues (in our database).

Similarly, E_{pair} is calculated based on a pairwise energy parameter $e_{pair}(i,j)$, that is, $E_{pair} = \Sigma_{i\le j}\, e_{pair}(i,j)$. In PROSPECT, the cutoff distance for $e_{pair}(i,j)$ is 7.0 Å between C_β atoms, which accounts for most of the important inter-residue interactions. We only consider the residue pairs that are separated by at least three amino acids in the protein sequence[2]. $e_{pair}(i,j)$ is derived from the frequency of the inter-residue pairs, that is,

$$e_{pair}(I,j) \;=\; -\log\frac{M(i,j)}{M_E(i,j)},\quad M_E(i,j) = \frac{M(i)\,M(j)}{M} \qquad ..(9.4)$$

where $M(i,j)$ is the number of pairs between residues of types *i*, and *j* in the database; $M_E(I,J)$ is the average of the reference state, the estimated number of i-j pairs assuming that residues of types *i* and *j* form pairs without any preference. $M(k) = \Sigma_s\, M(k,s)\ (k = i,j)$. *M* is the total number of pairs—$M = \Sigma_{ij} M(i,j)$. In the calculation, the number of pairs between residues of types *i* and *j* are partitioned equally in $M(i,j)$ and $M(j,i)$ such that $M(i,j) = M(j,i)$. All statistics are collected from our protein database.

Alignment Algorithm

If we do not consider the pairwise energy, a threading problem is essentially the same as a sequence alignment problem. Such a problem can be solved efficiently by a dynamic programming approach. There are a number of computer programs that use dynamic programming for their threading problem, such as 123D TOPITS, SAS and the UCLA-DOE Structure Prediction Server. An advantage of a threading algorithm without considering pairwise energy is its speed. However, without pairwise interactions, the threading accuracy is severely compromised. Threading with pairwise terms and variable-length gaps is generally considered to be a very difficult problem.

Under the assumption that pairwise interactions between every pair of amino acids in a protein structure should be considered in threading, the problem was proved to be NP-hard (meaning that the problem's intrinsic computational complexity is too high for an efficient rigorous solution). Two previous existing threading programs with rigorous solutions both have exponential computational complexity. To overcome the computational difficulty, several approaches have been proposed. One approach is to employ the "frozen approximation," that is, when calculating a pairwise contact potential for a residue it is assumed that its contacting residues are the same in the template and in the query structure. Such an assumption has reduced the general threading problem to a problem solvable by dynamic programming but at the expense of threading accuracy.

Another approach attacks the problem through statistical sampling. Generally this type of approach does not guarantee to find the globally optimal threading alignment, though they typically work better than the frozen approximation approach. We have developed a threading algorithm that guarantees to solve the globally optimal threading problem efficiently under the assumption that residue-residue contact potentials need to be considered only between spatially close pairs in threading. The detailed algorithm will be discussed in the following section.

Assessment of Threading Results

A threading score between a query sequence and a template structure may not provide enough information about whether the template is the *"correct"* fold. This is because the scores are generally not normalized to the same scale. Consider a query protein and a false template. It is conceivable that their threading score would be improved if we artificially increased the length of the template sequence by appending some irrelevant protein sequence to it without actually changing the *"significance"* of the threading alignment. So from the threading scores between a query and a pool of templates, we generally cannot tell if the query's correct fold template is in the pool, nor can we always tell which is the correct fold even if it is there.

There have been a number of attempts to "normalize" threading scores so that they can be compared with each other. An early attempt was to use *s-jcore*.

For the energy E resulting from a particular alignment, the z-score of E is defined as

$$z = \frac{E - \bar{E}}{\sigma}$$

where $\bar{E}$ and σ are, respectively, the average and the standard deviation of the energy distribution resulting from the same alignment after re-shuffling the amino acids of the query sequence. However, it has been shown that the z-score is clearly not effective.

There have also been attempts to use the P-value scheme as a way to assign a meaning to a threading score. P-value, which estimates the probability of having a particular alignment score between two random sequences, have been successfully applied to sequence alignment, thanks to Karlin's seminal work on a rigorous model of gapless alignments. Due to the lack of a rigorous model for threading, the P-values are typically estimated through compiling a *"large"* number of threading scores between a query sequence and a template after randomly shuffling its residues.

Although some usefulness of the estimated P-value has been demonstrated, the problem of developing a rigorous P-value scheme for threading remains a challenging open problem. The recent work of Jones (1999) seems to have provided a practical way to "normalize" the threading scores. The idea is to feed a neural network the threading scores along with various normalization factors, such as sequence length, and to let the neural network learn to "optimally" combine these factors based on a training set. The result of the training is a mapping from a threading score to a real number between 0 and 1, with an estimated probability that this number corresponds to a correct fold recognition.

PROSPECT: A THREADING BASED PROTEIN STRUCTURE PREDICTION SYSTEM

PROSPECT (PROtein Structure Prediction and Evaluation Computer Toolkit) is a computer program that we have developed for protein structure prediction. The core of the program is a threading algorithm which guarantees to find the globally optimal alignment between a query sequence and a template structure for an energy function. Compared with other threading programs, two unique features of the algorithm are (1) that it finds the globally optimal alignment efficiently when using a widely accepted cutoff distance between residue-residue contacts, 7 Å between the C_β atoms; and (2) that it allows a user to incorporate experimental data as constraints in the threading process, and guarantees to find the globally optimal threading under the specified constraints.

Currently it allows following types of constraints: (1) disulfide bonds between specified residues; (2) active sites involving a specified set of residues; (3) long-range NOE (nuclear Overhauser effects) restraints from NMR experiments; (4) secondary structures predicted by computer programs like PHD or determined from chemical shift data of NMR experiments; and (5) position-dependent profiles based on multiple sequence alignment using SAM PROSPECT provides a confidence value for each of its predictions being a true fold-recognition. Recently, PROSPECT has been ported to a 184-node IBM/SP3 supercomputer at Oak Ridge National Laboratory. In this section, we will first give an informal description of the threading algorithm and explain how its computational efficiency is achieved. Then we will briefly explain how we assign a confidence value to each threading result, followed by PROSPECT'S prediction performance. This section concludes with a short description on the application side of the PROSPECT program.

Algorithm

The basic problem solved by PROSPECT's threading algorithm is to find an alignment between a query sequence and a template structure, which optimizes an energy function. To simplify the problem, we have applied two widely adopted assumptions: (1) no gap can occur within a core secondary structure; (2) any pairwise interaction involving a non-core secondary structure is ignored. Figure 9.1 shows a schematic of such an alignment. What makes the threading problem difficult is the consideration of pairwise contacts, which separate the problem from sequence alignment. When we are aligning the first part of the query sequence with the first core (and its loop regions), we do not have any knowledge about which residues of the query protein will be aligned to the third, fourth, and fifth cores of the template (in the final optimal alignment), and so do not have enough information for calculation of the corresponding pairwise contact potentials.

This illustrates why the dynamic programming scheme, which has been widely used for sequence alignment, does not work for the threading problem. The very basic idea of our solution to the threading problem follows the idea outlined above—considering all possible amino acid types when considering which type of amino acid is aligned to the template position at the other end *(e.g.,* the third, fourth, and fifth cores of the above example) of each arc, and calculating the pairwise contact potential for each amino acid type. Clearly doing this directly is not computationally practical because of the enormous number of combinations we need to consider. We have developed an efficient way to implement this idea, which we now explain. Detailed formal description can be found elsewhere. The algorithm employs a *divide and conquer* strategy to solve the optimal threading problem.

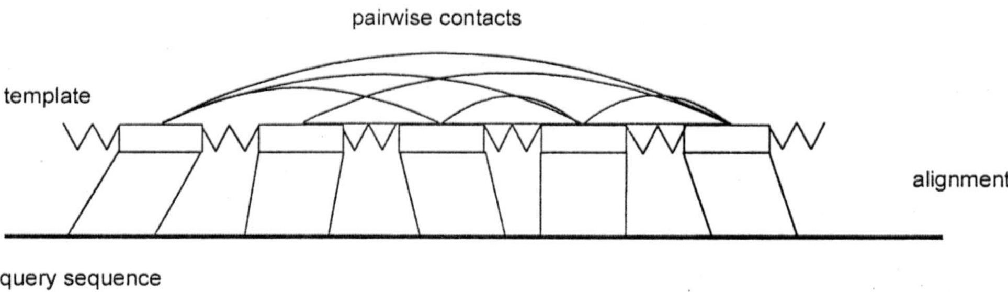

Fig. 9.1. A schematic sequence-structure alignment. Each box represents a core secondary structure; the connecting line between two boxes represents a loop region of the template structure. A core secondary structure, or a core, means an a-helix or a β-strand. An arc between two cores indicates that at least one pair of residues from the two cores have contacts *(i.e.,* their Cβ atoms are within our cutoff distance). The thick horizontal line represents a query sequence. Lines connecting the query sequence and the template represent a possible sequence-structure alignment.

For this purpose, we first pre-process the template by repeatedly dividing (bi-partitioning) it into substructures until each substructure contains only one core secondary structure. Dividing the template cuts an interaction between two cores into two *open links,* represented as an arc with one of its ends being a hollow circle. Our divide and conquer algorithm works correctly on any bi-partition of the template. However, the way a template is partitioned affects the computing time. algorithm solves the entire optimal alignment problem by recursively solving a series of subalignment problems between substructures and subsequences, under various constraints, and then combining these subalignments in a consistent and optimal way.

In this example, the substructure *AB* is partitioned into two cores, *A* and *B*. The interaction link between *A* and *B* in the partition is cut into two open links, that is, a_3 and b_1. The algorithm first calculates the alignment score between *A* (similarly *B)* and each sequence position (meaning that the leftmost residue of *A* is aligned with that position). Because we assume that there is no alignment gap within a core alignment, this score can be calculated by simply adding the singleton scores, E_{single}, of the aligned residues and structural positions, plus the E_{pair} scores. The calculation of E_{pair} is tricky because we do not know the sequence positions that the cores at the other ends of the open links a_1, a_2, and a_3 are aligned to. Once the other end of an open link is determined, all the pairwise interactions between the two cores connected by the open link can be subsequently calculated.

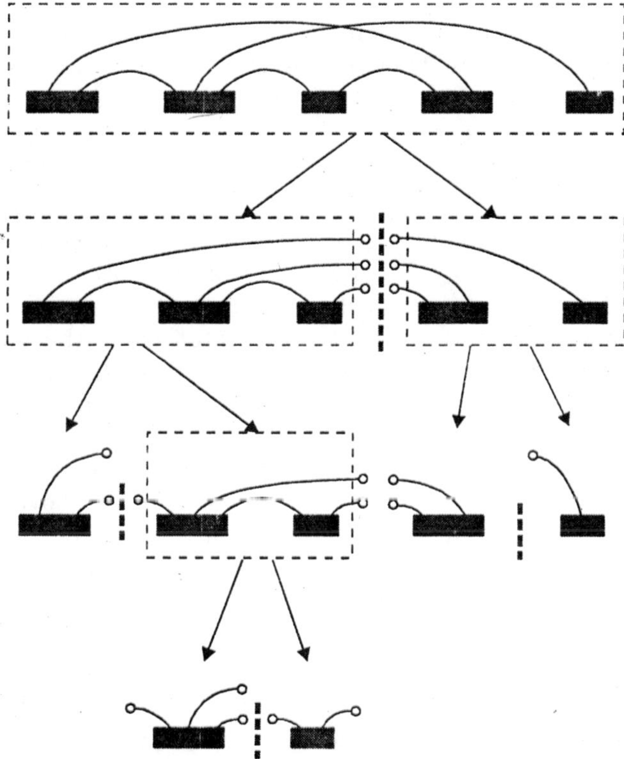

Fig. 9.2. A partition for a **template** structure. The first row shows the template with five core secondary structures. The second row shows a partition of the template into two substructures, one with three cores and the other with two cores. A broken arc ended with a circle is called an *open link.* The third and fourth rows show further partition of the template until each substructure contains only one core secondary structure. Note that a partition forms a tree structure as indicated by the arrows.

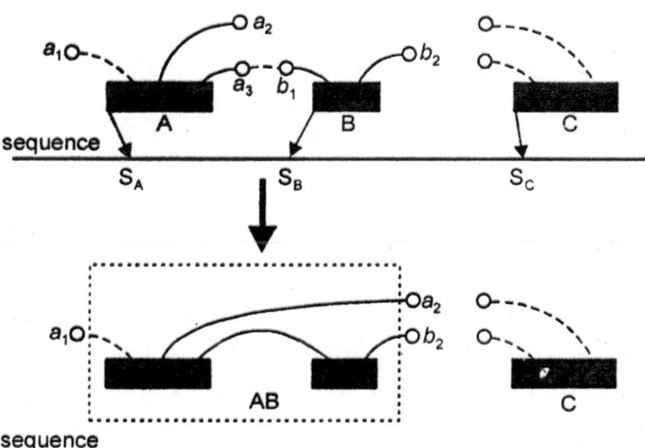

Fig. 9.3. A schematic example of the divide-and-conquer algorithm.

To overcome the missing information, we simply consider all possible *legal* alignments of these cores. Note that not every combination of the alignments of these cores makes a legal (overall) alignment as some of them may (1) violate the relative order of these cores *(e.g.,* the first core is aligned to a sequence position that is to the right of the aligned sequence position of the fourth core); (2) overlap each other; and (3) violate the allowed minimum and maximum length difference in a loop alignment. Though now we have to consider many possible aligned positions of the cores connected with a_1, a_2, a_3 (from now on, we simply call them *assignments* to a_1, a_2, a_3), we have enough information to calculate the pairwise contact term E_{pair} for each fixed assignment to the open links. To avoid double counting, we treat the two open links created from each interaction differently.

We use a dashed arc to represent one open link and a solid arc to represent the other. We only calculate the pair contact energy E_{pair} for the solid arc; both open links are assigned to the aligned position of the core with a dashed arc. Here only a_2 is used to calculate E_{pair}. The corresponding E_{pair} terms for a_1 and a_3 will be calculated when the alignment for the first and the third cores are calculated, respectively. In addition, we define $a_3 = b_1 = s_A$, where s_A is the aligned position of the leftmost residue of core *A* on the sequence. After the algorithm calculates the core alignment scores for each core under various assumptions that their open links are assigned to particular sequence positions, it calculates the alignment scores for larger substructures consisting of multiple cores. We continue to use the above example to illustrate the basic idea. We now want to calculate the optimal alignment score for *AB* under the assumption that a_1, a_2, and b_2 are assigned to particular sequence positions. Calculation of this optimal score consists of two parts: (1) the sum of the alignment scores for *A* and for *B* under the condition that the two partial alignments for *A* and *B* are consistent, and *(2)* the alignment score of the loop connecting *A* and *B*. The optimal alignment score for *AB* is the lowest combined score of (1) and (2) among all possible legal assignments to a_3 and b_1.

To find the lowest combined score, the algorithm goes through all the alignment scores for *A* and *B* under the conditions (1) that a_3 and b_1 have the same sequence-position assignment, and (2) that $a_2 = b_2$, because they point to the same core (i.e. core C). For each such assignment, the algorithm calculates the optimal loop alignment (now the aligned positions of both *A* and *B* are fixed) using the dynamic programming method and adds up the total alignment score for *AB*. The assignment to a_3 (b_1) that has the lowest combined score gives the optimal alignment score for *AB* under the specified assignment condition. This process continues for larger substructures (*e.g.*, the next step is substructure *ABC*) until the top level of the partition tree is reached, and the whole template is considered. Note that in the lower level calculations, the algorithm repeatedly solves the *constrained* alignment problem, such as finding the optimal alignment score for a substructure under the condition that its open links have particular assignments.

On the top level, the whole structure has no open links, and the optimal alignment obtained gives the final solution to our threading problem. We have found that the computational bottleneck of this algorithm is the consideration of all legal combinations of link assignments. More specifically, the dominating term in the algorithm's computational complexity (running time) increases exponentially with the maximum number of links among all division points. Fortunately, this (maximum) number is generally a small number if we only consider pairwise contacts between residues that are spatially close, given the optimal way to divide a template. We have previously demonstrated that the maximum number of links changes according to how a template structure is divided (*e.g.*, where to cut first, and then second, etc.), and presented an algorithm for finding a division scheme that minimizes the maximum link number among all division points.

The minimized maximum link number is called the *topological complexity* of a template. The topological complexity is a small number (≤ 8) in all the cases that we have studied when the cutoff distance between the C_β atoms is no more than 15 Å. Table 9.1 shows the distribution of the topological complexity versus the cutoff distance between C_β atoms for pairwise contacts. The statistics were compiled from 750 structures in a subset of the FSSP database.

Table 9.1. Distribution of topological complexity

TC	Cutoff (A) 5	6	7	8	9	10	11	12	13	14	15
0	52 (7%)	52 (7%)	52 (7%)	52 (7%)	52 (7%)	52 (7%)	52 (7%)	52 (7%)	52 (7%)	52 (7%)	52 (7%)
1	87 (12%)	87 (12%)	87 (12%)	87 (12%)	87 (12%)	87 (12%)	87 (12%)	87 (12%)	87 (12%)	87 (12%)	87 (12%)
2	218 (29%)	162 (22%)	135 (18%)	118 (16%)	112 (15%)	106 (14%)	102 (14%)	98 (13%)	96 (13%)	94 (13%)	92 (12%)
3	364 (49%)	318 (42%)	245 (33%)	189 (25%)	152 (20%)	131 (17%)	114 (15%)	104 (14%)	93 (12%)	83 (11%)	77 (10%)
4	29 (4%)	126 (17%)	216 (29%)	245 (33%)	218 (29%)	198 (26%)	185 (25%)	175 (23%)	161 (21%)	156 (21%)	154 (21%)
5	0 (0%)	5 (.7%)	15 (2%)	58 (8%)	121 (16%)	156 (21%)	162 (22%)	151 (20%)	136 (18%)	120 (16%)	105 (14%)
6	0 (0%)	0 (0%)	0 (0%)	1 (.1%)	8 (1%)	19 (3%)	47 (6%)	82 (11%)	117 (16%)	135 (18%)	141 (19%)
7	0 (0%)	0 (0%)	0 (0%)	0 (0%)	0 (0%)	1 (.1%)	1 (.1%)	1 (.1%)	8 (1%)	23 (3%)	38 (5%)
8	0 (0%)	0 (0%)	0 (0%)	0 (0%)	0 (0%)	0 (0%)	0 (0%)	0 (0%)	0 (0%)	0 (0%)	4 (.5%)

Each column represents a cutoff distance, and each row represents the number of proteins having a particular topological complexity (TC).

The lengths of these proteins range from 31 to 793; the numbers of their core secondary structures range from 1 to 34. By using a similar analysis to that of Xu et al. (1998a, b), it can be proved that this threading algorithm runs in a time proportional to $mn + Mn^{TC}N^{TC/2}$ where m and n represent the lengths of the template and query protein sequences, respectively; M is the number of core secondary structures of the template; N is the maximum allowed difference between the lengths of two aligned loops; and TC is the topological complexity of the template. This function shows how various parameters affect the running time of the algorithm. Using a recently improved version of PROSPECT, the actual threading time on a workstation is typically less than a minute.

Performance by Protect : In a recent study, we tested PROSPECT on 312 pairs of query-template proteins, each pair being from the same superfamily and having a sequence identity $\leq 30\%$. We randomly selected 175 pairs as the training set to obtain the weights between different energy terms, and another 137 pairs as the independent test set to verify the threading performance. Figure 9.4 gives a benchmark of CPU time for threading the 312 pairs on a Linux workstation with a Pentium III 550 MHz CPU.

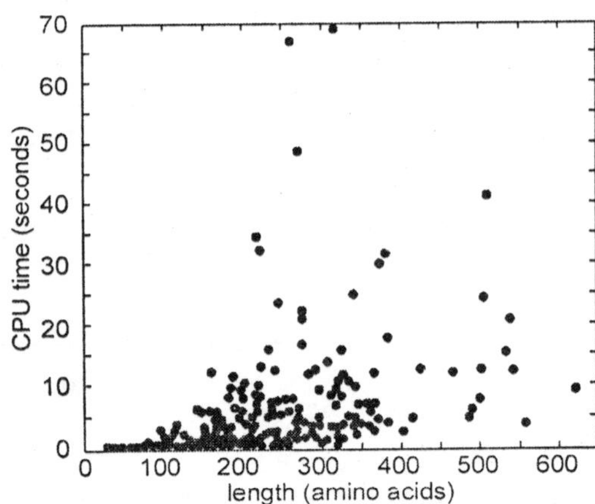

Fig. 9.4. CPU time of PROSPECT vs. the length of query sequence for threading each of 312 proteins on the template of its native folds using a *Linux* workstation with Pentium III 550 MHz. To show the majority of the data clearly, two extreme data points (421 amino acids, 117.9 seconds) and (907 amino acids, 8.4 seconds) are not included within the scale of the figure.

For each pair of proteins, we put one into the query set and the other into the template set (both proteins of each pair have solved structures). In the fold recognition test, we ran each protein from the query set against the whole template set. Table 9.2 shows the performance of threading accuracies. Overall, prospect recognizes 69 percent of the templates correctly (among top five in ranking) and aligns 66 percent of the structurally alignable residues correctly (defined as being aligned to within a four-residue shift from the SARF's structure-structure alignment position). Interestingly, these numbers may be compared with the 55 percent fold recognition and 64 percent alignment accuracy for the same test set without using the pairwise energy, indicating the significant contribution from the pairwise term, particularly for fold recognition. The fold recognition and alignment accuracy are further improved to 72 percent and 74 percent, respectively, when the secondary structure information predicted by the PHD program is used in scoring.

Normalization of Threading Scores

The goal of normalizing PROSPECT'S threading scores is to map all its scores to a fixed range, say [0, 1], so that we can assign a fixed meaning to each normalized score. We have trained the neural network to accomplish such a mapping. To train a neural network, we need to provide a list of training vectors, each of which consists of an array of input data and a desired output value. In our case, the input data for each training vector consists of the following values: the total threading score, the pair contact potential, the singleton potential, the mutation energy, the gap penalty, and a series of "normalization" factors, including the lengths of the query and the template, the sequence identity between the query and the template in threading, and the (threading) scoring distribution between the template and the protein sequences of all the representative entries in FSSP (the actual input to the neural net includes the two extreme values and the average value of the distribution). One way to define the desired output value is simply to use 1 or 0, representing whether the template is the correct fold of the query or not, respectively, as Jones did in his recent work (1999). We have used a different function to define a desired output value.

Table 9.2. Threading performance vs. sequence identity

Seq identity		1–6%	7–9%	9–12%	13–15%	16–18%	19–21%	22–24%	25–27%	28–30%	overall
train	total	(1)	(4)	(26)	(28)	(44)	(28)	(19)	(15)	(10)	(175)
	top-1	0	0	2 8%	9 32%	25 57%	25 89%	13 68%	15 100%	9 90%	98 56%
	top-5	0 0%	0	4 15%	15 54%	32 73%	28 100%	17 89%	15 100%	10 100%	121 69%
	align	25/88 28%	143/421 34%	797/2153 37%	1564/2786 56%	3127/4423 71%	2704/3122 87%	1589/1743 91%	1022/1221 84%	886/979 91%	11857/16936 70%
test	total	(1)	(3)	(24)	(36)	(19)	(15)	(18)	(16)	(5)	(137)
	top-1	0	0	6 25%	18 50%	7 37%	11 73%	17 94%	16 100%	4 80%	79 58%
	top-5	0	0	9 37%	24 67%	10 53%	12 80%	18 100%	16 100%	5 100%	94 69%
	align	0/88 0%	116/312 37%	866/3023 29%	3384/5650 60%	1426/2392 60%	1673/2106 79%	3022/3412 89%	2091/2253 93%	489/505 97%	13067/19723 66%

Each column represents a different range of identity level: *total* is the total number of pairs in a particular sequence-identity range; *top-1* and *top-5* represent the numbers of pairs whose templates are ranked in top one and top five, respectively; and *align* denotes the averaged alignment accuracy over all pair alignments in a particular sequence-identity range.

For each query-template pair, we calculate the number of structurally alignable residues between them (note that all query proteins in our training set have 3D experimental structures in PDB) defined by the SARF program. Then for each query-template pair, the two proteins are considered to be the "true" pair if their FSSP in dices share the same first digit that is, they belong to the same fold family. The frequency of the true pairs within a range of alignable residues provides a continuous function of how probable it may correspond to a true query-template pair if measured by structure-structure alignments alone. For example, on our dataset of 17,000+ false pairs and 708 true pairs, this number has the distribution shown in table 9.3. We found that using the relative frequencies listed as the desired outputs for neural net training has generated better normalization results (than the 0/1 output value) on our training set. This is probably because they provide more information than simply 0 or 1 to the neural net. Detailed discussion about the strength and properties of this training function is described by Xu et al. (2001). Our neural net consists of 14 input nodes, one output node, and one hidden layer with 20 nodes. Among the entire data set (more than 18,000 pairs in total), one-half is used as the training set and the remaining half as the (independent) testing set.

The goal of the training is to find a mathematical function that maps the input data to values in the range of [0,1], which are as close to the desired output values as possible. Similar prediction results are achieved on the training and the testing sets, both having an error rate of about 11 percent Table 18.4 summarizes the prediction performance on the combined set of training the testing. This normalization by neural net has clearly given a meaning to each (normalized) score. For example, if a (normalized) score is between 0.6 and 0.7, we know that the probability of having a true fold recognition is about 0.83. Our preliminary test shows that ranking templates using normalized scores improves the accuracy of fold recognition, compared with using raw scores.

Table 9.3. Relative-frequency of true query-template pairs

Number of alignables	*0–9*	*10–19*	*20–29*	*30–39*	*40–49*	*50–59*	*60–69*	*70–79*	*80–89*	*90–99*
Frequency being a true pair	0.000	0.001	0.003	0.020	0.150	0.500	0.750	0.900	0.990	1.000

Using Prospect for Structure Prediction

Prospect provides a suite of options for various applications. It also has an interface to modeller for generating atomic structures after a sequence-structure alignment has been constructed. It currently uses two template libraries that a user can choose: protein chains defined by the FSSP non-redundant set and compact domains denned by the DALI non-redundant domain library.

We now use the CASP-3 target t0057 (the CbiK protein) as a prediction example of applying the prospect program. To obtain a template for t0057, we ran prospect against the database of more than two thousand protein templates. The template lakl was identified. Then we refined the alignment between t0053 and the template lakl using the information of an active site. Through a detailed analysis, we found that lakl has an active site at His-183. Searching the block database, we identified His-145 of t0053 as the corresponding active site. We then re-ran prospect using the constraint that His-145 of t0053 should be aligned with His-183 of lakl. The final alignment between t0053 and lakl (with a sequence identities of 12.2 percent) is given in figure 9.5. The refined alignment agrees quite well with the structure-structure alignment, and it is significantly better than the original one without the constraint.

Table 9.4. Prediction performance on combined set of training and testing

Neural net output	*0–0.1*	*0.1–0.2*	*0.2–.03*	*0.3–0.4*	*0.4–0.5*	*0.5–0.6*	*0.6–0.7*	*0.7–0.8*	*0.8–0.9*	*0.9–0.10*
No. of false pairs	17279	474	143	56	25	7	7	2	0	0
No. of true pairs	314	135	60	51	44	38	34	14	12	8
Percent of true pairs	1.7%	22%	30%	48%	64%	84%	83%	87%	100%	100%

The full alignment with 264 residues in the query.

The structure using MODELLER based on the refined alignment is shown in figure detailed description of our prediction experience using PROSPECT in CASP-3 appears in Xu et al. 1999.

Application of Experimental Data as Threading Constraints

Though threading often provides valuable information, the computational structure models are still predictions. One way to significantly improve the reliability of predictions is through the incorporation of partial experimental data related to protein structures. Two classes of data have proven to be particularly useful in the pursuit of combined methods of experiments and computation for protein structure determination: (1) intramolecular cross-links from mass spectrometry, and (2) chemical shifts and long-range NOEs from NMR experiments. Potentially, these experimental procedures could produce structural data, not necessarily sufficient for an accurate structure determination, in a high-throughput mode. Our experience has shown that even a small number of distance restraints from NMR or cross-links could significantly reduce the search space size of *feasible* threading alignments, quickly ruling out the vast majority of the wrong folds and incorrect sequence-structure alignments.

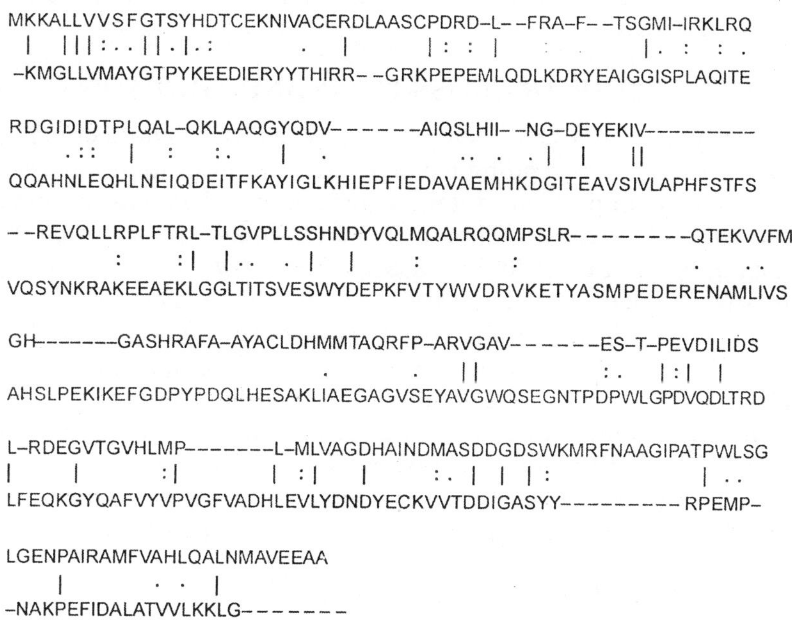

The full alignment with 264 residues in the query sequence

MKKALLVVSFGTSYHDTCEKNIVACERDLAASCPDRD–L– –FRA–F– –TSGMI–IRKLRQ
–KMGLLVMAYGTPYKEEDIERYYTHIRR– –GRKPEPEMLQDLKDRYEAIGGISPLAQITE

RDGIDIDTPLQAL–QKLAAQGYQDV– – – – – –AIQSLHII– –NG–DEYEKIV– – – – – – – – –
QQAHNLEQHLNEIQDEITFKAYIGLKHIEPFIEDAVAEMHKDGITEAVSIVLAPHFSTFS

– –REVQLLRPLFTRL–TLGVPLLSSHNDYVQLMQALRQQMPSLR– – – – – – – –QTEKVVFM
VQSYNKRAKEEAEKLGGLTITSVESWYDEPKFVTYWVDRVKETYASMPEDERENAMLIVS

GH– – – – – – –GASHRAFA–AYACLDHMMTAQRFP–ARVGAV– – – – – –ES–T–PEVDILIDS
AHSLPEKIKEFGDPYPDQLHESAKLIAEGAGVSEYAVGWQSEGNTPDPWLGPDVQDLTRD

L–RDEGVTGVHLMP– – – – – – –L–MLVAGDHAINDMASDDGDSWKMRFNAAGIPATPWLSG
LFEQKGYQAFVYVPVGFVADHLEVLYDNDYECKVVTDDIGASYY– – – – – – – – – RPEMP–

LGENPAIRAMFVAHLQALNMAVEEAA
–NAKPEFIDALATVVLKKLG– – – – – – –

Fig. 9.5. Sequence alignment between 10053 and lakl, where |, :, and . indicate the two aligned residues are identical, similar, as related, respectively.

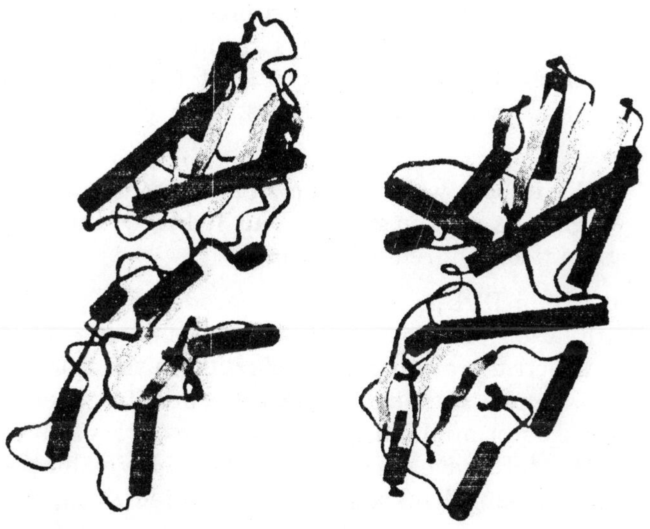

Fig. 9.6. A comparison between the predicted structure (left) based on the template lakl and the experimental one (right) for the CASP-3 target 10053. The target 10053 and the template lakl belong to the same super-family. The cylinders represent a-helices, the strands denote b-strands, the dark lines are for turns, and the thin lines represent loops.

As the number of such distance restraints increases, we have observed that the quality of the predicted structures rapidly approaches the ones of the experimental structures. We could foresee the emergence of a new paradigm for protein structure determination in the near future, which will fully utilize the structural information derived from both the experiments and sophisticated computational procedures, making the structure determination process much more efficient and cost effective.

INTRAMOLECULAR CROSS-LINKS AND THREADING

A recent study by Young et al. (2000) demonstrates the effectiveness of using lysine (Lys-Lys) cross-links as constraints on threading for high-throughput protein fold recognition. The basic idea of this approach is to probe a protein structure with a lysine-specific cross-linking agent, bis *(sulfosuccinimidyl)* suberate. Then a few cross-linked sites can be determined by tryptic peptide mapping using time-of-flight mass spectrometry. The distance between a pair of cross-linked lysines is bounded by the length of the cross-linking agent, that is, 24 Å between C_α-C_α atoms. The distance restraints are then used to assess the quality of threading models—how consistent the predicted structure models are with the obtained distance measures. The model that is most consistent with the cross-links is then chosen. Young et al. (2000) used 18 unique lysine cross-links, in conjunction with threading, and correctly identified FGF-2 as a member of the β-trefoil fold family. The structure model has a backbone error of 4.8 Å rmsd. This method is fast and uses a small amount of material. It is easy for automation and should be broadly applicable, so it has a potential for the high-throughput fold recognition.

Use of NMR Partial Data as Threading Constraints

The NMR method for protein structure determination is based on (1) proton and heteronuclear chemical shifts, (2) a network of distance restraints between spatially close (≤ 5 Å) hydrogen atoms derived from nuclear Overhauser effects (NOEs), (3) dihedral angle restraints calculated from scalar coupling constants, and so on. An NMR structure is generally determined through molecular dynamics simulation and energy minimization under the constraints specified by NMR restraints. It typically requires 15-20 NOE restraints per residue to obtain an accurate (mean) structure (equivalent to a ~2Å X-ray structure).

It is well known that NMR methods work only for small proteins and that the effectiveness goes down rapidly as the weight of a protein goes beyond 30 kD. Of the 1,901 NMR structures in PDB only 27 are larger than two hundred amino acids.

The largest NMR structure in PDB is intimin (PDB code linm), with 282 residues (30.1 kD). The main problems with larger proteins are the increased number of resonances and line broadening, which result in spectral crowding and reduction in the fraction of spectral peaks that can be identified and assigned. Although NMR experiments generally cannot produce sufficient data for 3D structure determination of large proteins, it is often possible to get some NMR data. These partial NMR data have turned out to be highly valuable as constraints on threading-based fold recognition and backbone structure prediction, as we have recently demonstrated. Particularly, we have found that (1) the topological information provided by just a few long-range NOEs (NOEs between residues that are not adjacent in the protein sequence) can quickly rule out certain false fold classes and alignments during threading, and (2) the geometric patterns of a group of NOEs, can rapidly narrow down the list of potential alignments between a query protein and a template structure in threading.

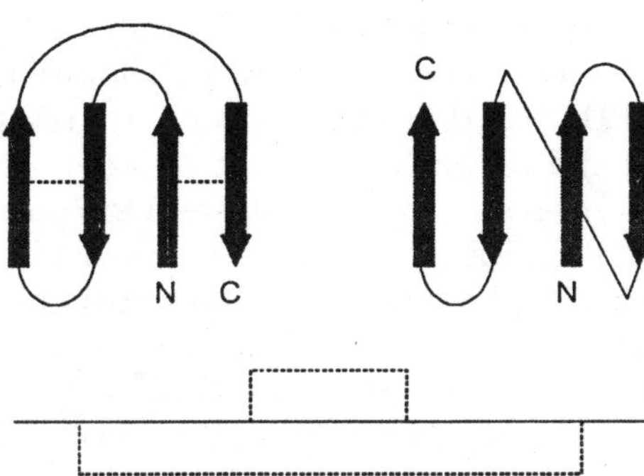

Fig. 9.7. Ruling out false topological classes of proteins by using a few NOEs. On the left is a Greek key substructure, and on the right is a 4-antiparallel (flat) β-strand motif. A partial sequence with two NOEs is given in the bottom. Although an alignment between the Greek key structure and the partial sequence may exist that does not violate the NOE restraints, there is no such alignment between the 4-antiparallel β-strand motif and the sequence because it has the "wrong" topology.

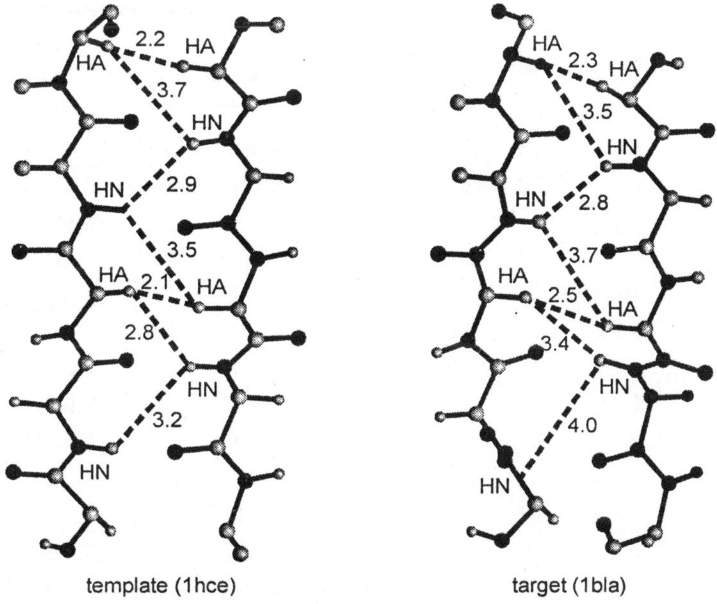

Fig. 9.8. Comparison between the structurally equivalent β-sheets in the two proteins of similar folds. Residues 74-78 and 103-107 for template Ihce, and residues 83-87 and 1 13-117 for the query protein Ibla, are shown. The dashed lines with the numbers give the distance in Å.

We can define an *NOE-constrained threading problem* as to find a globally optimal threading between a query sequence and a template structure, under the constraint that no NOEs are *"significantly"* violated; that is, two residues with an NOE between them should not be aligned to template positions that are *"too far"* from each other. This can be implemented using a penalty function that penalizes threading alignments with *"significant"* deviations from the NOE-specified distances. We can treat structural information from other types of NMR data in a similar fashion. The following gives a case study on how long-range NOEs affect the threading accuracy in both fold recognition and threading alignments.

For this study, we have selected from the FSSP database 17 proteins as the queries. These proteins were selected randomly under the following conditions: (1) the query proteins have experimental NMR data in PDB; (2) they should evenly represent three classes of proteins: all-α, all-β, and α and β mixed; and (3) each has a native-like structure in the template set, with sequence identity less than 35 percent. In this test, we first ran the NMR-constrained threading program, and then ran MODELLER, to generate detailed 3D structures based on threading alignments without using additional data. The fold recognition study is done against 667 unique folds. We tested how the amount of NOE information affects the threading performance, using 0, 0.5, 1.0, 1.5, 2.0, 2.5, and 3.0 NOEs (including both long- and short- NOEs) per residue, respectively. The NOEs are selected randomly and uniformly from the NMR data file of the corresponding protein. Table 9.5 summarizes the test results. The tabulated alignment accuracy between structurally equivalent residues in the query and template (C_α-RMSD) is the highest accuracy in ten runs. The fold recognition accuracy is based on a single run.

Table 9.5. Threading accuracy vs. the number of NOE restraints.

Class	*Query*	*Nres* *(a.a.)*	*temp*	*iden* *(%)*	*rmsd* *(Å)*	*RMSD (Å) / rank vs. NOE / a.a.* *0*	*0.5*	*1.0*	*1.5*	*2.0*	*2.5*	*3.0*
α	1bbn	133	Icntl	8	2.6	**16.2/29**	6.6/15	6.6/1	6.6/2	5.3/1	5.3/1	**5.3/1**
	1itl	130	3inkC	12	2.2	**13.1/7**	4.3/26	4.2/1	4.2/4	3.6/3	3.6/3	**3.6/3**
	1ner	74	Ilmb3	14	2.4	3.8/**57**	3.8/57	3.8/57	3.8/79	3.8/49	3.8/39	3.8/**39**
	1il6	166	lbgc	15	2.1	3.0/1	3.4/1	3.4/1	3.4/1	3.4/1	3.4/1	3.4/1
	1ocd	104	lcyj	28	2.0	3.4/1	3.4/1	3.4/1	3.4/1	3.4/1	3.4/1	3.4/1
β	1maj	113	lagdB	7	2.5	**8.8/37**	8.0/10	6.9/1	6.9/1	6.6/1	6.6/1	**6.6/1**
	1nct	98	lbec	12	2.5	**14.7**/1	10.8/1	5.4/1	5.4/1	5.4/1	5.4/1	**5.4**/1
	1bla	155	Ihce	14	2.1	7.7/1	4.0/1	4.0/1	4.0/1	4.0/1	4.0/1	**4.0**/1
	1vhp	117	Icd8	15	2.9	**4.5**/1	4.4/1	4.4/1	4.4/1	4.4/1	4.4/1	**4.4**/1
	1ghj	79	Iiyu	24	1.8	2.0/1	2.0/1	2.0/1	2.0/1	2.0/1	2.0/1	2.0/1
	1a 7i	60	Iqli	33	2.5	**2.8**/1	2.8/1	2.8/1	2.6/1	2.6/1	2.6/1	**2.6**/1
α/β	1afi	72	2acy	7	2.8	**6.8/376**	5.5/193	5.6/26	5.5/20	3.8/3	3.8/2	**3.8/2**
	3trx	105	Ia8y	12	2.7	**4.1**/1	3.1/1	3.1/1	3.1/1	3.2/1	3.0/1	3.0/1
	1ikm	69	IdokB	13	2.1	2.1/2	2.1/2	2.1/2	2.1/1	2.1/1	2.1/1	2.1/1
	3phy	125	Ibv6	15	2.3	**12.2/41**	8.8/25	3.6/31	3.6/5	3.6/6	3.6/6	**3.6/4**
	1crp	166	IbyuB	24	1.7	2.6/1	2.6/1	2.6/1	2.6/1	2.6/1	2.6/1	2.6/1
	1fht	116	2ula	34	2.1	4.5/1	4.5/1	4.5/1	4.5/1	4.5/1	4.5/1	**2.1**/1

Query and *temp* represent the query and template proteins, respectively. *Nres* is the number of alignable residues between the query and template. *Iden* denotes the sequence identity between the query and template sequences. The entry *rmsd* is the C_α-RMSD between the structurally equivalent residues of the query and template structures. *RMSD / ronk vs. NOE / a.a.* represents (a) the C_α-RMSD between the experimental structure and the predicted structure and (b) the rank of the correct template structure among all templates, respectively, when using 0, 0.5, 1.0, 1.5, 2.0, 2.5, 3.0 (long- and short-range) NOEs per residue as threading constraints.

In this test, NOEs improve the threading performance in 13 out of the 17 cases, with seven cases showing improvements in fold recognition and 11 cases showing improvements in alignment accuracy. One observation we have is that although the pure threading results on all three analogous pairs (lafi-2acy, 3phy-lbv6, and Ibla-Ihce, as classified by SCOP are generally poor and worse than those of homologous pairs, the prediction results with at least one NOE per residue for the analogous pairs and the remaining 14 homologous pairs are quite comparable. This seems to suggest the possibility of reliably extending the scope of threading from homologs to analogs, with the help of a few NOEs.

IMPLICATIONS OF TREADING FOR THE STRUCTURAL GCNOMICS INITIATIVE

Traditionally, protein structures were solved at a low-throughput mode, one protein at a time. Recent advances in synchrotron and high-resolution NMR have substantially accelerated the rate of protein structure determination. There is an overwhelming consensus in the structural biology community that protein structures can be solved massively (an effort called *"structural genomics")*, in a similar fashion to how DNA sequences are being sequenced. The goal of the recent NIH Structural Genomics Initiative is to determine the structures of a hundred thousand human proteins. The potential impact of such a project could be as significant as the Human Genome Project.

The basic strategy of the project is to first select a representative from each fold family and solve them using experimental means including X-ray crystallography and NMR, and then to computationally model the rest of the structures using the representative structures as templates. This would save a tremendous amount of time and resources if successful. Two key computational challenges in this ambitious initiative are (1) to identify which protein sequences represent novel folds, and (2) to computationally model the rest of the structures based on the representative structures. Clearly threading could play a key role in addressing these challenges. Several groups have employed threading tools to structurally model all coding sequences from a whole genome. These studies have provided timely analysis for the current genome sequencing efforts, and allow gene-hunting researchers to find valuable structure information quickly.

It has been found that about 17 per cent of the identified genes fall into the category of homology modeling. The sequence profile approach extends the genome-wide coverage from 17 percent to about 30-40 percent. The threading approach can predict about an additional 10-20 percent of identified genes with good confidence. For the rest of the proteins, either existing threading methods cannot recognize the templates with good confidence, or templates do not exist in the structure database (*i.e.,* novel folds). This raises two challenging issues—how to further improve the efficacy of threading methods, and how to reliably determine if a protein has a new fold? Among some of the proteins with new folds is the class of membrane proteins, which have not been touched by threading methods. Due to the technical difficulty of experimentally solving these structures and the lack of available templates for computational methods like threading, it may take a long time to solve all these proteins. However, with the advancement of new experimental techniques and computational methods, it is foreseeable to have the structures of all globular proteins solved within the next ten years, and hence put an end to the folding problem of globular proteins.

Challenging Issues and New Developments

Although about 90 per cent of new globular proteins may have native-like folds in the structure database, as we discussed in section 18.2, current threading results clearly fall short of these theoretical limits. Among all the prediction targets in CASP-3, 25 proteins were considered as "solvable" by protein threading and were put in the category of fold recognition. Only 13 of the 25 proteins had their correct folds recognized and had at least 50 per cent of the alignable residues aligned correctly by any participating team. Further improving the accuracy and enlarging the scope of protein threading are among the key challenging issues in computational structural biology.

Local Threading : Virtually all existing threading programs attempt to solve the "global" threading problem, finding an optimal alignment between the whole sequence of a query protein and the entire structure of a template protein. This type of threading method is not very sensitive in detecting templates when the percentage of structurally alignable residues is low (say, <50%)—existing threading methods work well only when the query protein has a template in PDB with at least 60-70 per cent of their residues structurally alignable.

This has clearly limited the applicability of protein threading. One way to solve this problem is through "local threading", locating good partial sequence-structure alignments (*e.g.,* a structure domain or a super-secondary structure). This will extend the scope of current threading methods to deal with a much larger class of structure prediction problems. Local threading may provide a better way to recognize function-related structural motifs (for example, the EF-hand motif frequently found in calcium binding proteins with various folds) than sequence-based motif search methods. However, as we know, many of the motifs do not have strong sequence characteristics, so sequence-based motif search methods may not work well.

Mini-Threading : Threading takes advantage of enormous structure information in PDB and provides more reliable structure predictions than ab initio methods. However, threading does not apply to the proteins that do not have templates in PDB or have templates that threading cannot recognize. On the other hand, ab initio prediction applies to all proteins. However, it generally requires a significant amount of computing resource, and its prediction results might be unreliable. Mini-threading attempts to combine the strengths of these two approaches. It first builds local structures through sequence comparison or local threading, and then assembles them into a global structure through ab initio methods.

Current mini-threading methods typically recognize structure segments in PDB that are most similar to a segment of the query sequence (with a typical size of nine amino acids) in terms of sequence similarity. Then a distribution profile is constructed for the ϕ and ψ angles of the query protein's backbones, based on the distributions in the structure segments. It is possible to use templates that are fine-tuned for small structure segments like *I-sites* to obtain more reliable local structure templates. Once local structures are more or less defined, assembling them based on their backbone distributions would require significantly fewer computational resources than pure ab initio methods. The optimization process is typically carried out using genetic algorithms or Monte Carlo simulations. Some success of mini-threading has been demonstrated in CASP-3. Mini-threading provides a promising method to generate low resolution structures for the proteins to which threading might not apply.

Using Multiple Sequence or Multiple-Structure Information

One of the problems in threading is the structural variations among proteins of the same fold, which is not well captured in the existing threading scoring functions. When two proteins share the same fold, the length of the structurally alignable secondary structures and the packing between the secondary structures may differ, particularly when the two proteins are structural analogs (in the same fold but not the same superfamily), where the variations in the secondary structures of peripheral elements as well as in some core elements are much larger than structural homologs.

The solvent accessibilities, the secondary structure types, and the C_β pairs of the template may not correspond exactly to those of the query protein in the aligned positions. One way to overcome the problem is to derive the basic statistics (used to calculate the scoring functions) from the "consensus" of the query sequence or template fold. One can generate a sequence profile for the query's protein family based on multiple-sequence alignments. Such a profile reflects common characteristics of the protein family, and hence it can enhance threading sensitivity while reducing noises.

Another way to use protein classification is to construct the sequence profile for the protein family containing the fold template or to build profile for the structures of the same fold through multiple structure alignments. By using the structure alignment, one can identify which positions (as well as solvent accessibilities, the secondary structure types, or the C_β pairs) are more conserved across different templates of the same fold. The advantage of using multiple templates instead of multiple sequences in the query's family is that the former covers the profile of proteins of the same fold (not necessarily of the same family or superfamily).

FUTURE OUTLOOK

Protein threading is becoming a major tool in computational biology. Its applicability will probably cover all globular proteins within the next decade. We will continue to see improvement in the prediction accuracy and computational efficiency of threading-based structure prediction methods. Genome-scale application of threading will become routine in the next few years. As technologies mature in high-throughput experimental and computational structural biology, we will see a merger of the two approaches. Information from structural experiments and template-based computation will be tightly combined for significantly more efficient and cost-effective ways of determining protein structures.

Index

❑❑❑